Random Vibration

Mechanical Vibration and Shock Analysis
Third edition – Volume 3

Random Vibration

Christian Lalanne

 WILEY

First edition published 2002 by Hermes Penton Ltd © Hermes Penton Ltd 2002
Second edition published 2009 in Great Britain and the United States by ISTE Ltd and John Wiley & Sons, Inc. © ISTE Ltd 2009
Third edition published 2014 in Great Britain and the United States by ISTE Ltd and John Wiley & Sons, Inc.

Apart from any fair dealing for the purposes of research or private study, or criticism or review, as permitted under the Copyright, Designs and Patents Act 1988, this publication may only be reproduced, stored or transmitted, in any form or by any means, with the prior permission in writing of the publishers, or in the case of reprographic reproduction in accordance with the terms and licenses issued by the CLA. Enquiries concerning reproduction outside these terms should be sent to the publishers at the undermentioned address:

ISTE Ltd
27-37 St George's Road
London SW19 4EU
UK

www.iste.co.uk

John Wiley & Sons, Inc.
111 River Street
Hoboken, NJ 07030
USA

www.wiley.com

© ISTE Ltd 2014
The rights of Christian Lalanne to be identified as the author of this work have been asserted by him in accordance with the Copyright, Designs and Patents Act 1988.

Library of Congress Control Number: 2014933739

British Library Cataloguing-in-Publication Data
A CIP record for this book is available from the British Library
ISBN 978-1-84821-643-3 (Set of 5 volumes)
ISBN 978-1-84821-646-4 (Volume 3)

Table of Contents

Foreword to Series xiii

Introduction xvii

List of Symbols xix

Chapter 1. Statistical Properties of a Random Process 1

    1.1. Definitions 1
        1.1.1. Random variable 1
        1.1.2. Random process 2
    1.2. Random vibration in real environments 2
    1.3. Random vibration in laboratory tests 3
    1.4. Methods of random vibration analysis 3
    1.5. Distribution of instantaneous values 5
        1.5.1. Probability density 5
        1.5.2. Distribution function 6
    1.6. Gaussian random process 7
    1.7. Rayleigh distribution 12
    1.8. Ensemble averages: through the process 12
        1.8.1. n order average 12
        1.8.2. Centered moments 14
        1.8.3. Variance 14
        1.8.4. Standard deviation 15
        1.8.5. Autocorrelation function 16
        1.8.6. Cross-correlation function 16
        1.8.7. Autocovariance 17
        1.8.8. Covariance 17
        1.8.9. Stationarity 17
    1.9. Temporal averages: along the process 23
        1.9.1. Mean 23

1.9.2. Quadratic mean – rms value 25
1.9.3. Moments of order n. 27
1.9.4. Variance – standard deviation 28
1.9.5. Skewness. 29
1.9.6. Kurtosis. 30
1.9.7. Crest Factor 33
1.9.8. Temporal autocorrelation function. 33
1.9.9. Properties of the autocorrelation function 39
1.9.10. Correlation duration 41
1.9.11. Cross-correlation 47
1.9.12. Cross-correlation coefficient 50
1.9.13. Ergodicity 50
1.10. Significance of the statistical analysis (ensemble or temporal). 52
1.11. Stationary and pseudo-stationary signals. 52
1.12. Summary chart of main definitions 53
1.13. Sliding mean 54
1.14. Test of stationarity 58
1.14.1. The reverse arrangements test (RAT) 58
1.14.2. The runs test 61
1.15 Identification of shocks and/or signal problems 65
1.16. Breakdown of vibratory signal into "events": choice of signal samples 68
1.17. Interpretation and taking into account of environment variation 75

Chapter 2. Random Vibration Properties in the Frequency Domain 79

2.1. Fourier transform 79
2.2. Power spectral density 81
  2.2.1. Need 81
  2.2.2. Definition 82
2.3. Amplitude Spectral Density 89
2.4. Cross-power spectral density 89
2.5. Power spectral density of a random process. 90
2.6. Cross-power spectral density of two processes 91
2.7. Relationship between the PSD and correlation function of a process. . 93
2.8. Quadspectrum – cospectrum. 93
2.9. Definitions 94
  2.9.1. Broadband process 94
  2.9.2. White noise 95
  2.9.3. Band-limited white noise 95
  2.9.4. Narrow band process. 96
  2.9.5. Colors of noise 97
2.10. Autocorrelation function of white noise 98
2.11. Autocorrelation function of band-limited white noise 99

2.12. Peak factor	101
2.13. Effects of truncation of peaks of acceleration signal on the PSD	101
2.14. Standardized PSD/density of probability analogy	105
2.15. Spectral density as a function of time	106
2.16. Sum of two random processes	106
2.17. Relationship between the PSD of the excitation and the response of a linear system	108
2.18. Relationship between the PSD of the excitation and the cross-power spectral density of the response of a linear system	111
2.19. Coherence function	112
2.20. Transfer function calculation from random vibration measurements	114
2.20.1. Theoretical relations	114
2.20.2. Presence of noise on the input	116
2.20.3. Presence of noise on the response	118
2.20.4. Presence of noise on the input and response	120
2.20.5. Choice of transfer function	121

Chapter 3. Rms Value of Random Vibration 127

3.1. Rms value of a signal as a function of its PSD	127
3.2. Relationships between the PSD of acceleration, velocity and displacement	131
3.3. Graphical representation of the PSD	133
3.4. Practical calculation of acceleration, velocity and displacement rms values	135
3.4.1. General expressions	135
3.4.2. Constant PSD in frequency interval	135
3.4.3. PSD comprising several horizontal straight line segments	137
3.4.4. PSD defined by a linear segment of arbitrary slope	137
3.4.5. PSD comprising several segments of arbitrary slopes	147
3.5. Rms value according to the frequency	147
3.6. Case of periodic signals	149
3.7. Case of a periodic signal superimposed onto random noise	151

Chapter 4. Practical Calculation of the Power Spectral Density 153

4.1. Sampling of signal	153
4.2. PSD calculation methods	158
4.2.1. Use of the autocorrelation function	158
4.2.2. Calculation of the PSD from the rms value of a filtered signal	158
4.2.3. Calculation of PSD starting from a Fourier transform	159
4.3. PSD calculation steps	160
4.3.1. Maximum frequency	160
4.3.2. Extraction of sample of duration T	160

4.3.3. Averaging 167
4.3.4. Addition of zeros 170
4.4. FFT. 175
4.5. Particular case of a periodic excitation. 177
4.6. Statistical error. 178
   4.6.1. Origin. 178
   4.6.2. Definition 180
4.7. Statistical error calculation. 180
   4.7.1. Distribution of the measured PSD 180
   4.7.2. Variance of the measured PSD 183
   4.7.3. Statistical error 183
   4.7.4. Relationship between number of degrees of freedom,
   duration and bandwidth of analysis 184
   4.7.5. Confidence interval. 190
   4.7.6. Expression for statistical error in decibels 202
   4.7.7. Statistical error calculation from digitized signal 204
4.8. Influence of duration and frequency step on the PSD 212
   4.8.1. Influence of duration 212
   4.8.2. Influence of the frequency step. 213
   4.8.3. Influence of duration and of constant statistical error
   frequency step. 214
4.9. Overlapping 216
   4.9.1. Utility. 216
   4.9.2. Influence on the number of degrees of freedom 217
   4.9.3. Influence on statistical error. 218
   4.9.4. Choice of overlapping rate 221
4.10. Information to provide with a PSD 222
4.11. Difference between rms values calculated from a signal
according to time and from its PSD 222
4.12. Calculation of a PSD from a Fourier transform 223
4.13. Amplitude based on frequency: relationship with the PSD 227
4.14. Calculation of the PSD for given statistical error. 228
   4.14.1. Case study: digitization of a signal is to be carried out 228
   4.14.2. Case study: only one sample of an already
   digitized signal is available. 230
4.15. Choice of filter bandwidth 231
   4.15.1. Rules 231
   4.15.2. Bias error 233
   4.15.3. Maximum statistical error 238
   4.15.4. Optimum bandwidth 240
4.16. Probability that the measured PSD lies
between ± one standard deviation 243
4.17. Statistical error: other quantities. 245

4.18. Peak hold spectrum 250
4.19. Generation of random signal of given PSD 252
    4.19.1. Random phase sinusoid sum method. 252
    4.19.2. Inverse Fourier transform method 255
4.20. Using a window during the creation of a random signal from a
PSD 256

Chapter 5. Statistical Properties of Random Vibration in the Time Domain 259

    5.1. Distribution of instantaneous values 259
    5.2. Properties of derivative process 260
    5.3. Number of threshold crossings per unit time 264
    5.4. Average frequency 269
    5.5. Threshold level crossing curves 272
    5.6. Moments 279
    5.7. Average frequency of PSD defined by straight line segments 282
        5.7.1. Linear-linear scales 282
        5.7.2. Linear-logarithmic scales 284
        5.7.3. Logarithmic-linear scales 285
        5.7.4. Logarithmic-logarithmic scales. 286
    5.8. Fourth moment of PSD defined by straight line segments 288
        5.8.1. Linear-linear scales 288
        5.8.2. Linear-logarithmic scales 289
        5.8.3. Logarithmic-linear scales 290
        5.8.4. Logarithmic-logarithmic scales. 291
    5.9. Generalization: moment of order n 292
        5.9.1. Linear-linear scales 292
        5.9.2. Linear-logarithmic scales 292
        5.9.3. Logarithmic-linear scales 292
        5.9.4. Logarithmic-logarithmic scales. 293

Chapter 6. Probability Distribution of Maxima of Random Vibration . . . 295

    6.1. Probability density of maxima 295
    6.2. Moments of the maxima probability distribution 303
    6.3. Expected number of maxima per unit time 304
    6.4. Average time interval between two successive maxima 307
    6.5. Average correlation between two successive maxima 308
    6.6. Properties of the irregularity factor 309
        6.6.1. Variation interval 309
        6.6.2. Calculation of irregularity factor for band-limited white noise . . . 313
        6.6.3. Calculation of irregularity factor for noise of
form $G = Const.f^b$ 316

6.6.4. Case study: variations of irregularity factor for two
narrowband signals 320
6.7. Error related to the use of Rayleigh's law instead of a complete
probability density function. 321
6.8. Peak distribution function 323
   6.8.1. General case 323
   6.8.2. Particular case of narrowband Gaussian process 325
6.9. Mean number of maxima greater than the given
threshold (by unit time) 328
6.10. Mean number of maxima above given threshold between two times . 331
6.11. Mean time interval between two successive maxima 331
6.12. Mean number of maxima above given level reached by signal
excursion above this threshold 332
6.13. Time during which the signal is above a given value 335
6.14. Probability that a maximum is positive or negative 337
6.15. Probability density of the positive maxima 337
6.16. Probability that the positive maxima is lower than a given threshold . 338
6.17. Average number of positive maxima per unit of time 338
6.18. Average amplitude jump between two successive extrema 339
6.19. Average number of inflection points per unit of time 341

Chapter 7. Statistics of Extreme Values 343

7.1. Probability density of maxima greater than a given value 343
7.2. Return period. 344
7.3. Peak ℓ_p expected among N_p peaks 344
7.4. Logarithmic rise 345
7.5. Average maximum of N_p peaks. 346
7.6. Variance of maximum 346
7.7. Mode (most probable maximum value) 346
7.8. Maximum value exceeded with risk α 346
7.9. Application to the case of a centered narrowband normal process . . . 346
   7.9.1. Distribution function of largest peaks over duration T 346
   7.9.2. Probability that one peak at least exceeds a given threshold. 349
   7.9.3. Probability density of the largest maxima over duration T. 350
   7.9.4. Average of highest peaks 353
   7.9.5. Mean value probability 355
   7.9.6. Standard deviation of highest peaks 356
   7.9.7. Variation coefficient 357
   7.9.8. Most probable value 358
   7.9.9. Median 358
   7.9.10. Value of density at mode. 360
   7.9.11. Value of distribution function at mode 361

7.9.12. Expected maximum. 361
7.9.13. Maximum exceeded with given risk α. 361
7.10. Wideband centered normal process 363
7.10.1. Average of largest peaks 363
7.10.2. Variance of the largest peaks 366
7.10.3. Variation coefficient 367
7.11. Asymptotic laws 368
7.11.1. Gumbel asymptote 368
7.11.2. Case study: Rayleigh peak distribution 369
7.11.3. Expressions for large values of N_p 370
7.12. Choice of type of analysis 371
7.13. Study of the envelope of a narrowband process. 374
7.13.1. Probability density of the maxima of the envelope 374
7.13.2. Distribution of maxima of envelope 379
7.13.3. Average frequency of envelope of narrowband noise. 381

**Chapter 8. Response of a One-Degree-of-Freedom Linear System
to Random Vibration** 385

8.1. Average value of the response of a linear system 385
8.2. Response of perfect bandpass filter to random vibration 386
8.3. The PSD of the response of a one-dof linear system. 388
8.4. Rms value of response to white noise 389
8.5. Rms value of response of a linear one-degree of freedom system
subjected to bands of random noise 395
    8.5.1. Case where the excitation is a PSD defined by a straight line
    segment in logarithmic scales 395
    8.5.2. Case where the vibration has a PSD defined by a straight line
    segment of arbitrary slope in linear scales 401
    8.5.3. Case where the vibration has a constant PSD between
    two frequencies. 404
    8.5.4. Excitation defined by an absolute displacement 409
    8.5.5. Case where the excitation is defined by PSD comprising
    n straight line segments 411
8.6. Rms value of the absolute acceleration of the response 414
8.7. Transitory response of a dynamic system under stationary
random excitation 415
8.8. Transitory response of a dynamic system under amplitude
modulated white noise excitation. 423

xii   Random Vibration

Chapter 9. Characteristics of the Response of a One-Degree-of-Freedom Linear System to Random Vibration   427

9.1. Moments of response of a one-degree-of-freedom linear system: irregularity factor of response   427
   9.1.1. Moments   427
   9.1.2. Irregularity factor of response to noise of a constant PSD   431
   9.1.3. Characteristics of irregularity factor of response   433
   9.1.4. Case of a band-limited noise   444
9.2. Autocorrelation function of response displacement   445
9.3. Average numbers of maxima and minima per second   446
9.4. Equivalence between the transfer functions of a bandpass filter and a one-degree-of-freedom linear system   449
   9.4.1. Equivalence suggested by D.M. Aspinwall   449
   9.4.2. Equivalence suggested by K.W. Smith   451
   9.4.3. Rms value of signal filtered by the equivalent bandpass filter . . .   453

Chapter 10. First Passage at a Given Level of Response of a One-Degree-of-Freedom Linear System to a Random Vibration   455

10.1. Assumptions   455
10.2. Definitions   459
10.3. Statistically independent threshold crossings   460
10.4. Statistically independent response maxima   468
10.5. Independent threshold crossings by the envelope of maxima   472
10.6. Independent envelope peaks   476
   10.6.1. S.H. Crandall method   476
   10.6.2. D.M. Aspinwall method   479
10.7. Markov process assumption   486
   10.7.1. W.D. Mark assumption   486
   10.7.2. J.N. Yang and M. Shinozuka approximation   493
10.8. E.H. Vanmarcke model   494
   10.8.1. Assumption of a two state Markov process   494
   10.8.2. Approximation based on the mean clump size   500

Appendix   511

Bibliography   571

Index   591

Summary of Other Volumes in the Series   597

Foreword to Series

In the course of their lifetime simple items in everyday use such as mobile telephones, wristwatches, electronic components in cars or more specific items such as satellite equipment or flight systems in aircraft, can be subjected to various conditions of temperature and humidity, and more particularly to mechanical shock and vibrations, which form the subject of this work. They must therefore be designed in such a way that they can withstand the effects of the environmental conditions to which they are exposed without being damaged. Their design must be verified using a prototype or by calculations and/or significant laboratory testing.

Sizing, and later, testing are performed on the basis of specifications taken from national or international standards. The initial standards, drawn up in the 1940s, were blanket specifications, often extremely stringent, consisting of a sinusoidal vibration, the frequency of which was set to the resonance of the equipment. They were essentially designed to demonstrate a certain standard resistance of the equipment, with the implicit hypothesis that if the equipment survived the particular environment it would withstand, undamaged, the vibrations to which it would be subjected in service. Sometimes with a delay due to a certain conservatism, the evolution of these standards followed that of the testing facilities: the possibility of producing swept sine tests, the production of narrowband random vibrations swept over a wide range and finally the generation of wideband random vibrations. At the end of the 1970s, it was felt that there was a basic need to reduce the weight and cost of on-board equipment and to produce specifications closer to the real conditions of use. This evolution was taken into account between 1980 and 1985 concerning American standards (MIL-STD 810), French standards (GAM EG 13) or international standards (NATO), which all recommended the *tailoring of tests*. Current preference is to talk of the *tailoring of the product to its environment* in order to assert more clearly that the environment must be taken into account from the very start of the project, rather than to check the behavior of the material *a posteriori*. These concepts, originating with the military, are currently being increasingly echoed in the civil field.

Tailoring is based on an analysis of the life profile of the equipment, on the measurement of the environmental conditions associated with each condition of use and on the synthesis of all the data into a simple specification, which should be of the same severity as the actual environment.

This approach presupposes a proper understanding of the mechanical systems subjected to dynamic loads and knowledge of the most frequent failure modes.

Generally speaking, a good assessment of the stresses in a system subjected to vibration is possible only on the basis of a finite element model and relatively complex calculations. Such calculations can only be undertaken at a relatively advanced stage of the project once the structure has been sufficiently defined for such a model to be established.

Considerable work on the environment must be performed independently of the equipment concerned either at the very beginning of the project, at a time where there are no drawings available, or at the qualification stage, in order to define the test conditions.

In the absence of a precise and validated model of the structure, the simplest possible mechanical system is frequently used consisting of mass, stiffness and damping (a linear system with one degree of freedom), especially for:

– the comparison of the severity of several shocks (shock response spectrum) or of several vibrations (extreme response and fatigue damage spectra);

– the drafting of specifications: determining a vibration which produces the same effects on the model as the real environment, with the underlying hypothesis that the equivalent value will remain valid on the real, more complex structure;

– the calculations for pre-sizing at the start of the project;

– the establishment of rules for analysis of the vibrations (choice of the number of calculation points of a power spectral density) or for the definition of the tests (choice of the sweep rate of a swept sine test).

This explains the importance given to this simple model in this work of five volumes on "Mechanical Vibration and Shock Analysis".

Volume 1 of this series is devoted to *sinusoidal vibration*. After several reminders about the main vibratory environments which can affect materials during their working life and also about the methods used to take them into account, following several fundamental mechanical concepts, the responses (relative and absolute) of a mechanical one-degree-of-freedom system to an arbitrary excitation are considered, and its transfer function in various forms are defined. By placing the properties of sinusoidal vibrations in the contexts of the real environment and of laboratory tests, the transitory and steady state response of a single-degree-of-freedom system with viscous and then with non-linear damping is evolved. The various sinusoidal modes of sweeping with their properties are described, and then, starting from the response of a one-degree-of-freedom system, the consequences of an unsuitable choice of sweep rate are shown and a rule for choice of this rate is deduced from it.

Volume 2 deals with *mechanical shock*. This volume presents the shock response spectrum (SRS) with its different definitions, its properties and the precautions to be taken in calculating it. The shock shapes most widely used with the usual test facilities are presented with their characteristics, with indications how to establish test specifications of the same severity as the real, measured environment. A demonstration is then given on how these specifications can be made with classic laboratory equipment: shock machines, electrodynamic exciters driven by a time signal or by a response spectrum, indicating the limits, advantages and disadvantages of each solution.

Volume 3 examines the analysis of *random vibration* which encompasses the vast majority of the vibrations encountered in the real environment. This volume describes the properties of the process, enabling simplification of the analysis, before presenting the analysis of the signal in the frequency domain. The definition of the power spectral density is reviewed, as well as the precautions to be taken in calculating it, together with the processes used to improve results (windowing, overlapping). A complementary third approach consists of analyzing the statistical properties of the time signal. In particular, this study makes it possible to determine the distribution law of the maxima of a random Gaussian signal and to simplify the calculations of fatigue damage by avoiding direct counting of the peaks (Volumes 4 and 5). The relationships that provide the response of a one-degree-of-freedom linear system to a random vibration are established.

Volume 4 is devoted to the calculation of *damage fatigue*. It presents the hypotheses adopted to describe the behavior of a material subjected to fatigue, the laws of damage accumulation and the methods for counting the peaks of the response (used to establish a histogram when it is impossible to use the probability density of the peaks obtained with a Gaussian signal). The expressions of mean damage and its standard deviation are established. A few cases are then examined using other hypotheses (mean not equal to zero, taking account of the fatigue limit, non-linear accumulation law, etc.). The main laws governing low cycle fatigue and fracture mechanics are also presented.

Volume 5 is dedicated to presenting the method of *specification development* according to the principle of tailoring. The extreme response and fatigue damage spectra are defined for each type of stress (sinusoidal vibrations, swept sine, shocks, random vibrations, etc.). The process for establishing a specification as from the lifecycle profile of the equipment is then detailed taking into account the uncertainty factor (uncertainties related to the dispersion of the real environment and of the mechanical strength) and the test factor (function of the number of tests performed to demonstrate the resistance of the equipment).

First and foremost, this work is intended for engineers and technicians working in design teams responsible for sizing equipment, for project teams given the task of writing the various sizing and testing specifications (validation, qualification, certification, etc.) and for laboratories in charge of defining the tests and their performance following the choice of the most suitable simulation means.

Introduction

The vibratory environment found in the majority of vehicles essentially consists of random vibrations. Each recording of the same phenomenon results in a signal different from the previous ones. The characterization of a random environment therefore requires an infinite number of measurements to cover all the possibilities. Such vibrations can only be analyzed statistically.

The first stage consists of defining the properties of the processes comprising all the measurements, making it possible to reduce the study to the more realistic measurement of single or several short samples. This means evidencing the stationary character of the process, making it possible to demonstrate that its statistical properties are conserved in time, thus its ergodicity, with each recording representative of the entire process. As a result, only a small sample consisting of one recording has to be analyzed (Chapter 1).

The value of this sample gives an overall idea of the severity of the vibration, but the vibration has a continuous frequency spectrum that must be determined in order to understand its effects on a structure. This frequency analysis is performed using the power spectral density (PSD) (Chapter 2) which is the ideal tool for describing random vibrations. This spectrum, a basic element for many other treatments, has numerous applications, the first being the calculation of the rms (root mean square) value of the vibration in a given frequency band (Chapter 3).

The practical calculation of the PSD, completed on a small signal sample, provides only an estimate of its mean value, with a statistical error that must be evaluated. Chapter 4 shows how this error can be evaluated according to the analysis conditions and how it can be reduced, before providing rules for the determination of the PSD.

The majority of signals measured in the real environment have a Gaussian distribution of instantaneous values. The study of the properties of such a signal is extremely rich in content (Chapter 5). For example, knowledge of the PSD alone gives access, without having to count the peaks, to the distribution of the maxima of a random signal (Chapter 6), and to the law of distribution of the largest peaks, in itself useful information for the pre-sizing of a structure (Chapter 7).

It is also used to determine the response of a system with one degree-of-freedom (Chapters 8 and 9), which is necessary to calculate the fatigue damage caused by the vibration in question (Volume 4).

The study of the first crossing of a given response threshold for a one-degree-of-freedom system can also be useful in estimating the greatest stress value over a given duration. Different methods are presented (Chapter 10).

List of Symbols

The list below gives the most frequent definition of the main symbols used in this book. Some of the symbols can have another meaning which will be defined in the text to avoid any confusion.

a	Threshold value of $\ell(t)$ or maximum of $\ell(t)$	$G(\ )$	Power spectral density for $0 \leq f \leq \infty$
A	Maximum of $A(t)$	$\hat{G}(\ )$	Measured value of $G(\ )$
$A(t)$	Envelope of a signal	$G_{\ell u}(\ )$	Cross-power spectral density
b	Exponent	h	Interval (f/f_0) or f_2/f_1
c	Viscous damping constant	$h(t)$	Impulse response
$e(t)$	Narrow band white noise	$H(\ )$	Transfer function
$E(\ )$	Expectation of...	i	$\sqrt{-1}$
$E_1(\ )$	First definition of error function	k	Stiffness
$E_2(\ )$	Second definition of error function	K	Number of subsamples
		ℓ	Value of $\ell(t)$
Erf	Error function	$\bar{\ell}$	Mean value of $\ell(t)$
$E(\ )$	Expected function of ...	$\bar{\ell}_N$	Average maximum of N_p peaks
f	Frequency of excitation		
$f_{samp.}$	Sampling frequency	ℓ_{rms}	Rms value of $\ell(t)$
f_{max}	Maximum frequency	$\ddot{\ell}_{rms}$	Rms value of $\ddot{\ell}(t)$
f_0	Natural frequency	$\ell(t)$	Generalized excitation (displacement)
g	Acceleration due to gravity	$\dot{\ell}(\ )$	First derivative of $\ell(t)$
G	Particular value of power spectral density	$\ddot{\ell}(t)$	Second derivative of $\ell(t)$
		L	Given value of $\ell(t)$
		L_{rms}	Rms value of filtered signal

$L(\Omega)$	Fourier transform of $\ell(t)$	N_p^+	Average number of positive maxima for given length of time
$\dot{L}(\Omega)$	Fourier transform of $\dot{\ell}(t)$		
m	Mean	$p(\ )$	Probability density
M	Number of points of PSD	$p_N(\ )$	Probability density of largest maximum over given duration
M_a	Average number of maxima which exceeds threshold per unit time		
M_n	Moment of order n	P	Probability
n	Order of moment or number of degrees of freedom	PSD	Power spectral density
		q	$\sqrt{1-r^2}$
		q_E	$\dot{R}_{rms}/\dot{u}_{rms}$
n_a	Average number of crossings of threshold a per unit time	q_{max}^+	Probability that a maximum is positive
n_a^+	Average number of crossings of threshold **a** with positive slope per unit time	q_{max}^-	Probability that a maximum is negative
		$q(\ )$	Probability density of maxima of $\ell(t)$
n_0	Average number of zero-crossings per unit time	$q(\theta)$	Reduced response
		$\dot{q}(\theta)$	First derivative of $q(\theta)$
n_0^+	Average number of zero-crossings with positive slope per second (average frequency)	$\ddot{q}(\theta)$	Second derivative of $q(\theta)$
		Q	Q factor (quality factor)
		$Q(\ )$	Distribution function of maxima of $\ell(t)$
n_p^+	Average number of maxima per unit time	$Q(u)$	Probability that a maximum is higher than a given threshold
N	Number of curves or Number of points of signal or Numbers of dB		
		r	Irregularity factor
		rms	Root mean square (value)
N_p	Number of peaks	$r(t)$	Temporal window
N_a^+	Average number of crossings of threshold **a** with positive slope for given length of time	R	Slope in dB/octave or Ratio of the number of minima to the number of maxima
		$R_E(\ )$	Auto-correlation function of envelope
N_0^+	Average number of zero-crossings with positive slope for given length of time	$R_{\ell u}$	Cross-correlation function between $\ell(t)$ and $u(t)$
		$R(f)$	Fourier transform of $r(t)$
		$R(t)$	Envelope of maxima of $u(t)$

Symbol	Description
$\dot{R}(t)$	First derivative of R(t)
$R(\tau)$	Auto-correlation function
s	Standard deviation
S_0	Value of constant PSD
$S(\ )$	Power spectral density for $-\infty \leq f \leq +\infty$
t	Time
T	Duration of sample of signal or duration of vibration
T_a	Average time between two successive maxima
u	Ratio of threshold a to rms value ℓ_{rms} of $\ell(t)$
u_0	Initial value of $u(t)$
$\dot{u}_0$	Initial value of $\dot{u}(t)$
$\bar{u}_0$	Average of highest peaks
u_{rms}	Rms value of $u(t)$
$\dot{u}_{rms}$	Rms value of $\dot{u}(t)$
$\ddot{u}_{rms}$	Rms value of $\ddot{u}(t)$
$u(t)$	Generalized response
$\dot{u}(t)$	First derivative of $u(t)$
$\ddot{u}(t)$	Second derivative of $u(t)$
v	Ratio a/u_{rms}
v_{rms}	Rms value of $\dot{x}(t)$
x_{rms}	Rms value of $x(t)$
$\ddot{x}(t)$	Absolute acceleration of base of one-degree-of-freedom system
$\ddot{x}_{rms}$	Rms value of $\ddot{x}(t)$
$\ddot{x}_m$	Maximum value of $\ddot{x}(t)$
y_{rms}	Rms value of $y(t)$
$\dot{y}_{rms}$	Rms value of $\dot{y}(t)$
$\ddot{y}_{rms}$	Rms value of $\ddot{y}(t)$
z_{rms}	Rms value of $z(t)$
$\dot{z}_{rms}$	Rms value of $\dot{z}(t)$
$\ddot{z}_{rms}$	Rms value of $\ddot{z}(t)$
α	Time-constant of the probability density of the first passage of a threshold or Risk of up-crossing or $2\sqrt{1-\xi^2}$
β	$2\left(1-2\xi^2\right)$
χ_n^2	Variable of chi-square with n degrees of freedom
δt	Time step
$\delta(\ )$	Dirac delta function
$\Delta\tau$	Effective time interval
Δf	Frequency interval between half-power points or frequency step of the PSD
ΔF	Bandwidth of analysis filter
$\Delta\ell$	Interval of amplitude of $\ell(t)$
Δt	Time interval
ε	Statistical error or Euler's constant (0.57721566490…)
$\gamma_{\ell u}$	Coherence function between $\ell(t)$ and $u(t)$
φ	Phase
$\Phi(t)$	Gauss complementary distribution function
$\lambda(\ )$	Reduced excitation
π	3.14159265…
θ	Reduced time ($\omega_0\, t$)
μ_n	Central moment of order n
μ'_n	Reduced central moment of order n
π	3.14159265 …
ρ	Correlation coefficient
τ	Delay

τ_m Average time between two successive maxima

ω_0 Natural pulsation ($2\pi f_0$)

Ω Pulsation of excitation ($2\pi f$)

ξ Damping factor

ψ Phase

Chapter 1

Statistical Properties of a Random Process

1.1. Definitions

1.1.1. *Random variable*

A random variable is a quantity whose instantaneous value cannot be predicted. Knowledge of the values of the variable before time t does not make it possible to deduce the value at the time t from it.

Example: the Brownian movement of a particle.

If a vibration was perfectly random, its analysis would be impossible. The points that define the signal would have an amplitude that varied in a completely unpredictable way. Thankfully, in practice, it is possible to associate with all the points that characterize the signal a probability law which will enable a statistical analysis [AND 11].

The principal characteristic of a random vibration is to simultaneously excite all the frequencies of a structure [TUS 67]. In contrast to sinusoidal functions, random vibrations are made up of a continuous range of frequencies, the amplitude of the signal and its phase varying with respect to time in a random fashion [TIP 77] [TUS 79]. Thus, the random vibrations are also called *noise*.

Random functions are sometimes defined as a continuous distribution of sinusoids of all frequencies whose amplitudes and phases vary randomly with time [CUR 64], [CUR 88].

2   Random Vibration

1.1.2. *Random process*

Let us consider, as an example, the acceleration recorded at a given point on the dial of a truck traveling on a good road between two cities A and B. For a journey, the recorded acceleration obeys the definition of a random variable. The vibration characterized by this acceleration is said to be *random* or *stochastic*.

Complexity of the analysis

Even in the most simple hypothesis where a vehicle runs at a constant speed on a straight road in the same state, each vibration measure $^i\ell(t)$ at one point of the vehicle is different from the other. An infinity of measures to completely characterize the trip should be completed *a priori*.

We define as a *random process* or *stochastic process* the ensemble of the time functions $\{^i\ell(t)\}$ for t included between $-\infty$ and $+\infty$, this ensemble being able to be defined by statistical properties [JAM 47].

By their very nature, the study of vibrations would be intensive if we did not have the tools to limit the complete process analysis, made up of a large number of signals according to time, with a very long duration, to that of a very restricted number of samples of reasonable duration. Fortunately, random movements are not erratic in the common sense, but instead follow well-defined statistical laws. The study of statistical process properties, with averages in particular, will enable the simplification of the analysis from two very useful notions for this objective: stationarity and ergodicity.

1.2. Random vibration in real environments

By its nature, the real vibratory environment is random [BEN 61a]. These vibrations are encountered:

– on road vehicles (irregularities of the roads);

– on aircraft (noise of the engines, aerodynamic turbulent flow around the wings and fuselage, creating non-stationary pressures, etc.) [PRE 56a];

– on ships (engine, swell, etc.);

– on missiles. The majority of vibrations encountered by military equipment, and in particular by the internal components of guided missiles, are random with respect to time and have a continuous spectrum [MOR 55]: the gas jet emitted with a large velocity creates important turbulences resulting in acoustic noise which attacks the skin of the missile until its velocity exceeds Mach 1 approximately (or until it leaves the Earth's atmosphere) [ELD 61], [RUB 64], [TUS 79];

– in mechanical assemblies (ball bearings, gears, etc.).

1.3. Random vibration in laboratory tests

Tests using random vibrations first appeared around 1955 as a result of the inability of sine tests to correctly excite equipment exhibiting several resonances [DUB 59], [TUS 73]. The tendency in standards is thus to replace the old swept sine tests which excite resonances one after the other by a random vibration whose effects are nearer to those of the real environment.

Random vibration tests are also used in a much more marginal way:

– to identify the structures (research of the resonance frequencies and measurement of Q factors), their advantage being that of shorter test duration;

– to simulate the effects of shocks containing high frequencies and which are difficult to replace by shocks of simple form.

1.4. Methods of random vibration analysis

The first mathematical analysis of random vibrations was carried out by A. Einstein [EIN 05], [PAE 11] for the study of Brownian motion of particles in a liquid medium. The theoretical works of S.O. Rice [RIC 39], [RIC 44] on the mathematical analysis of random noise and the distribution of the peaks of random signals quickly became the basis for different studies such as the effect of turbulence on airplanes [DIE 56], [FUN 53], [FUN 55], action of swell on marine structures [CAR 56], [LON 52] or that of wind on civil engineering structures [DAV 64], then for the calculation of the response of structures to the noise of propulsion engines in airplanes and missile thrusters [CRA 58], [CRA 63].

Taking into account their randomness and their frequency contents, these vibrations can be studied only using statistical methods applied to the signals with respect to time or using curves plotted in the frequency domain (spectra).

4   Random Vibration

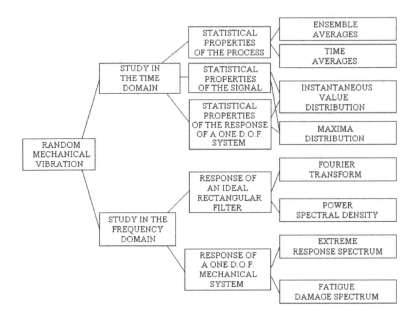

Figure 1.1. *Analysis possibilities for random vibration*

We can schematically distinguish four ways of approaching the analysis of random vibrations [CUR 64], [RAP 69]:

– analysis of the ensemble statistical properties of the process;

– methods of correlation;

– spectral analysis;

– analysis of statistical properties of the signal with respect to time.

The block diagram (Figure 1.1) summarizes the main possibilities which will be considered in turn in what follows.

The parameters most frequently used in practice are:

– the rms (root mean square) value of the signal and, if it is the case, its variation as a function of time;

– the distribution of instantaneous accelerations of the signal with respect to time;

– the PSD.

1.5. Distribution of instantaneous values

1.5.1. *Probability density*

One of the objectives of the analysis of a random process is to determine the probability of finding extreme or peak values, or of determining the percentage of time that a random variable (acceleration, displacement, etc.) exceeds a given value [RUD 75]. Figure 1.2 shows a sample of a random signal with respect to time defined over duration T.

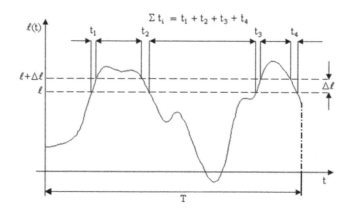

Figure 1.2. *Sample of random signal*

The probability that this function $\ell(t)$ is in the interval ℓ ($\ell + \Delta\ell$ being equal to the percentage of time during which it has values in this interval) is expressed mathematically as:

$$\text{prob}[\ell < \ell(t) < \ell + \Delta\ell] = \sum_i \frac{t_i}{T} \qquad [1.1]$$

If this interval $\Delta\ell$ is small, a density function probability $p(\ell)$ is defined by:

$$\text{prob}[\ell < \ell(t) < \ell + \Delta\ell] = p(\ell)\Delta\ell \qquad [1.2]$$

where:

$$p(\ell) = \frac{1}{T} \frac{\sum t_i}{\Delta \ell} \qquad [1.3]$$

To precisely define $p(\ell)$, it is necessary to consider very small intervals $\Delta \ell$ and of very long duration T, so that, mathematically, the probability density function is defined by:

$$p(\ell) = \lim_{\Delta \ell \to 0} \left[\lim_{T \to \infty} \left(\frac{1}{T} \frac{\sum t_i}{\Delta \ell} \right) \right] \qquad [1.4]$$

1.5.2. *Distribution function*

Owing to the fact that $p(\ell)$ was given for the field of values of $\ell(t)$, the probability that the signal is inside the limits $a < \ell(t) < b$ is obtained by integration from [1.2]:

$$\text{prob}[a < \ell(t) < b] = \int_a^b p(\ell)\, d\ell \qquad [1.5]$$

Since the probability that $\ell(t)$ within the limits $-\infty, +\infty$ is equal to 1 (absolutely certain event), it follows that

$$\int_{-\infty}^{+\infty} p(\ell)\, d\ell = 1 \qquad [1.6]$$

and the probability that ℓ exceeds a given level L is simply

$$\text{prob}[L \leq \ell(t)] = 1 - \int_L^\infty p(\ell)\, d\ell \qquad [1.7]$$

There are electronic equipment and calculation programs that make it possible to determine either *the distribution function* or *the probability density function* of the instantaneous values of a real random signal $\ell(t)$.

Among the mathematical laws representing the usual probability densities, we distinguish two that are particularly important in the field of random vibrations: Gauss's law (or Normal law) and Rayleigh's law.

1.6. Gaussian random process

A *Gaussian random process* $\ell(t)$ is such that the ensemble of the instantaneous values of $\ell(t)$ obeys a law of the form:

$$p[\ell(t)] = \frac{1}{s\sqrt{2\pi}} \exp\left\{-\frac{[\ell(t) - m]^2}{2s^2}\right\} \quad [1.8]$$

where m and s are constants. The utility of the Gaussian law lies in the central limit theorem, which establishes that the sum of independent random variables follows a roughly Gaussian distribution whatever the basic distribution.

This is the case for many physical phenomena, for quantities which result from a large number of independent and comparable fluctuating sources, and in particular the majority of vibratory random signals encountered in the real environment [BAN 78], [CRE 56], [PRE 56a].

A Gaussian process is fully determined by knowledge of the mean value m (generally zero in the case of vibratory phenomena) and of the standard deviation s.

Moreover, it is shown that:

– if the excitation is a Gaussian process, the response of a linear time-invariant system is also a Gaussian process [CRA 83], [DER 80];

– the vibration, in part excited at resonance, tends to be Gaussian.

For a strongly resonant system subjected to broadband excitation, the central limit theorem makes it possible to establish that the response tends to be Gaussian even if the input is not. This applies when the excitation is not a white noise, provided that it is a broadband process covering the resonance peak [NEW 75] (provided that the probability density of the instantaneous values of the excitation does not have too significant an asymmetry [MAZ 54] and that the structure is not very strongly damped [BAN 78], [MOR 55]).

In many practical cases, we are thus led to conclude that the vibration is stationary and Gaussian, which simplifies the problem of calculation of the response of a mechanical system (Chapter 9) and, consequently, the simplification of fatigue damage calculations, the possibility of simply evaluating the probability of exceeding a given value, etc.

The reduced variable $t = \dfrac{\ell(t) - m}{s}$ is sometimes used. The distribution function makes it possible to calculate the probability for the amplitude to be lower than a given value. For a Gaussian distribution, it is equal to:

$$F(L) = P(\ell < L) = \dfrac{1}{s\sqrt{2\pi}} \int_{-\infty}^{L} e^{-\frac{1}{2}\left(\frac{\ell - m}{s}\right)^2} d\ell \qquad [1.9]$$

It can also be written as:

$$F(T) = \dfrac{1}{2}\left[1 + E_1\left(\dfrac{T}{\sqrt{2}}\right)\right] \qquad [1.10]$$

if we say $T = \dfrac{L - m}{s}$ and if E_1 is the error function defined by:

$$E_1(x) = \dfrac{2}{\sqrt{\pi}} \int_0^L e^{-t^2} dt \qquad [1.11]$$

The interest of the relation between the distribution function and error function lies in the calculation possibility from a development in series, thus enabling us to avoid making an integration (Appendix, section A4.1).

Figure 1.3 provides an image of the way to estimate the probability density and distribution function of a signal $\ell(t)$. Signal amplitudes are divided into small $\Delta\ell$ intervals. In each one, we count the number of times the signal is located in each one of these intervals for all its duration. This number is transferred to a curve based on the average amplitude of interval $\Delta\ell$. The resulting diagram is no more than the histogram of instantaneous values of signal. When the amplitude interval becomes very small, this histogram tends toward the probability density of instantaneous values, which is generally close to a Gaussian distribution.

Statistical Properties of a Random Process 9

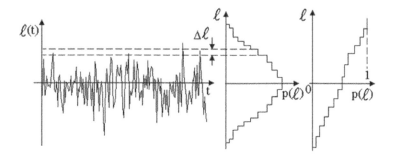

Figure 1.3. *Distribution of the instantaneous values of the signal*

Example 1.1.

Let us take a time history vibratory signal having a constant PSD equal to 1 $(m/s^2)^2/Hz$ between 0 and 2,000 Hz (Figure 1.4.).

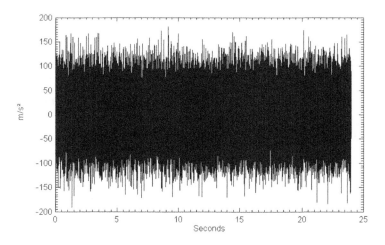

Figure 1.4. *Random vibration signal according to time (96,000 points, rms value: 44.72 m/s^2)*

The histogram of the instantaneaous values, representing the number of values counted in a given amplitude class, is close to a Gaussian distribution, the difference being due to statistic dispersion; the result is all the closer as the duration of the signal sample analyzed is greater (24 seconds in this case). We verify that the skewness (0.005) (see section 1.9.5) and kurtosis (3.0) (see section 1.9.6) are respectively close to zero and three (Figure 1.5).

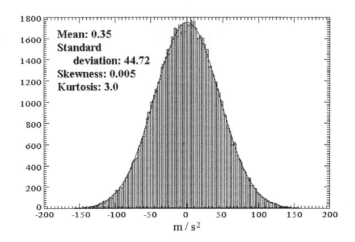

Figure 1.5. *Histogram of instantaneous values of signal of Figure 1.4 compared to the density of probability of a Gaussian distribution with the same mean and standard deviation*

Most vibrations (stationary phase) measured in the real environment are Gaussian.

NOTE.– *A signal with a varying rms value over time (and thus non-stationary) cannot be Gaussian. The standard deviation of instantaneous values (or rms value) is then seen as a constant in the definition of probability density of a Gaussian distribution.*

Example 1.2.– Signal presenting a transitory phase between two stationary phases

Consider a random vibration made up of three phases with the same duration (8 s), one stationary with an rms value equal to 44.72 m/s2, followed by a transitory phase during which the rms value decreases, in a linear way, until it reaches a second stationary phase with an rms value of 11.18 m/s2 (Figure 1.6).

Statistical Properties of a Random Process 11

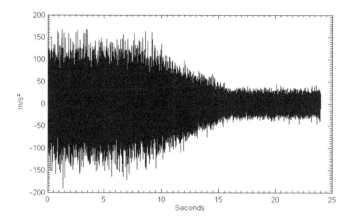

Figure 1.6. *Signal presenting a transitory phase between two stationary phases*

The instantaneous values histogram of this signal is traced in Figure 1.7, superposed to the probability density of a Gaussian distribution where the standard deviation is equal to the rms value of the complete signal (zero mean). This density is far from the histogram.

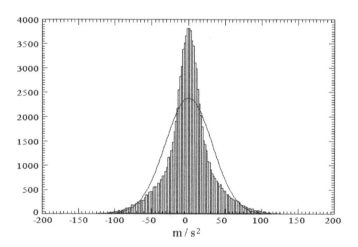

Figure 1.7. *Histogram of instantaneous values of signal of Figure 1.6 and Gaussian probability density calculated from the global rms value of this same signal (31.62 m/s²)*

1.7. Rayleigh distribution

Rayleigh distribution, of which the probability has the form

$$p(\ell) = \frac{\ell}{s^2} e^{-\frac{\ell^2}{2s^2}} \qquad [1.12]$$

$(\ell \geq 0)$ is also an important law in the field of vibration for the representation of:

– variations in the instantaneous value of the envelope of a narrowband Gaussian random process;

– peak distribution in a narrowband Gaussian process.

Because of its very nature, the study of vibration would be very difficult if we did not have tools enabling the limitation of analysis of the complete process, which comprises a great number of signals varying with time and of very great duration, using a very restricted number of samples of reasonable duration. The study of statistical properties of the process will make it possible to define two very useful concepts with this objective in mind: stationarity and ergodicity.

1.8. Ensemble averages: through the process

1.8.1. *n order average*

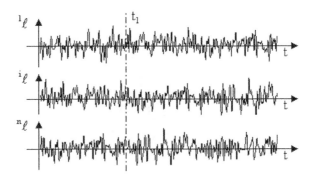

Figure 1.8. *"Through the process" study*

Let us consider N recordings of a random phenomenon varying with time $^i\ell(t)$ $[i \in (1, N)]$ for t varying from 0 to T (Figure 1.8). The ensemble of the curves $^i\ell(t)$

constitutes the process $\{^i\ell(t)\}$. A first possibility may consist of studying the distribution of the values of ℓ for $t = t_1$ given [JAM 47].

If we have (N) records of the phenomenon, we can calculate, for a given t_1, the mean [BEN 62], [BEN 63], [DAV 58], [JEN 68]:

$$\overline{\ell(t)} = \frac{^1\ell(t_1) + {^2\ell(t_1)} + \cdots + {^n\ell(t_1)}}{N} \qquad [1.13]$$

If the values $^i\ell(t)$ belong to an infinite discrete ensemble, the *moment of order n* is defined by:

$$E[\ell^n(t_1)] = \lim_{N \to \infty} \sum_{i=1}^{N} \frac{^i\ell^n(t_1)}{N} \qquad [1.14]$$

(E[ ] = mathematical expectation). By considering the ensemble of the samples at the moment t_1, the statistical nature of $\ell(t_1)$ can be specified by its probability density [LEL 73]:

$$p[\ell(t_1)] = \lim_{\Delta\ell \to 0} \frac{\text{Prob}[\ell \leq \ell(t_1) \leq \ell + \Delta\ell]}{\Delta\ell} \qquad [1.15]$$

and by the moments of the distribution:

$$E[\ell^n(t_1)] = \int_{-\infty}^{\infty} \ell^n(t_1)\, p[\ell(t_1)]\, d\ell(t_1) \qquad [1.16]$$

if the density $p[\ell(t_1)]$ exists and is continuous (or the distribution function). The moment of order 1 is the *mean* or *expected value*; the moment of order 2 is the *quadratic mean*.

For two random variables

The joint probability density is written:

$$p(\ell_1, t_1; \ell_2, t_2) = \lim_{\substack{\Delta\ell_1 \to 0 \\ \Delta\ell_2 \to 0}} \frac{\text{Prob}[\ell_1 \leq \ell(t_1) \leq \ell_1 + \Delta\ell_1;\ \ell_2 \leq \ell(t_2) \leq \ell_2 + \Delta\ell_2]}{\Delta\ell_1\, \Delta\ell_2}$$

$$[1.17]$$

14 Random Vibration

and joint moments:

$$E[\ell(t_1)\,\ell(t_2)] = \int_{-\infty}^{\infty}\int_{-\infty}^{\infty} {}^i\ell(t_1)\,{}^j\ell(t_2)\,p[\ell(t_1),\ell(t_2)]\,d\ell(t_1)\,d\ell(t_2) \qquad [1.18]$$

1.8.2. Centered moments

The *central moment of order n* (with regard to the mean) is the quantity:

$$E\{[\ell(t_1) - m]^n\} = \lim_{N \to \infty} \frac{1}{N}\sum_{i=1}^{N}[{}^i\ell(t_1) - m]^n \qquad [1.19]$$

in the case of a discrete ensemble and, for $p(\ell)$ continuous:

$$E\{[\ell(t_1) - m]^n\} = \int_{-\infty}^{\infty}[\ell(t_1) - m]^n\,p[\ell(t_1)]\,d\ell(t_1) \qquad [1.20]$$

1.8.3. Variance

The *variance* is the centered moment of order 2

$$s^2_{\ell(t_1)} = E\{[\ell(t_1) - m]^2\} \qquad [1.21]$$

By definition:

$$s^2_{\ell(t_1)} = \int_{-\infty}^{\infty}[\ell(t_1) - m]^2\,p[\ell(t_1)]\,d\ell(t_1) \qquad [1.22]$$

$$s^2_{\ell(t_1)} = \int_{-\infty}^{\infty}\ell^2(t_1)\,p[\ell(t_1)]\,d\ell(t_1)$$

$$-2m\underbrace{\int_{-\infty}^{\infty}\ell(t_1)\,p[\ell(t_1)]\,d\ell(t_1)}_{m} + m^2\underbrace{\int_{-\infty}^{\infty}p[\ell(t_1)]\,d\ell(t_1)}_{1}$$

$$s^2_{\ell(t_1)} = E\{[\ell(t_1)]^2\} - 2m^2 + m^2$$

$$s^2_{\ell(t_1)} = E\left\{\left[\ell(t_1)\right]^2\right\} - m^2 \qquad [1.23]$$

1.8.4. *Standard deviation*

The quantity $s_{\ell(t_1)}$ is called the *standard deviation*. If the mean is zero,

$$s^2_{\ell(t_1)} = E\left\{\left[\ell(t_1)\right]^2\right\} \qquad [1.24]$$

When the mean m is known, an absolutely unbiased estimator of s^2 is $\sum \dfrac{\left(^i\ell - m\right)^2}{N}$. When m is unknown, the estimator of s^2 is $\sum \dfrac{\left(^i\ell - m'\right)^2}{N-1}$ where $m' = \dfrac{1}{N}\sum {}^i\ell$.

Example 1.3.

Let us consider 5 samples of a random vibration $\ell(t)$ and the values of ℓ at a given time $t = t_1$ (Figure 1.9).

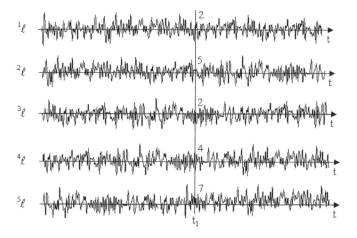

Figure 1.9. *Example of a stochastic process*

If the exact mean m is known (m = 4.2 m/s² for example), the variance is estimated from:

$$s^2 = \frac{(2-4.2)^2 + (5-4.2)^2 + (2-4.2)^2 + (4-4.2)^2 + (7-4.2)^2}{5} \ (m/s^2)^2$$

$$s^2 = \frac{18.2}{5} = 3.64 \ (m/s^2)^2$$

If the mean m is unknown, it can be evaluated from

$$m' = \frac{1}{N} \sum_i {}^i \ell(t_1) = \frac{2+5+2+4+7}{5} = \frac{20}{5} = 4 \ m/s^2$$

$$s^2 = \frac{18}{4} = 4.50 \ (m/s^2)^2$$

1.8.5. *Autocorrelation function*

Given a random process ${}^i \ell(t)$, the *autocorrelation function* is the function defined, in the discrete case, by:

$$R(t_1, t_1 + \tau) = \lim_{N \to \infty} \frac{1}{N} \sum_i {}^i \ell(t_1) \ {}^i \ell(t_1 + \tau) \qquad [1.25]$$

$$R(t_1, t_1 + \tau) = E\left[x(t_1) \cdot x(t_1 + \tau)\right] \qquad [1.26]$$

or, for a continuous process, by:

$$R(\tau) = \int_{-\infty}^{\infty} x(t_1) \, x(t_1 + \tau) \, p[x(t_1)] \, dx(t_1) \qquad [1.27]$$

1.8.6. *Cross-correlation function*

Given the two processes $\{\ell(t)\}$ and $\{u(t)\}$ (for example, the excitation and the response of a mechanical system), the *cross-correlation function* is the function:

$$R_{\ell u}(t_1, t_1 + \tau) = E[\ell(t_1) \cdot u(t_1 + \tau)] \quad [1.28]$$

or

$$R(\tau) = \lim_{N \to \infty} \frac{1}{N} \sum_i {}^i\ell(t_1) \cdot {}^i u(t_1 + \tau) \quad [1.29]$$

The *correlation* is a number measuring the degree of resemblance or similarity between two functions of the same parameter (time generally) [BOD 72].

1.8.7. *Autocovariance*

Autocovariance is the quantity:

$$C(t_1, t_1 + \tau) = E\left\{\left[\ell(t_1) - \overline{\ell(t_1)}\right]\left[\ell(t_1 + \tau) - \overline{\ell(t_1 + \tau)}\right]\right\} \quad [1.30]$$

$$C(t_1, t_1 + \tau) = R(t_1, t_1 + \tau) - \overline{\ell(t_1)}\,\overline{\ell(t_1 + \tau)} \quad [1.31]$$

$C(t_1, t_1 + \tau) = R(t_1, t_1 + \tau)$ if the mean values are zero.

We have in addition:

$$R(t_1, t_2) = R(t_2, t_1) \quad [1.32]$$

1.8.8. *Covariance*

We define *covariance* as the quantity:

$$C_{\ell u} = E\left\{\left[\ell(t_1) - \overline{\ell(t_1)}\right]\left[u(t_1 + \tau) - \overline{u(t_1 + \tau)}\right]\right\} \quad [1.33]$$

1.8.9. *Stationarity*

A phenomenon is *strictly stationary* if every moment of all orders and all the correlations are invariable with time t_1 [CRA 67], [JAM 47], [MIX 69], [PRE 90] [RAP 69], [STE 67].

The phenomenon is *wide-sense (or weakly) stationary* if only the mean, the mean square value and the autocorrelation are independent of time t_1 [BEN 58], [BEN 61b], [SVE 80].

18   Random Vibration

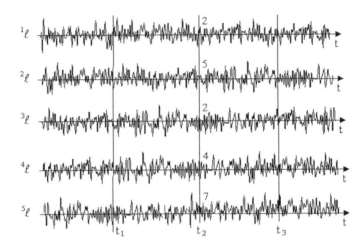

Figure 1.10. *Calculation of ensemble averages at different times*

Interest of stationarity

Since statistical process properties do not evolve over time in a stationary process, it is not necessary to record the signals for a long period of time. This time, however, must be long enough to subsequently enable a significant frequency analysis. In Chapter 4, we will see which rule must be respected.

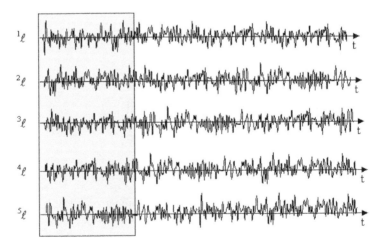

Figure 1.11. *Calculation of ensemble averages on a short duration if the process is stationary*

Example 1.4.

The following cases are completely unrealistic and are only illustrated for learning purposes.

CASE 1.–*The process is described by 5 measurements of a constant value signal equal to 1.*

Since all averages are equal regardless of moment t used for calculation, the process represented in Figure 1.12 is stationary.

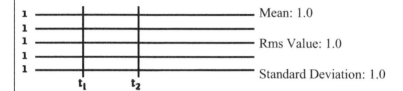

Figure 1.12. *Example of a stationary process*

CASE 2.– *The process is described by 5 measurements of signals with different constant values*

Means are independent of moment t chosen: the process is stationary (Figure 1.13).

Figure 1.13. *Example of a stationary process*

CASE 3.– *The process is described by 4 signals with a constant value equal to 1 and by a sine curve*

The means in the process in Figure 1.14 vary according to the moment of calculation: the process is not stationary.

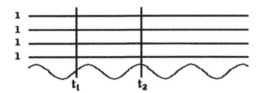

Figure 1.14. *Example of a non-stationary process*

Autostationarity

If only one recording of the phenomenon $\ell(t)$ is available, we sometimes define the autostationarity of the signal by studying the stationarity with n samples taken at various moments of the recording, by regarding them as samples obtained independently during n measurements (Figure 1.15).

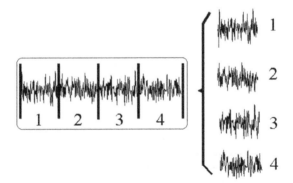

Figure 1.15. *Study of autostationarity*

We can also define strong autostationarity and weak autostationarity.

For a stationary process, the autocorrelation function is written:

$$R(\tau) = E[\ell(0)\,\ell(\tau)]$$

$$R(\tau) = \lim_{N \to \infty} \frac{1}{N} \sum_{i=1}^{N} {}^i \ell(0) \, {}^i \ell(\tau)$$ [1.34]

NOTES.–

Based on this assumption, we have:

$$R(-\tau) = E\{\ell(0)\ell(-\tau)\}$$

$$R(-\tau) = E\{\ell(\tau)\ell(0)\}$$

$$R(-\tau) = R(\tau)$$ [1.35]

(R is an even function of τ) [PRE 90].

$$R(0) = E\{\ell(0)\ell(0)\} = E\{\ell^2(t)\}$$ [1.36]

$R(0)$ *is the ensemble mean square value at the arbitrary time t.*

– $R(0) \geq |R(\tau)|$

We have

$$E\left\{\left[\ell(0) \pm \ell(\tau)\right]^2\right\} \geq 0$$

yielding

$$E\{\ell^2(0)\} \pm 2 E\{\ell(0)\ell(\tau)\} + E\{\ell^2(\tau)\} \geq 0$$

$$R(0) \pm 2 R(\tau) + R(0) \geq 0$$

and

$$R(0) \geq |R(\tau)|$$ [1.37]

As for the cross-correlation function, it becomes, for a stationary process,

$$R_{\ell u}(\tau) = E\{\ell(0)\,\ell(\tau)\} = \lim_{N\to\infty} \frac{\sum_{i=1}^{N} {}^i\ell(0)\,{}^i\ell(\tau)}{N} \qquad [1.38]$$

Properties

1. $\quad R_{\ell u}(-\tau) = R_{u\ell}(\tau)$ [1.39]

Indeed

$$R_{\ell u}(-\tau) = E\{\ell(0)\,u(-\tau)\}$$

$$R_{\ell u}(-\tau) = E\{\ell(\tau)\,u(0)\}$$

$$R_{\ell u}(-\tau) = E\{u(0)\,\ell(\tau)\}$$

$$R_{\ell u}(-\tau) = R_{u\ell}(\tau)$$

2. Whatever τ

$$R_{\ell u}(\tau) \le \sqrt{R_\ell(0)\,R_u(0)} \qquad [1.40]$$

Cyclostationary process

A random process $\{x(t)\}$ is *cyclostationary* if its random properties repeat periodically over the course of time. At order 2, this means that the correlation

$$R(\tau, t) = E\big[\ell(t)\,\ell(t-\tau)\big]$$

is periodic in t: $R(\tau, t) = R(\tau, t + T)$.

An important example is that of a signal modulated by a linear modulation. If T is the period, the average power of the process is defined by

$$P_{moy} = \lim_{T\to\infty} \frac{1}{2T} E\left[\int_{-T}^{T} |x(t)|^2\,dt\right] \qquad [1.41]$$

As the process is cyclostationary, $E\left[|x(t)|^2\right] = R(0,t)$ is periodic of period T, and thus

$$P_{moy} = \frac{1}{T}\int_{(T)} E\left[|x(t)|^2\right] dt \qquad [1.42]$$

where (T) designates any interval of period T. The average power is thus obtained by averaging $E\left[|x(t)|^2\right]$ over a period.

1.9. Temporal averages: along the process

1.9.1. *Mean*

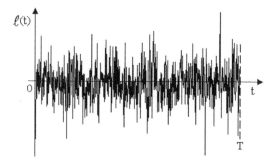

Figure 1.16. *Sample of a random signal*

Let us consider a sample $\ell(t)$ of duration T of a recording. It can be interesting to study the statistical properties of the instantaneous values of the function $\ell(t)$. The first possibility is to consider the temporal mean of the instantaneous values of the recording.

We have:

$$\overline{\ell(t)} = \lim_{T \to \infty} \frac{1}{2T} \int_{-T}^{T} \ell(t) dt \qquad [1.43]$$

24 Random Vibration

if this limit exists. This limit may very well not exist for some or all the samples and, if it exists, it may depend on the selected sample $\ell(t)$; however, it does not depend on time[1].

For practical reasons, we in fact calculate the mean value of the signal $\ell(t)$ over one finite duration T:

$$\overline{\ell(t)} = \frac{1}{T}\int_0^T \ell(t)\,dt \qquad [1.44]$$

The mean value is related to the difference between the positive and negative areas ranging between the curve $\ell(t)$ and the time axis [GRE 81].

The mean m of a centered signal is zero, so this parameter cannot be used by itself to correctly evaluate the severity of the excitation.

Figure 1.17. *Random vibration with non-zero mean*

The mean value is equal to the absolute value of the parallel shift of the Ot axis necessary to cancel out this difference. A signal $\ell(t)$ of mean m can be written:

$$\ell(t) = m + \ell^*(t) \qquad [1.45]$$

where $\ell^*(t)$ is a centered signal. This mean value is generally a static component which can be due to the weight of the structure, to the maneuverings of an aircraft,

---

1. We also define $\overline{x(t)}$ from:

$$\lim_{T\to\infty} \frac{1}{T}\int_0^T x(t)\,dt \quad \text{or} \quad \lim_{T\to\infty} \frac{1}{T}\int_{-T/2}^{T/2} x(t)\,dt$$

Statistical Properties of a Random Process 25

to the thrust of a missile in phase propulsion, etc. In practice, we often consider this mean to be zero.

1.9.2. *Quadratic mean – rms value*

The vibration $\ell(t)$ generally results in an oscillation of the mechanical system around its equilibrium position, so that the arithmetic mean of the instantaneous values can be zero if the positive and negative values are compensated. The arithmetic mean represents the signal poorly [RAP 69], [STE 67]. Therefore, it is sometimes preferred to calculate the mean value of the absolute value of the signal

$$\overline{|\ell(t)|} = \frac{1}{T}\int_0^T |\ell(t)|\, dt \qquad [1.46]$$

and, much more generally, by analogy with the measurement of the rms value of an electrical quantity, the *quadratic mean* (or *mean square value*) of the instantaneous values of the signal of which the square root is the *rms value*.

The rms value $\ell_{rms} = \sqrt{\overline{\ell^2(t)}}$ is the simplest statistical characteristic to obtain. It is also most significant since it provides an order of magnitude of the intensity of the random variable.

If we can analyze the curve $\ell(t)$ by dividing the sample of duration T into N intervals of duration Δt_i $(i \in [1, N])$, and if ℓ_i is the value of the variable during the time interval Δt_i, the mean quadratic value is written:

$$\overline{\ell^2} = \frac{\ell_1^2\, \Delta t_1 + \cdots + \ell_i^2\, \Delta t_i + \cdots + \ell_N^2\, \Delta t_N}{T} \qquad [1.47]$$

with $T = \sum_{i=1}^{N} \Delta t_i$. If the intervals of time are equal to (Δt) and if N is the number of points characterizing the signal, $T = N\, \Delta t$ and:

$$\ell_{rms} = \frac{1}{N}\sum_i \ell_i^2 \qquad [1.48]$$

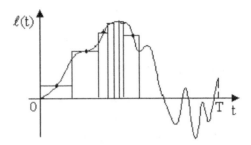

Figure 1.18. *Approximation of the signal*

If all Δt_i tend towards zero and if $N \to \infty$, the quadratic mean is defined by [BEN 63]:

$$\overline{\ell^2(t)} = \frac{1}{T}\int_0^T \ell^2(t)\,dt \qquad [1.49]$$

(or by $\dfrac{1}{2T}\int_{-T}^{T}\ell^2(t)\,dt$).

Two signals having very different frequency contents, corresponding to very dissimilar temporal forms, can have the same mean quadratic value. In this expression, the rms value takes into account the totality of the frequencies of the signal.

Example 1.5.

1. Consider a signal is defined by 11 points (Figure 1.19).

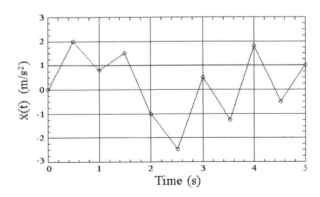

Figure 1.19. *Signal of acceleration*

The use of relation [1.48] leads to:

$$\ddot{x}_{rms} = \sqrt{\frac{0^2 + 2^2 + 0.8^2 + 1.5^2 + (-1)^2 + (-2.5)^2 + 0.5^2 + (-1.25)^2 + 1.8^2 + (-0.5)^2 + 1^2}{11}}$$

or: $\ddot{x}_{rms} \approx 1.36$.

2. Let us consider a sinusoid $\ddot{x}(t) = \ddot{x}_m \sin(\Omega t + \varphi)$

$$\overline{\ddot{x}(t)} = 0$$

$$s^2 = \overline{\ddot{x}^2(t)} = \frac{\ddot{x}_m^2}{2}$$

$$\overline{|\ddot{x}(t)|} = \frac{2}{\pi} \ddot{x}_m \left(= \frac{2}{\pi}\sqrt{2}\, s \approx 0.9\, s \right)$$

(for a normal distribution, $\overline{|\ddot{x}(t)|} \approx 0.798\, s$).

1.9.3. *Moments of order n*

As in the preceding section, we also define:

– moments of an order higher than 2; the *moment of order n* is expressed:

$$E\{\ell^n(t)\} = \overline{\ell^n(t)} = \lim_{T \to \infty} \frac{1}{2T} \int_{-T}^{T} \ell^n(t)\, dt \qquad [1.50]$$

– *centered moments*: measured vibratory signals usually have a zero mean value. We then call the signal *centered*. When that is not the case, the rms value is still calculated as in section 1.9.2. We can also look at centered moments of order n defined by:

$$\mu_n = E\left\{[\ell(t) - \overline{\ell(t)}]^n\right\} = \lim_{T \to \infty} \frac{1}{2T} \int_{-T}^{T} [\ell(t) - \overline{\ell(t)}]^n\, dt \qquad [1.51]$$

For a signal made up of N points of mean $\bar{\ell}$:

$$\mu_n = \frac{1}{N} \sum_{i=1}^{N} (\ell_i - \bar{\ell})^n \qquad [1.52]$$

1.9.4. *Variance – standard deviation*

The rms value is calculated from the mean quadratic value of the instantaneous values of the signal. The centered moment of order 2 is the *variance*, denoted by s_ℓ^2:

$$s_\ell^2 = E\left[(\ell - \bar{\ell})^2\right] = \overline{\ell^2(t)} - \overline{\ell(t)}^2 \qquad [1.53]$$

s_ℓ is called the *standard deviation*. This parameter characterizes the dispersion of the signal around its mean.

For a signal defined by N points:

$$s_\ell^2 = \frac{1}{N} \sum_{i=1}^{N} (\ell_i - \bar{\ell})^2 \qquad [1.54]$$

If the mean m is zero, the standard deviation s is equal to the rms value of the signal $\ell(t)$.

NOTE.– *On the assumption of zero mean, we can however note a difference between the standard deviation and the rms value when the latter is calculated starting from the PSD, which does not necessarily cover all of the frequency contents of the signal, in particular beyond 2,000 Hz (a value often selected as the upper limit of the analysis band). The rms value is then lower than the standard deviation. The comparison of the two values makes it possible to evaluate the importance of the neglected range.*

Non biased estimator

Relationship [1.54] gives a biased measurement of the variance. The bias can be corrected using the relation:

$$s_\ell^2 = \frac{1}{N-1} \sum_{i=1}^{N} (\ell_i - \bar{\ell})^2 \qquad [1.55]$$

Example 1.6.

Using the values of Example 1.5, the mean m value is equal to:

$$m = \sqrt{\frac{0+2+0.8+1.5-1-2.5+0.5-1.25+1.8-0.5+1}{11}}$$

$m \approx 0.214$

The variance is given by:

$$V = \sqrt{\frac{(0-0.214)^2 + (2-0.214)^2 + (0.8-0.214)^2 + \cdots + (-0.5-0.214)^2 + (1-0.214)^2}{11}}$$

$V \approx 1.813$

Standard deviation: $s = \sqrt{V} \approx 1.346$. We can verify that $\ddot{x}_{rms}^2 = s^2 + m^2$.

Non-biased standard deviation: ≈ 1.412.

1.9.5. Skewness

The centered moment of order 3, denoted by μ_3, is sometimes reduced by division by s_ℓ^3:

$$\mu_3' = \frac{E\left\{\left[\ell(t) - \overline{\ell(t)}\right]^3\right\}}{s_\ell^3} \qquad [1.56]$$

We can show [GMU 68] that μ_3' is characteristic of the symmetry of the probability density law $p(\ell)$ with regard to the mean $\overline{\ell(t)}$; for this reason, μ_3' is sometimes called *skewness*.

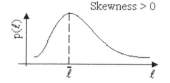
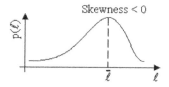

Figure 1.20. *Probability densities with non-zero skewness*

For a signal that is made up of N points:

$$\mu'_3 = \frac{\sum_{i=1}^{N}(\ell_i - \bar{\ell})^3}{N\, s_\ell^3} \qquad [1.57]$$

$\mu'_3 = 0$ characterizes a normal process.

For $\mu'_3 > 0$, the probability density curve presents a peak towards the left and for $\mu'_3 < 0$, the peak of the curve is shifted towards the right.

Non-biased estimator of skewness

Definition [1.57] is a biased measurement of the asymmetry of the population. A non-biased estimator of the asymmetry is given by the relation:

$$\mu'_3 = \frac{N}{(N-1)(N-2)} \frac{\sum_{i=1}^{N}(\ell_i - \bar{\ell})^3}{s_\ell^3} \qquad [1.58]$$

where $\bar{\ell}$ and s_ℓ are the non-biased estimators of the average and the standard deviation.

1.9.6. *Kurtosis*

The centered moment of order 4, reduced by division by s_ℓ^4, is also sometimes considered, as it makes it possible to estimate the flatness of the probability density curve. This is often called *kurtosis* [GUE 80].

$$\mu'_4 = \frac{E\left\{\left[\ell(t) - \overline{\ell(t)}\right]^4\right\}}{s_\ell^4} \qquad [1.59]$$

For a signal made up of N points:

$$\mu'_4 = \frac{\sum\limits_{i=1}^{N}(\ell_i - \bar{\ell})^4}{N\, s_\ell^4} \qquad [1.60]$$

The kurtosis characterizes the relative importance of the major distribution values in relations to the values close to zero:

$\mu'_4 = 3$ for a normal process.

$\mu'_4 < 3$ characteristic of a truncated signal or existence of a sinusoidal component ($\mu'_4 = 1.5$ for a pure sine).

$\mu'_4 > 3$ presence of peaks of high value (more than in the normal case).

We sometimes use a normalized expression of the kurtosis in the form

$$\gamma_4 = \frac{\sum\limits_{i=1}^{N}(\ell_i - \bar{\ell})^4}{N\, s_\ell^4} - 3 \qquad [1.61]$$

The quantity γ_4 is called "excess kurtosis". γ_4 is positive if the probability density is more acute than that of a normal law and γ_4 is negative if it is flatter.

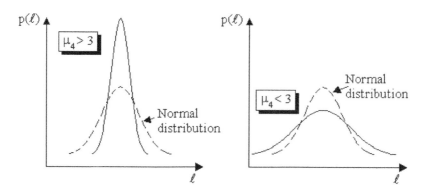

Figure 1.21. *Kurtosis influence on probability density*

Non-biased estimator of kurtosis

Definition [1.60] is a biased measurement of the flattening of the population. A non-biased estimator of this parameter is given by the relation:

$$\mu'_4 = \frac{(N+1)N}{(N-1)(N-2)(N-3)} \frac{\sum_{i=1}^{N}(\ell_i - \bar{\ell})^4}{s_\ell^4} + 3\frac{(N-2)(N-3)-(N-1)^2}{(N-2)(N-3)} \quad [1.62]$$

where $\bar{\ell}$ and s_ℓ are the non-biased estimators of the average and the standard deviation.

The normalized form of the kurtosis can be estimated without bias by:

$$\mu'_4 = \frac{(N+1)N}{(N-1)(N-2)(N-3)} \frac{\sum_{i=1}^{N}(\ell_i - \bar{\ell})^4}{s_\ell^4} - 3\frac{(N-1)^2}{(N-2)(N-3)} \quad [1.63]$$

Example 1.7.

With the information from Example 1.5, we obtain:

$$As = \frac{1}{(1.346)^3} \frac{(0-0.214)^3 + (2-0.214)^3 + \cdots + (-0.5-0.214)^3 + (1-0.214)^3}{11}$$

$$Ap = \frac{1}{(1.346)^4} \frac{(0-0.214)^4 + (2-0.214)^4 + \cdots + (-0.5-0.214)^4 + (1-0.214)^4}{11}$$

yielding $As \approx -0.475$ and $Ap \approx 2.241$.

Non-biased skewness: ≈ -0.553

Non-biased kurtosis: ≈ 2.568.

The advantage of these parameters

The rms value provides information on the global severity of the vibration. This information is useful, but not quite useful enough since it does not indicate the energy distribution in the frequency domain and thus potential vibration effects based on the natural frequencies of a mechanical structure.

Since measured vibration averages are mostly zero, the standard deviation is generally equal to the rms value. Comparing these two parameters makes it possible to verify this condition and to evaluate the value of the average when it is not zero.

Skewness and kurtosis are two parameters for verifying that the analyzed signal has a Gaussian instantaneous value distribution. When that is the case (the most common), skewness is theoretically equal to zero and kurtosis to 3. Strong variations of these parameters (particularly of kurtosis) calculated from sliding means also make it possible to detect the presence of mechanical shocks and signal problems (see section 1.16).

1.9.7. Crest Factor

The *crest factor* (or *peak-to-average* ratio) is equal to the ratio between the largest peak of the signal (in absolute value) and its rms value.

$$c_f = \frac{\max\left[|\ell(t)|\right]}{\sigma_\ell}$$

For a sinusoid, the crest factor is equal to $\sqrt{2}$. Random signals have an undefined crest factor which is extremely large, which may be expected on rare occasions very much larger than its rms value.

Each peak is associated with a probability. It is thus always possible, in theory, to find a very large peak whose probability of occurrence is very small. In practice, if the signal is Gaussian, we can observe crest factor values in the order of 5 to 6.

1.9.8. Temporal autocorrelation function

We define in the time domain the autocorrelation function $R_\ell(\tau)$ of the calculated signal, for a given τ delay, of the product $\ell(t)\,\ell(t+\tau)$ [BEA 72], [BEN 58], [BEN 63], [BEN 80], [BOD 72], [JAM 47], [MAX 65], [RAC 69], [SVE 80].

Figure 1.22. *Sample of a random signal*

34 Random Vibration

$$R_\ell(\tau) = E[\ell(t)\,\ell(t+\tau)] \qquad [1.64]$$

$$R_\ell(\tau) = \lim_{T\to\infty} \frac{1}{2T} \int_{-T}^{T} \ell(t)\,\ell(t+\tau)\,dt \qquad [1.65]$$

The result is independent of the selected signal sample i. The *delay* τ being given, we thus create, for each value of t, the product $\ell(t)$ and $\ell(t+\tau)$ and we calculate the mean of all the products thus obtained. The function $R_\ell(\tau)$ indicates the influence of the value of ℓ at time t on the value of the function ℓ at time $t+\tau$. Indeed let us consider the mean square of the variation between $\ell(t)$ and $\ell(t+\tau)$, i.e. $E\{[\ell(t)-\ell(t+\tau)]^2\}$, equal to:

$$E\{[\ell(t)-\ell(t+\tau)]^2\} = E[\ell^2(t)] + E[\ell^2(t+\tau)] - 2\,E[\ell(t).\ell(t+\tau)]$$

$$E\{[\ell(t)-\ell(t+\tau)]^2\} = 2\,R_\ell(0) - 2\,R_\ell(\tau) \qquad [1.66]$$

We note that the weaker the autocorrelation $R_\ell(\tau)$, the greater the mean square of the difference $[\ell(t)-\ell(t+\tau)]$ and, consequently, the less $\ell(t)$ and $\ell(t+\tau)$ resemble each other.

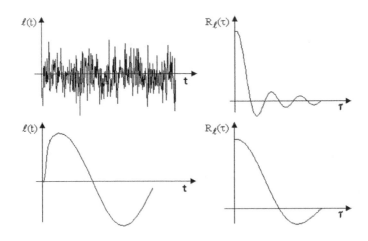

Figure 1.23. *Examples of autocorrelation functions*

The autocorrelation function measures the correlation between two values of $\ell(t)$ considered at different times t. If R_ℓ tends towards zero quickly when τ becomes large, the random signal probably fluctuates quickly and contains high frequency components.

If R_ℓ tends slowly towards zero, the changes in the random function are probably very slow [BEN 63], [BEN 80], [RAC 69].

R_ℓ is thus a measurement of the degree of random fluctuation of a signal.

Autocovariance

When the signal studied has a non-zero $\bar{\ddot{x}}$ average, we sometimes use the *autocovariance function* defined in an analog way by:

$$C_{\ddot{x}}(\tau) = \lim_{T \to \infty} \frac{1}{2T} \int_{-T}^{T} [\ddot{x}(t) - \bar{\ddot{x}}][\ddot{x}(t+\tau) - \bar{\ddot{x}}] dt \qquad [1.67]$$

Autocovariance is connected to autocorrelation by:

$$C_{\ddot{x}}(\tau) = R_{\ddot{x}}(\tau) - \bar{\ddot{x}}^2 \qquad [1.68]$$

Discrete form

The autocorrelation function calculated for a sample of signal digitized with N points separated by Δt is equal, for $\tau = m \, \Delta t$, to [BEA 72]:

$$R_\ell(\tau) = \frac{1}{N-m} \sum_{i=1}^{N-m} \ell_i \cdot \ell_{i+m} \qquad [1.69]$$

Catalogs of correlograms exist allowing typological study and facilitating the identification of the parameters characteristic of a vibratory phenomenon [VIN 72]. Their use makes it possible to analyze, with some care, the composition of a signal (white noise, narrowband noise, sinusoids, etc.).

36 Random Vibration

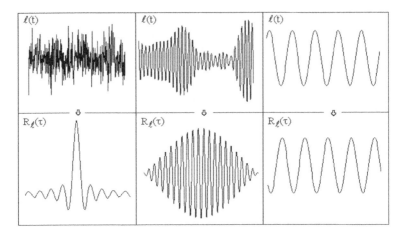

Figure 1.24. *Examples of autocorrelation functions*

Example 1.8.

Wideband noise autocorrelation

In the examples that follow, we often use a random vibration defined by a signal lasting 1 s (2,600 points), by a 31.6 m/s² rms value and by constant PSD equal to 1 (m/s²)²/Hz between 1 Hz and 1,000 Hz (Figure 1.25).

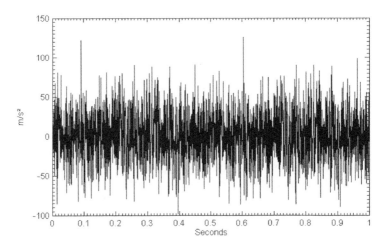

Figure 1.25. *Wideband random vibration*

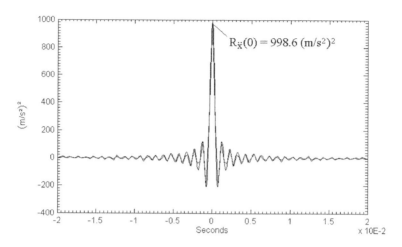

Figure 1.26. *Autocorrelation function of wideband noise between 1 Hz and 1,000 Hz, calculated from the signal according to time and its PSD*

Figure 1.26 illustrates the autocorrelation function of this vibration calculated from the signal and its PSD.

The two curves are very close, the difference being due to the number of points used for its line. This number is equal to that of the original signal in the first case, whereas it can be greater in the second case, leading to a smoother curve.

Autocorrelation shows a peak at origin with an amplitude that is equal to the square of the vibration rms value ($31.6^2 = 998.6$ $(m/s^2)^2$). Since the noise is wideband, the curve quickly tends toward zero.

Example 1.9.

Narrowband noise autocorrelation

This same time history signal was used to calculate the response of a linear one-degree-of-freedom system ($f_0 = 300$ Hz, $Q = 10$) and its autocorrelation was calculated.

The response is a narrowband noise with an rms value equal to 65.55 m/s^2. The autocorrelation of the response does not tend toward zero as quickly as a wideband noise. The peak at origin remains equal to the square of the signal's rms value.

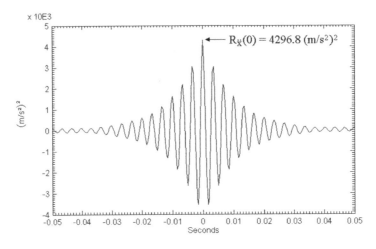

Figure 1.27. *Autocorrelation of the response of a linear one-degree-of-freedom system (300 Hz, Q = 10)*

Calculation of the autocorrelation function of a sinusoid

$$\ell(t) = \ell_m \sin(\Omega t) \qquad [1.70]$$

$$R_\ell(\tau) = \frac{1}{T} \int_0^T \ell_m \sin \Omega t \, \sin \Omega(t+\tau) \, dt$$

$$R_\ell(\tau) = \frac{\ell_m^2}{2} \cos \Omega \tau \qquad [1.71]$$

The correlation function of a sinusoid of amplitude ℓ_m and angular frequency Ω is a cosine of amplitude $\dfrac{\ell_m^2}{2}$ and pulsation Ω. The amplitude of the sinusoid can thus, conversely, be deduced from the autocorrelation function:

$$\ell_m = \sqrt{2} \left[R_\ell(\tau) \right]_{max} \qquad [1.72]$$

Example 1.10.

Autocorrelation of a sinusoid

Let us take a sinusoidal vibration with 20 m/s² amplitude and 300 Hz frequency. The autocorrelation of this signal (Figure 1.28) is a 200 (m/s²)² amplitude and 300 Hz frequency sinusoid.

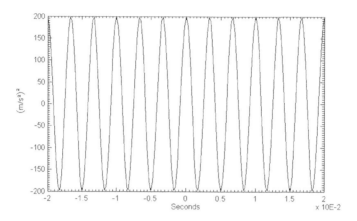

Figure 1.28. *Autocorrelation of a sinusoidal vibration*

1.9.9. Properties of the autocorrelation function

1. $R_\ell(0) = E[\ell^2(t)] = \overline{\ell^2(t)}$ = quadratic mean

$$R_\ell(0) = s^2 + \overline{\ell}^2 \qquad [1.73]$$

For a centered signal $(\overline{\ell} = 0)$, the ordinate at the origin of the autocorrelation function is equal to the variance of the signal.

2. The autocorrelation function is even [BEN 63], [BEN 80], [RAC 69]:

$$R_\ell(\tau) = R_\ell(-\tau) \qquad [1.74]$$

3. $|R_\ell(\tau)| < R_\ell(0) \quad \forall \, \tau \qquad [1.75]$

If the signal is centered, $R_\ell(\tau) \to 0$ when $\tau \to \infty$. If the signal is not centered, $R_\ell(\tau) \to \bar{\ell}^2$ when $\tau \to \infty$.

4. It is shown that:

$$\frac{dR_\ell(\tau)}{d\tau} = E\left[\ell(t-\tau)\,\dot{\ell}(t)\right] \qquad [1.76]$$

$$\frac{d^2 R_\ell(\tau)}{d\tau^2} = -E\left[\dot{\ell}(t)\,\dot{\ell}(t+\tau)\right] \qquad [1.77]$$

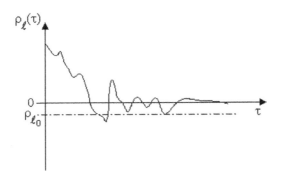

Figure 1.29. *Correlation coefficient*

NOTES.–

1. *The autocorrelation function, normalized autocorrelation function [BOD 72] or correlation coefficient is sometimes expressed in the reduced form:*

$$\rho_\ell(\tau) = \frac{R_\ell(\tau)}{R_\ell(0)} \qquad [1.78]$$

$\rho_\ell(\tau)$ *varies between –1 and +1,*

$\rho_\ell = 1$ *if the signals are identical (superimposable),*

$\rho_\ell = -1$ *if the signals are identical in absolute value and of opposite sign.*

2. *If the mean m is not zero, the correlation coefficient is given by*

$$\rho_\ell(\tau) = \frac{R_\ell(\tau) - m^2}{R_\ell(0)} \qquad [1.79]$$

1.9.10. *Correlation duration*

Correlation duration of a signal is the term given to the value τ_0 of τ from which the reduced autocorrelation function ρ_ℓ is always lower, in absolute value, than a certain value ρ_{ℓ_0}.

The correlation duration of:

– a wideband noise is weak;

– a narrowband noise is large; in extreme cases, a sinusoidal signal, which is thus deterministic, has an infinite correlation duration.

This last remark is sometimes used to detect in a signal $\ell(t)$ a sinusoidal wave $s(t) = S \sin \Omega t$ embedded in a random noise $b(t)$:

$$\ell(t) = s(t) + b(t) \qquad [1.80]$$

The autocorrelation is written:

$$R_\ell(\tau) = R_S(\tau) + R_b(\tau) \qquad [1.81]$$

If the signal is centered, for sufficiently large τ, $R_b(\tau)$ becomes negligible so that:

$$R_\ell(\tau) = R_S(\tau) = \frac{S^2}{2} \cos \Omega \tau \qquad [1.82]$$

This calculation makes it possible to detect a sinusoidal wave of low amplitude embedded in a very significant noise [SHI 70a].

Examples of application of the correlation method are as follows [MAX 69]:

– determination of the dynamic characteristics of a system;

– extraction of a periodic signal embedded in a noise;

42   Random Vibration

– detection of periodic vibrations of a vibratory phenomenon;

– study of transmission of vibrations (cross-correlation between two points of a structure);

– study of turbulences;

– calculation of PSDs [FAU 69];

– more generally, applications in the field of signal processing, in particular in medicine, astrophysics, geophysics, etc. [JEN 68].

Example 1.11.

A sinusoidal vibration with a frequency of 300 Hz and rms value of 10 m/s^2 was superimposed on the random vibration in Figure 1.25 (rms value 31.6 m/s^2). Because of the low rms value of the sinusoid, drowning in random noise and almost undetectable in the global signal (Figure 1.28), the rms value of the composite signal is not much different (33.16 m/s^2).

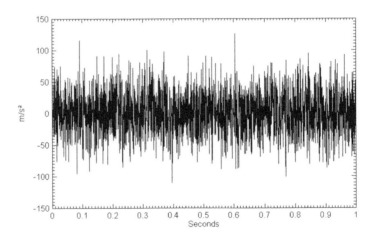

Figure 1.30. *Sine 300 Hz superimposed to random vibration in Figure 1.25*

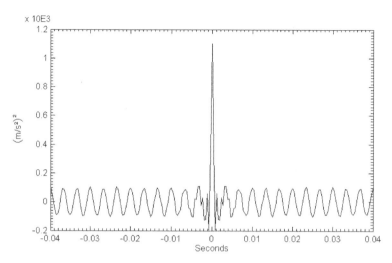

Figure 1.31. *Autocorrelation of the vibration made up of the random wideband vibration and the rms value 10 m/s² at 300 Hz sinusoid*

The autocorrelation function of this signal (Figure 1.31) clearly highlights the presence of this sinusoid and makes it possible to find its frequency and amplitude. The highest peak is equal to the square of the global signal rms value (random plus sine).

With this method, we can detect and characterize a low amplitude sinusoidal component. The threshold of detection, defined by the ratio of the rms values of the sine and of the PSD in the frequency interval between two consecutive points (frequency step Δf), is approximately 4.

Example 1.12.

Let us take a PSD with a 1 $(m/s^2)^2/Hz$ amplitude, calculated with a $\Delta f = 1.27$ Hz frequency step.

The calculation of the autocorrelation function can also be done from this PSD on which a 300 Hz frequency line was superimposed 0. The amplitude G of the composite PSD at this frequency is such that $G = \dfrac{\ddot{x}_{rms\ sine}^2}{\Delta f}$ ($\ddot{x}_{rms\ sine}$ = sinusoid rms value, i.e. 10 m/s²) (Figure 1.32).

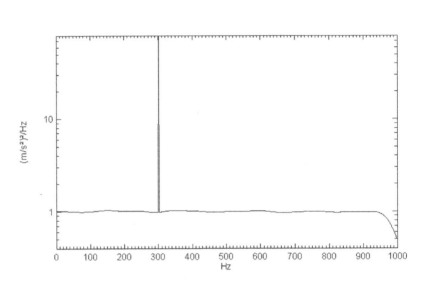

Figure 1.32. *PSD of wideband noise with 300 Hz sinusoidal line*

The sinusoid can be identified if its rms value is higher than approximately $4\sqrt{G \, \Delta f} = 4\sqrt{1.27} \approx 4.5$ m/s^2.

The resulting autocorrelation function can be superimposed over that in Figure 1.30. In order for this to occur, the line added to the PSD must be defined in a single frequency point. Otherwise, the autocorrelation function is transformed and tends toward that of a narrowband noise when this number increases.

Example 1.13.

Consider the wideband noise PSD in Figure 1.25 on which a 300 Hz, 20 m/s^2 rms value sinusoidal line is superimposed.

Figures 1.33 to 1.36 illustrate the autocorrelation function obtained when the sinusoidal line is defined consecutively by 1, 2, 4 and 9 points.

Statistical Properties of a Random Process   45

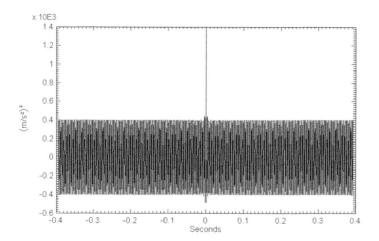

Figure 1.33. *Autocorrelation calculated from the wideband noise PSD with a sinusoidal line defined by a single point*

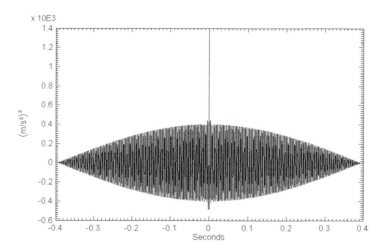

Figure 1.34. *Autocorrelation calculated from the wideband noise PSD with a sinusoidal line defined by two points*

46   Random Vibration

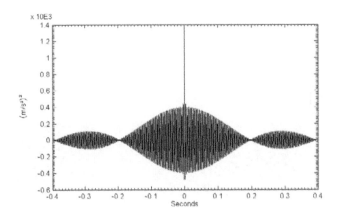

Figure 1.35. *Autocorrelation calculated from the wideband noise PSD with sinusoidal line defined by four points*

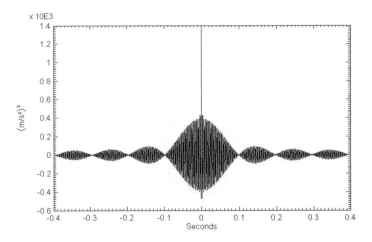

Figure 1.36. *Autocorrelation calculated from the wideband noise PSD with a sinusoidal line defined by eight points*

1.9.11. Cross-correlation

Let us consider two random functions $\ell(t)$ and $u(t)$; the *cross-correlation function* is defined by:

$$R_{\ell u}(\tau) = E[\ell(t)\, u(t+\tau)] = \lim_{T \to \infty} \frac{1}{2T} \int_{-T}^{T} \ell(t)\, u(t+\tau)\, dt \qquad [1.83]$$

The cross-correlation function makes it possible to establish the degree of resemblance between two functions of the same variable (time in general).

Discrete form [BEA 72]

If N is the number of sampled points and τ is a delay such that $\tau = m\, \Delta t$, where Δt is the temporal step between two successive points, the cross-correlation between two signals ℓ and u is given by

$$R_{\ell u}(\tau) = \frac{1}{N - m} \sum_{i=1}^{N-m} \ell_i\, u_{i+m} \qquad [1.84]$$

Covariance

The *covariance* of vibrations $\ddot{x}(t)$ and $\ddot{y}(t)$ is equal to the autocorrelation function of these centered signals. If $\overline{\ddot{x}}$ and $\overline{\ddot{y}}$ are the mean values of $\ddot{x}(t)$ and $\ddot{y}(t)$, covariance is defined by

$$C_{\ddot{x}\ddot{y}}(\tau) = \lim_{T \to \infty} \frac{1}{2T} \int_{-T}^{T} \left[\ddot{x}(t) - \overline{\ddot{x}}\right] \left[\ddot{y}(t+\tau) - \overline{\ddot{y}}\right] dt \qquad [1.85]$$

It is easy to show that:

$$C_{\ddot{x}\ddot{y}}(\tau) = R_{\ddot{x}\ddot{y}}(\tau) - \overline{\ddot{x}}\, \overline{\ddot{y}} \qquad [1.86]$$

If one of the mean values is zero, then the covariance is equal to the intercorrelation.

Example 1.14.

Consider a wideband random vibration and the response of a one-degree-of-freedom linear mechanical system with a natural frequency of 300 Hz and a Q factor of 10 for this noise.

Figure 1.37 shows the intercorrelation function between these two vibrations.

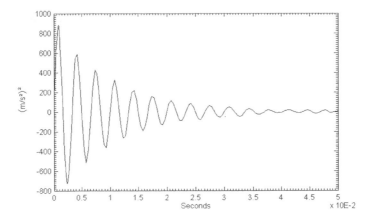

Figure 1.37. *Intercorrelation of wideband random vibration response from a linear one-degree-of-freedom system (300 Hz, Q = 10)*

Application: measure of the impulse response of a structure

If vibrations are stationary with zero mean, we can show that the intercorrelation of the vibration and response of a mechanical system is directly linked to the impulse response of the system.

Let h(t) be the impulse response of the system, $\ddot{x}(t)$ the random "excitation" vibration and $\ddot{y}(t)$ the response. This response can be calculated from the following relation:

$$\ddot{y}(t) = \int_0^t h(\alpha)\, \ddot{x}(t - \alpha)\, d\alpha \qquad [1.87]$$

The intercorrelation is equal to:

$$R_{\ddot{x}\ddot{y}}(\tau) = \int_0^\infty \ddot{x}(t - \tau)\, \ddot{y}(t)\, dt \qquad [1.88]$$

so, by considering [1.87]:

$$R_{\ddot{x}\ddot{y}}(\tau) = \int_0^\infty \ddot{x}(t-\tau) \int_0^\infty h(\alpha) \ddot{x}(t-\alpha) \, d\alpha \, dt \qquad [1.89]$$

$$R_{\ddot{x}\ddot{y}}(\tau) = \int_0^\infty \int_0^\infty h(\alpha) \ddot{x}(t-\tau) \ddot{x}(t-\alpha) \, d\alpha \, dt \qquad [1.90]$$

The "excitation" noise being wideband:

$$R_{\ddot{x}\ddot{y}}(\tau) = h(\tau) R_{\ddot{x}}(0) \qquad [1.91]$$

This method makes it possible to evaluate the impulse response of a system with a lower amplitude excitation than by impulsion, thus remaining in the linear domain.

Example 1.15.

The impulse response of the one-degree-of-freedom system (f_0 = 300 Hz, Q = 10) of Example 1.14 can be calculated (Figure 1.38) by dividing the intercorrelation in Figure 1.37 by $R_{\ddot{x}}(0)$, i.e. by the square of the rms value of the "excitation" wideband noise (31.6 m/s²).

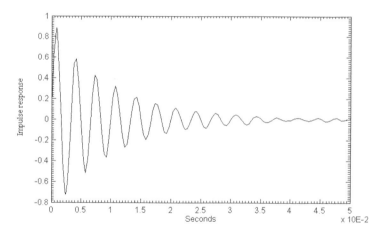

Figure 1.38. *Impulse response of the one-degree-of-freedom system (300 Hz, Q = 10) calculated from the intercorrelation function*

1.9.12. *Cross-correlation coefficient*

The *cross-correlation coefficient* $\rho_{\ell u}(\tau)$ or *normalized cross-correlation function* or *normalized covariance* is the quantity [JEN 68]

$$\rho_{\ell u}(\tau) = \frac{R_{\ell u}(\tau)}{\sqrt{R_\ell(0)\,R_u(0)}} \qquad [1.92]$$

It is shown that:

$$-1 \leq \rho_{\ell u}(\tau) \leq 1$$

If $\ell(t)$ is a random signal *input* of a system and $u(t)$ is the signal *response* at a point of this system, $\rho_{\ell u}(\tau)$ is characteristic of the degree of linear dependence of the signal u with respect to ℓ. At the limit, if $\ell(t)$ and $u(t)$ are independent, $\rho_{\ell u}(\tau) = 0$.

If the joint probability density of the random variables $\ell(t)$ and $u(t)$ is equal to $p(\ell, u)$, we can show that the cross-correlation coefficient $\rho_{\ell,u}$ can be written in the form:

$$\rho_{\ell u} = \frac{E\left[(\ell - m_\ell)(u - m_u)\right]}{s_\ell\, s_u} \qquad [1.93]$$

where m_ℓ, m_u, s_ℓ and s_u are respectively the mean values and the standard deviations of $\ell(t)$ and $u(t)$.

1.9.13. *Ergodicity*

A process is known as *ergodic* if all the temporal averages exist and have the same value as the corresponding ensemble averages calculated at an arbitrary given moment [BEN 58], [CRA 67], [JAM 47], [SVE 80].

Example 1.16.

In section 1.8.9, example 1.4, the process of case number 1 is ergodic and case number 2 is stationary, not ergodic. The question of ergodicity of process number 3 is irrelevant as it is not stationary.

An ergodic process is thus necessarily stationary [POU 02]. We dispose in general of only a very restricted number of records not permitting experimental evaluation of the ensemble averages. In practice, we simply calculate the temporal averages by making the assumption that the process is stationary and ergodic [ELD 61].

The concept of ergodicity is thus particularly important. Each particular realization of the random function makes it possible to consider the statistical properties of the whole ensemble of the particular realizations.

If a process is ergodic, we can limit ourselves to the frequency analysis of a short sample chosen over a single process record.

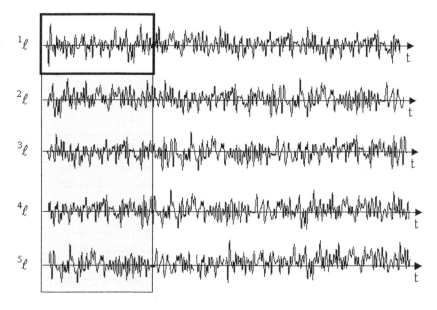

Figure 1.39. *Ergodic process: the analysis can involve a single sample*

NOTE.– *A necessary and sufficient condition such that a stationary random vibration $\ell(t)$ is ergodic is that its correlation function satisfies the condition [SVE 80]*

$$\lim_{T \to \infty} \frac{1}{T} \int_0^T \left(1 - \frac{\tau}{T}\right) R_\ell(\tau) \, d\tau = 0 \qquad [1.94]$$

where $R_\ell(\tau)$ is the autocorrelation function calculated from the centered variable $\ell(t) - m$.

1.10. Significance of the statistical analysis (ensemble or temporal)

The checking of stationarity and ergodicity should in theory be carried out before any analysis of a vibratory mechanical environment, in order to ensure that the consideration of only one sample is representative of the whole process. Very often, as a result of a lack of experimental data and to save time, we make these assumptions without checking (which is regrettable) [MIX 69], [RAC 69], [SVE 80].

1.11. Stationary and pseudo-stationary signals

We saw that the signal is known as stationary if the rms value as well as the other statistical properties remain constant over long periods of time.

In the real environment, this is not the case. The rms value of the load varies in a continuous or discrete way and gives the shape of signal known as *random pseudo-stationary*. For a road vehicle, for example, variations are due to the changes in road roughness, to changes of velocity of the vehicle, to mass transfers during turns, to wind effect, etc.

The temporal random function $\ell(t)$ is known as *quasi-stationary* if it can be divided into intervals of duration T that are sufficiently long compared with the characteristic correlation time, but sufficiently short to allow treatment in each interval as if the signal were stationary. Thus, the quasi-stationary random function is a function having characteristics which vary sufficiently slowly [BOL 84].

The study of the stationarity and ergodicity is an important stage in the analysis of vibration, but it is not generally sufficient by itself; in fact, it does not make it possible to answer the most frequently encountered problems, for example the estimate of the severity of a vibration or the comparison of several stresses of this nature.

1.12. Summary chart of main definitions

	Through the process (ensemble averages)	Along the process (temporal averages)
Moment of order n	$E[\ell^n(t_1)] = \lim_{N \to \infty} \frac{1}{N} \sum_{i=1}^{N} {}^i\ell^n(t_1)$ $E[x^n(t_1)] = \int_{-\infty}^{+\infty} \ell^n(t_1)\, p[\ell(t_1)]\, d\ell(t_1)$	$E[\ell^n(t)] = \lim_{T \to \infty} \frac{1}{T} \int_0^T \ell^n(t)\, dt$
Centered moment of order n	$E\left\{[\ell(t_1) - \overline{\ell(t_1)}]^n\right\} = \lim_{N \to \infty} \frac{1}{N} \sum_{i=1}^{N} [{}^i\ell(t_1) - \overline{\ell(t_1)}]^n$ $E\left\{[\ell(t_1) - \overline{\ell(t_1)}]^n\right\} = \int_{-\infty}^{+\infty} [\ell(t_1) - \overline{\ell}]^n\, p[\ell(t_1)]\, d\ell(t_1)$	$E\left\{[\ell(t) - \overline{\ell}]^n\right\} = \lim_{T \to \infty} \frac{1}{T} \int_0^T [\ell(t) - \overline{\ell}]^n\, dt$
Variance	$s^2 = E\left\{[\ell(t_1) - \overline{\ell(t_1)}]^2\right\}$	$s^2 = E\left\{[\ell(t) - \overline{\ell}]^2\right\} = \overline{\ell^2} - \overline{\ell}^2$
Autocorrelation	$R_\ell(t_1, t_1 + \tau) = \lim_{N \to \infty} \frac{1}{N} \sum_{i=1}^{N} {}^i\ell(t_1)\, {}^i\ell(t_1 + \tau)$ $R_\ell(\tau) = \int_{-\infty}^{+\infty} \ell(t_1)\, \ell(t_1 + \tau)\, p[\ell(t_1)]\, d\ell(t_1)$	$R_\ell(\tau) = \lim_{T \to \infty} \frac{1}{T} \int_0^T \ell(t)\, \ell(t + \tau)\, dt$
Cross-correlation	$R_{\ell u}(t_1, t_1 + \tau) = \lim_{N \to \infty} \frac{1}{N} \sum_{i=1}^{N} {}^i\ell(t_1)\, {}^i u(t_1 + \tau)$ $R_{\ell u}(\tau) = \int_{-\infty}^{+\infty} \ell(t_1)\, u(t_1 + \tau)\, p[\ell(t_1)]\, d\ell(t_1)$	$R_{\ell u}(\tau) = \lim_{T \to \infty} \frac{1}{T} \int_0^T \ell(t)\, u(t + \tau)\, dt$
	Stationarity if all the averages of order n are independent of the selected time t_1.	Ergodicity if the temporal averages are equal to the ensemble averages.

Table 1.1. *Main definitions*

Autocorrelation	$R_\ell(\tau) = \lim_{T\to\infty} \dfrac{1}{2T}\int_{-T}^{T} \ell(t)\,\ell(t+\tau)\,dt$		$R_\ell(\tau) = \dfrac{1}{N-m} \sum_{i=1}^{N-m} \ell_i \cdot \ell_{i+m}$
Normalized autocorrelation function or correlation coefficient		$\rho_\ell(\tau) = \dfrac{R_\ell(\tau)}{R_\ell(0)}$	
Autocovariance	$C_\ell(\tau) = \lim_{T\to\infty} \dfrac{1}{2T}\int_{-T}^{T} [\ell(t)-\bar\ell][\ell(t+\tau)-\bar\ell]\,dt$		$C_\ell(\tau) = \dfrac{1}{N-m} \sum_{i=1}^{N-m} (\ell_i-\bar\ell)(\ell_{i+m}-\bar\ell)$
Normalized autocovariance		$\rho_\ell(\tau) = \dfrac{C_\ell(\tau)}{C_\ell(0)}$	
Intercorrelation function	$R_{\ell u}(\tau) = \lim_{T\to\infty} \dfrac{1}{2T}\int_{-T}^{T} \ell(t)\,u(t+\tau)\,dt$		$R_{\ell u}(\tau) = \dfrac{1}{N-m} \sum_{i=1}^{N-m} \ell_i\,u_{i+m}$
Intercorrelation coefficient or normalized intercorrelation function		$\rho_{\ell u}(\tau) = \dfrac{R_{\ell u}(\tau)}{\sqrt{R_\ell(0)\,R_u(0)}}$	
Covariance	$C_{\ell u}(\tau) = \lim_{T\to\infty} \dfrac{1}{2T}\int_{-T}^{T}[\ell(t)-\bar\ell]\,[u(t+\tau)-\bar u]\,dt$		$C_{\ell u}(\tau) = \dfrac{1}{N-m}\sum_{i=1}^{N-m}(\ell_i-\bar\ell)(u_{i+m}-\bar u)$
Normalized covariance		$\rho_{\ell u}(\tau) = \dfrac{C_{\ell u}(\tau)}{\sqrt{C_\ell(0)\,C_u(0)}}$	

Table 1.2. *Some temporal averages with their discrete form*

1.13. Sliding mean

Instead of calculating the amplitude average of a signal over all available points, we can focus on a small number of consecutive points. This "block" is shifted by one point at each calculation (Figure 1.40). Each average thus obtained is attributed to the moment corresponding to the middle point in the block. All averages determined for the signal then make it possible to draw a curve based on time called the *sliding mean*.

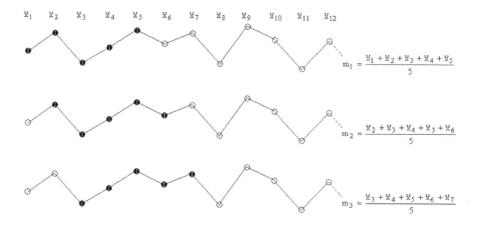

Figure 1.40. *Calculation process of the sliding mean*

There are several strategies to calculate the averages closer to the extremities. For example, we can determine the average of the first n/2 points and attribute it to the first point in the sliding mean curve, the average of the first n/2 + 1 points and the second point, etc., until we have all n points. The same methodology can be applied symmetrically for the extremity of the signal.

Similarly, this type of calculation can also be applied to the quadratic mean (for rms value, standard deviation, etc.) or a higher average (skewness, kurtosis, etc.).

The calculation speed of all these sliding means can be greatly accelerated with the use of the previous point calculation results for each point.

Example 1.17.

Figure 1.41 shows the rms value, skewness and kurtosis calculated according to time from a Gaussian signal lasting 5 seconds (sliding means over 1,000 points). We can verify that skewness and kurtosis are close to 0 and 3 respectively.

56   Random Vibration

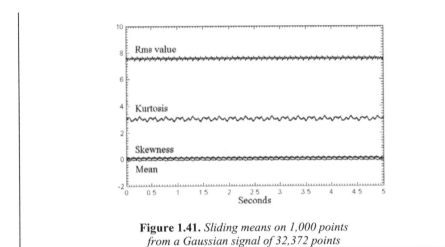

Figure 1.41. *Sliding means on 1,000 points from a Gaussian signal of 32,372 points*

Example 1.18.

Consider the random vibration taken from the platform of a truck (Figure 1.42).

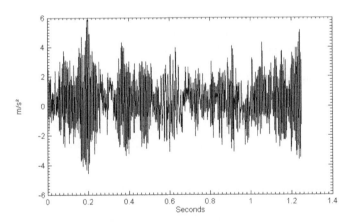

Figure 1.42. *"Truck" vibration*

Figures 1.43, 1.44 and 1.45 show these same sliding means calculated on 100, 500 and 1,000 points. The rms value, skewness and kurtosis vary with time (non-stationary vibration). The curves are smoother the higher the number of calculation points for each average.

Statistical Properties of a Random Process 57

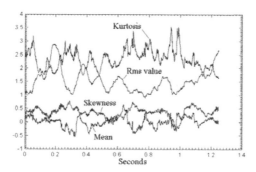

Figure 1.43. *"Truck" vibration – sliding means on 100 points*

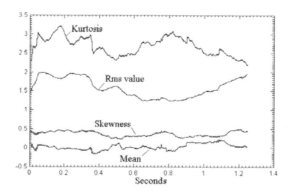

Figure 1.44. *"Truck" vibration – sliding means on 500 points*

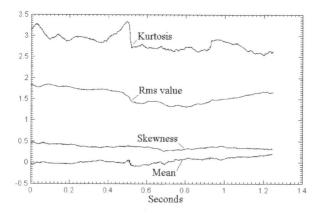

Figure 1.45. *"Truck" vibration – sliding means on 1,000 points*

1.14. Test of stationarity

The choice of analysis method of a random vibration in the frequency range is dependent on the stationary or non-stationary character of this vibration. We will see in particular that the calculation of the power spectral density only has meaning if the signal is stationary. Most often, the test simply consists of verifying that, in the time interval considered, the rms value calculated using a sliding mean varies little as a function of time.

Other more specific methods of a statistical nature have been developed, calling upon hypothesis tests.

1.14.1. *The reverse arrangements test (RAT)*

From a signal made up of N points, this method consists of [BEN 10], [BEN 65], [BRA 11], [HIM 06]:

– Cutting the signal into m samples of size equal to n points. H. Himelblau *et al.* propose to choose the integer m as a function of the signal duration (Table 1.3):

Duration	m
< 20 s	20
≥ 20 s and ≤ 100 s	Integer greater than the duration, so as to obtain samples with a duration close to 1 s
≥ 100 s	100

Table 1.3. *Choice of number of samples as a function of the signal duration*

m should be at least greater than 10, and preferably greater than 20.

– Calculating for each sample the parameter that will be used for the test: rms value in general, but another parameter could be used, such as the skewness or the kurtosis. This parameter will be denoted r_i.

– Calculating a function h such that

$$h_{ij} = \begin{cases} 1 & \text{if } r_i > r_j \\ 0 & \text{otherwise} \end{cases} \qquad [1.95]$$

where $i = 1, 2, \ldots, n-1$

$j = i+1, i+2, \ldots, n$

– Calculating $A_i = \sum_{j=i+1}^{n} h_{ij}$ [1.96]

and

$$A = \sum_{i=1}^{n-1} A_i$$ [1.97]

(total number of reverse arrangements).

– Choosing a significance level α as a function of the signal duration studied. H. Himelblau suggests the following values (Table 1.4):

m	α
< 20	0.10
20 to 40	0.05
> 40	0.02

Table 1.4. *Significance level α versus m*

– If parameter r is stationary, number A is a random variable having as a mean and standard deviation [HIM 06], [KEN 61]:

$$\mu_A = \frac{m(m-1)}{4}$$ [1.98]

$$\sigma_A = \sqrt{\frac{m(2m+5)(m-1)}{72}}$$ [1.99]

– For a chosen significance level α, the area of acceptation of the stationarity hypothesis is given by

$$A_{m;1-\alpha/2} < A < A_{m;\alpha/2}$$ [1.100]

where

$$A_{m;1-\alpha/2} = \mu_A - \sigma_A \, E_2^{-1}(\frac{1-\alpha}{2})$$ [1.101]

$$A_{m;\alpha/2} = \mu_A + \sigma_A \, E_2^{-1}(\frac{1-\alpha}{2})$$ [1.102]

$E_2^{-1}()$ being the inverse error function of $E_2(x)$ defined in Appendix A4.1.2:

$$E_2(x) = \frac{1}{\sqrt{2\pi}} \int_0^x e^{-\frac{t^2}{2}} dt$$

NOTE.– *The inverse error function of* $E_1(x) = \frac{2}{\sqrt{\pi}} \int_0^x e^{-t^2} dt$ *can also be used from*

$$A_{m;1-\alpha/2} = \mu_A - \sigma_A \sqrt{2} \, E_1^{-1}(1-\alpha)$$ [1.103]

$$A_{m;\alpha/2} = \mu_A + \sigma_A \sqrt{2} \, E_1^{-1}(1-\alpha)$$ [1.104]

There are, for example, as a reference [NEW 93], tables giving the probability for there being fewer than c arrangements of N of terms of a series whose order is random.

m	$A_{m;1-\alpha/2}$			$A_{m;\alpha/2}$		
	$1-\alpha/2 = 0.99$	$1-\alpha/2 = 0.975$	$1-\alpha/2 = 0.95$	$\alpha/2 = 0.05$	$\alpha/2 = 0.025$	$\alpha/2 = 0.01$
10	9	11	13	31	33	35
20	59	64	69	120	125	130
30	152	162	171	263	272	282
40	290	305	319	460	474	489
50	473	495	514	710	729	751
60	702	731	756	1013	1038	1067

Table 1.5. *Values of limits* $A_{m;1-\alpha/2}$ *and* $A_{m;\alpha/2}$ *as a function of the number of samples m and of the significance level* α

Example 1.19.

Let us consider the following rms values, calculated from a signal cut into m = 20 samples.

59.4	61.4	37.6	41.8	48.5	54.6	52.5	54.1	43.2	58.0
45.8	47.3	56.0	52.5	49.9	47.3	61.4	57.4	58.2	46.4

To a significance level $\alpha = 0.05$, Table 1.5 gives, for m = 20, the limits of the acceptance region $A_{20;\ 1-0.975} = 64$ and $A_{20;\ 1-0.025} = 125$. From relations [1.95] to [1.97], we obtain A = 85, value in the interval [64, 125]: the signal can be considered as stationary.

1.14.2. *The runs test*

The *runs test* (or *Wald–Wolfowitz test*) is a non-parametric statistical test that checks a randomness hypothesis for a two-valued data sequence. It is used to test the hypothesis that a series of numbers is random [BRA 68], [HOG 97], [SHE 04], [STE 39], [WAL 40].

To simplify computations, the data are first centered about or their rms (or mean or median) value.

Let us consider N observations. Each observation is classed exclusively in one of the three following categories, identified by the symbols + and – [BEN 63], [SIE 56]:

– (+) if the value is greater than the rms value;

– (–) if it is smaller than the rms value.

A *run* is defined as a sequence with the same sign ('+' or '–'). A positive run is a sequence of values greater than the rms value, and a negative run is a sequence of values less than the rms value [BEN 65], [BRA 11], [BRA 68], [YAM 87]. As an example, let us consider n = 25 observations of which n_1 are greater than the rms value and n_2 are lower (Figure 1.46).

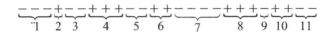

Figure 1.46. *Example of sequence of observations*

The first step is to count the number of runs in the data sequence. In this example, the n observations that lead to a total of runs $R = 11$ runs.

We can then test if the number of positive and negative runs are distributed equally in time.

W.L. Stevens [STE 39], then F.S. Swed and Eisenhart [SWE 43] have shown that the probability of the number of runs r being less than or equal to R is

$$P(r \leq R) = \frac{1}{C_n^{n_1}} \sum_{r=2}^{R} f_r \qquad [1.105]$$

where, by choosing $n_1 \leq n_2$,

if $r = 2k$

$f_r = 2\, C_{n_1-1}^{k-1}\, C_{n_2-2}^{k-1}$ [1.106]

if $r = 2k - 1$

$f_r = C_{n_1-1}^{k-1}\, C_{n_2-1}^{k-2} + C_{n_1-1}^{k-2}\, C_{n_2-1}^{k-1}$ [1.107]

with $k = 1, 2, 3, \cdots, m+1$.

From these relations, it is possible to search digitally by iterations the approximate limit values of the confidence interval corresponding to a given significance level α for a given value of n.

Table 1.6 gives several values thus obtained for $\alpha = 0.01$, $\alpha = 0.05$ and $\alpha = 0.10$ for $n_1 = n_2$. The values differ very little when this is not the case. As r varies by integer values, it is not possible to find a value that corresponds exactly to the desired probability. We have retained the value of r that leads to the probability closest to the one desired.

N	α = 0.01 Lower limit	α = 0.01 Upper limit	α = 0.05 Lower limit	α = 0.05 Upper limit	α = 0.10 Lower limit	α = 0.10 Upper limit
12	2	11	3	10	4	9
14	3	12	4	11	4	11
16	3	14	5	12	5	12
18	4	15	5	14	6	13
20	5	16	6	15	7	14
25	7	19	8	18	9	17
30	9	22	10	21	11	20
35	10	25	12	24	13	23
40	12	29	14	27	15	26
45	14	32	16	30	17	28
50	16	35	19	32	20	31
60	21	40	23	38	24	37
70	25	46	27	44	29	42
80	29	52	32	49	33	48
90	33	58	36	55	38	53
100	38	63	41	60	42	59
120	46	75	50	71	52	69
140	55	86	54	77	61	80
160	64	97	68	93	70	91
180	73	108	77	104	79	102
200	82	119	87	114	89	112

Table 1.6. *Upper and lower limits as a function of observations N and significance level α*

There are tables giving the bounds of this interval as a function of n_1 and n_2 in [MEN 86] and [SWE 43].

The signal is considered to be stationary if R is between the two bounds calculated for n and α given.

These parameters do not assume that the positive and negative values have the same probability of occurrence, but only assume that these values are independent and identically distributed.

If the number of runs is significantly more or less than that expected, the hypothesis of a statistical independence of observations can be rejected.

Example 1.20.

Let us consider the following 20 observations:

+	+	−	−	−	+	+	+	−	+
59.4	61.4	37.6	41.8	48.5	54.6	52.5	54.1	43.2	58.0

−	−	+	+	−	−	+	+	+	−
45.8	47.3	56.0	52.5	49.9	47.3	61.4	57.4	58.2	46.4

A (+) or (−) sign was attributed to each observation according to whether it is greater or lower than the rms value of the set of observations (52.09).

We count $n_1 = 9$ values with a (−) sign and $n_2 = 11$ with a (+) sign. The number of runs R is equal to 10.

For a significance level of 0.05, the limits calculated in these conditions are equal to 6 and 15. R being equal to 10, the hypothesis of a stationary signal can be accepted with this significance level.

For a large-sample runs test (where $n_1 > 10$ and $n_2 > 10$), the number of runs R is an approximately normally distributed random variable. It is thus possible to calculate the normal random variable Z as follows [SHE 04]:

$$Z = \frac{R - \overline{R}}{\sigma_R} \qquad [1.108]$$

The mean value $\overline{R}$ and variance σ_R^2 of this distribution are [BRO 84]:

$$\overline{R} = \frac{2 n_1 n_2}{n} + 1 \qquad [1.109]$$

$$\sigma_R^2 = \frac{2 n_1 n_2 (2 n_1 n_2 - n)}{n^2 (n-1)} \qquad [1.110]$$

The runs test rejects the hypothesis if $|Z| > Z_{1-\alpha/2}$ where α is the chosen significance level. It is thus necessary to verify that

$$\overline{R} - Z_{1-\alpha/2}\, \sigma_R < R < \overline{R} + Z_{\alpha/2}\, \sigma_R \qquad [1.111]$$

If there is a slowly varying periodicity in the data, with the rms value periodically increasing and dercreasing, the sequence of rms values will, of course, not be random.

The reverse arrangements test is based on rearranging the observations and it is thus not very sensitive to periodic components present in the signal. To highlight such components, which make the signal non-stationary, it is preferable to use a runs test [BRO 84], [SHE 04].

1.15 Identification of shocks and/or signal problems

The analysis of kurtosis according to time, drawn with a sliding mean, enables us to easily detect the presence of a local irregularity of the signal, whether it is due to a shock that is present in the real environment or to a signal problem that does not correspond to any physical mechanical reality (for example, a brief telemetry loss).

Kurtosis is actually very sensitive to "abnormal" signal values moving away from the distribution controlling the points.

Example 1.21.

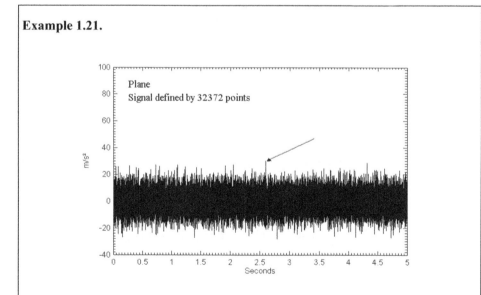

Figure 1.47. *Sample of vibratory signal measure on a plane*

66 Random Vibration

In order to show the sensitivity of this parameter, consider the stationary vibratory signal measured on a plane, lasting 5 s and defined by 32,372 points (Figure 1.47). This Gaussian signal presents a few high peaks whose amplitude can be about 5 times the signal rms value in accordance with what is expected with a Gaussian distribution.

Kurtosis according to time, calculated with a sliding mean over 500 points (Figure 1.48), remains approximately constant during the measure period. We can verify that its value is close to 3 (Gaussian signal).

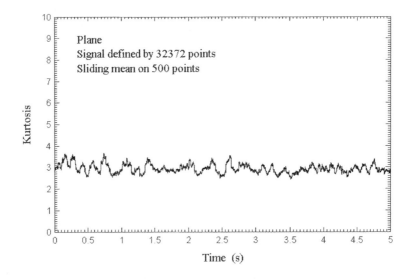

Figure 1.48. *Kurtosis versus time of vibration of Figure 1.47*

Figure 1.49 shows the signal of Figure 1.47 after multiplication of a single peak by a factor equal to 3.

Statistical Properties of a Random Process   67

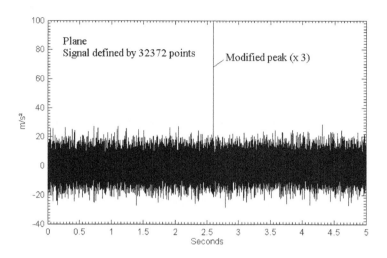

Figure 1.49. *Vibration of Figure 1.47 modified (multiplication by 3 of one peak)*

The effect on kurtosis of this modification is very important. We can observe at the peak moment a kurtosis value exceeding 40, which is thus much higher than 3 and very easily detectable (Figure 1.50).

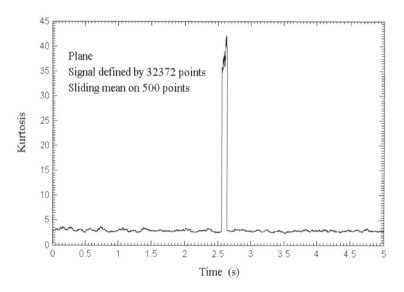

Figure 1.50. *Kurtosis of the signal modified in Figure 1.49*

> If this was a mechanical shock present within the random signal, there would be more modified points and kurtosis could reach, depending on the amplitude of the shock in relation to the noise, much greater values, by a few hundred for example.

1.16. Breakdown of vibratory signal into "events": choice of signal samples

If we only have a single record, it is sufficient to ensure that it is slightly autostationary to be able to characterize it with the help of PSD. If the distribution of instantaneous instant values of samples of a slightly autostationary record follows a Gaussian distribution, the record is *strongly autostationary*.

The mean value is generally zero (it can be brought down to zero if that is not the case). If the mean square value is constant in relation to time (the rms value), we can reasonably expect that the autocorrelation will be constant.

It is sufficient to verify that the rms value does not change in relation to time.

Practical problems

Each record is made up of several different sequences (also called *events*). Take the example of transportation by place: a record will include the runway haul, take-off, climb, cruising altitude, changes of direction, air brake phases, etc. to descent and landing. The process made up with these types of records is not stationary.

If we had a large number of records, we would have as many measures of each phase and we could imagine studying their stationarity and ergodicity. However, only rarely can we have more than a few measures. To reduce the cost of the measures campaign, we only carry out some records of the phenomenon in the best of cases. The phase studies may also be too short for these statistical tests in significant conditions.

If we carry out several instrumented flights for better environment understanding, we will probably not obtain the same rms value for each sequence (if only because weather conditions can change). In order for the process to be ergodic, all the conditions would have to be strictly the same, which would necessitate the characterization of all the processes corresponding to all possible conditions.

Clearly, it never works that way. Still, a reasonable methodology consists of carrying out several measures to be able to use a small statistic for estimating the most severe conditions.

Statistical Properties of a Random Process 69

In practice, for each record, we draw the rms value in relation to time, as well as the skewness and kurtosis (sliding means on the whole signal).

The first of these curves locates temporal ranges in which the rms value does not change much and which we will consider as stationary.

Each signal is separated into sections representing a specific phase of the flight, generating a vibratory or particular shock environment, with common characteristics from one flight to another: shock linked to landing, cruising flight without turbulence, etc.

Example 1.22.

Let us take a vibratory environment on an aircraft, represented by acceleration as a function of time:

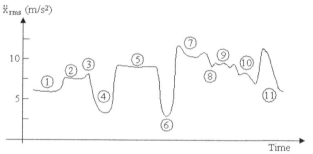

1. Taxi	7. Maximum velocity at low altitude
2. Take-off	8. Climbing turn
3. Climb	9. Deceleration
4. Cruise at high altitude	10. Landing approach
5. Maximum velocity at high altitude	11. Touchdown
6. Cruise at low altitude	

Figure 1.51. *Rms acceleration recorded on an aircraft during flight [KAT 65]*

70   Random Vibration

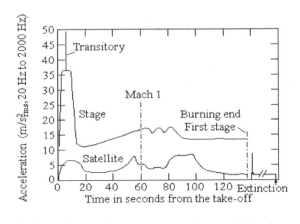

Figure 1.52. *Rms value of vibrations measured on a satellite during launch*

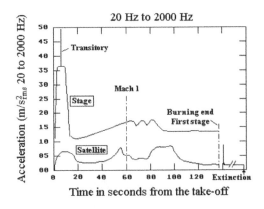

Figure 1.53. *Rms value of vibrations taken from a satellite during launch*

Skewness and kurtosis give an indication about the Gaussian character of the signal in these time intervals (close to 3 and 0 respectively in this hypothesis).

A quick and very important variation of the kurtosis generally indicated the presence of a shock. However, it is important that we consider the signal according to time to make sure it is not a problem with the signal (loss of telemetry, etc.), affecting kurtosis in a very similar way.

Example 1.23.

The chosen vibratory signal, created for this example, represents a measure that could have been carried out on a satellite launcher. It is represented in Figure 1.54.

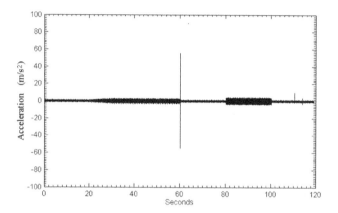

Figure 1.54. *Acceleration versus time*

The rms value, skewness and kurtosis according to time, calculated with the help of a sliding mean on 500 points, are given respectively in Figures 1.55, 1.56 and 1.57.

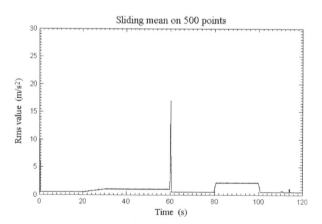

Figure 1.55. *Rms value versus time calculated from the signal in Figure 1.54*

72   Random Vibration

Figure 1.55 highlights the presence of several consecutive phases:

– a shock (to analyze in a more finite way to distinguish it from a signal fault);

– a stationary phase on the first 20 s (constant rms value);

– a transitory phase lasting 10 s;

– a new stationary phase between 30 s and 60 s;

– a shock (at approximately 60 s);

– a stationary phase approximately between 60 s and 80 s;

– a very brief transitory phase;

– a stationary phase between 80 and 100 s;

– a very brief transitory phase;

– a stationary phase between 100 and 120 s approximately, during which the rms value shows 2 peaks.

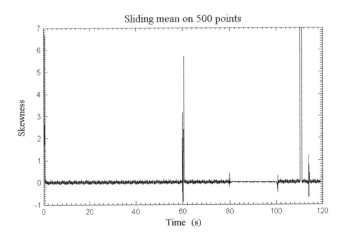

Figure 1.56. *Skewness versus time calculated from the signal in Figure 1.54*

Statistical Properties of a Random Process 73

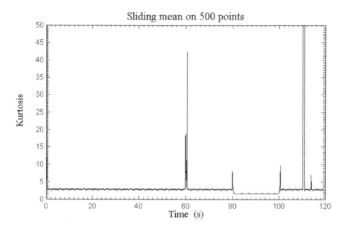

Figure 1.57. *Kurtosis versus time calculated from the signal in Figure 1.54*

Skewness (Figure 1.56) and kurtosis (Figure 1.57) show that the signal relative to stationary phases is Gaussian (skewness equal to 0 and kurtosis equal to 3), except between 80 s and 100 s (constant kurtosis lower than 3).

The shock at 60 s and the 2 peaks observed in the last stationary phase, particularly the first peak, appear clearly on the skewness and kurtosis.

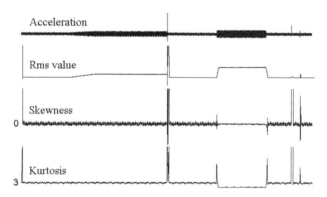

Figure 1.58. *Juxtaposition of acceleration, rms value, skewness and kurtosis according to time*

The juxtaposition of these curves (Figure 1.58) and the previous comments help us reach some conclusions. The vibratory signal is representative of a succession of events which can be described consecutively by (Figure 1.59):

– a mechanical shock (after analysis of the acceleration signal based on time), characterized by its shock response spectrum;

– a stationary random vibration, characterized by a PSD (Chapter 4), which is used to calculate an extreme response spectrum[2] (ERS) and a fatigue damage spectrum (FDS) (Volume 5);

– a non-stationary random vibration (we can use it to calculate an ERS and an FDS directly from the signal);

– a stationary random vibration;

– a shock;

– a stationary random vibration;

– between 80 s and 100 s, a periodic vibration to analyze in a more finite way to determine its components (kurtosis different from 3, and after analysis of the acceleration signal in this zone);

– a stationary phase in which we observe a signal fault (a very high peak that is extremely small and that cannot be linked to a specific phenomenon of the life profile);

– a low amplitude shock.

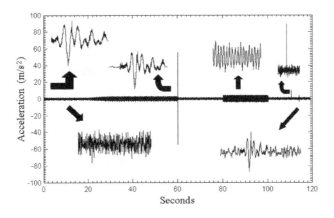

Figure 1.59. *Identification and extraction of the different events making up the vibratory signal*

If we have several logs, we should analyze each one of them.

---

2 Also called "maximum response spectrum" (MRS).

It is very useful to plot the variations of the rms value against time (sliding mean on n points) in order to:

– choose the time intervals over which the rms value varies little: each corresponding phase can then be characterized by a PSD;

– study the very short duration phenomena (non-stationary phenomena). The analysis for example measures the number of times that the rms value crosses a given threshold with respect to the amplitude of this threshold (rms value of the total signal or of the response of a one-degree-of-freedom mechanical system of constant Q factor, generally equal to 10, whose natural frequency varies in the useful frequency band) [KEL 61].

The variation of the rms value with time has also been used as a monitoring tool for the correct operation of rotating machinery based on a statistical study of their vibratory behavior [ALL 82].

1.17. Interpretation and taking into account of environment variation

Consider a stationary phase in a vibratory environment, for example, the cruising phase of air transport. The rms value is often different from one flight to another even though it is constant during this phase for each flight (slightly different weather conditions, etc.). This actually represents the non-ergodicity of the process.

This variation can be considered:

– by calculating the statistical PSD or, preferably, ERS and FDS (for example, average + 3 standard deviations) to describe the event with a crossing risk lower than 0.135%;

– by applying an uncertainty coefficient during the development of specifications (use of variation coefficient of environment, standard deviation and average ratio).

Example 1.24.

Let us assume that vibrations relative to air transport were measured during 6 different flights. It seems difficult to study the ensemble averages of this process with only 6 logs. Instead, each signal is analyzed to detect (from the rms value line according to time obtained with with a sliding mean) time intervals during which the signal is autostationary (slight variation of rms value).

76   Random Vibration

Each of these intervals is identified for a specific event (cruising altitude phase for example) in the situation involved ("air transport") and can be characterized by calculating a PSD on part of the corresponding signal or on its whole, depending on duration. The same procedure is applied to each of the other 6 logs. At the end of these analyses, we then have 6 PSDs representative of the cruising phase (Figure 1.60).

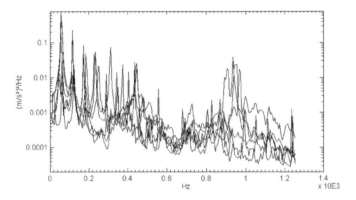

Figure 1.60. *Power spectral densities (6) characterizing the "cruising phase" random process of air transport*

The rms values of these 6 PSDs are equal to 1.5 m/s^2, 1.36 m/s^2, 1.71 m/s^2, 2.82 m/s^2, 3.16 m/s^2 and 1.65 m/s^2 respectively.

If the process was completely ergodic, all these PSDs would be extremely close with the same rms value (the differences are due to the short time of the signal samples used and the resulting statistical error).

We can observe that the PSDs are relatively dispersed, mainly in amplitude, since the dispersion is caused by the conditions of each flight (turbulences which can be important).

The random process is not really ergodic. In order to treat this general problem, we consider, from one flight to another, that the levels are random and we evaluate:

– an average spectrum (PSD) and a standard deviation spectrum (at each frequency, average and standard deviation of PSD values) if the number of available spectrums permits it (at least 4 or 5 measures);

– or, on the contrary, simply a spectrum "envelope" (greatest PSD value at each frequency).

These two spectrums can be used:

– to determine a PSD with a low probability of being exceeded (for example, average + 3 standard deviations), which can possibly be used by an engineering firm to size a mechanical part,

– or, by dividing the standard deviation by the mean at each frequency, to calculate the coefficient of variation that will characterize the dispersion of the environment involved.

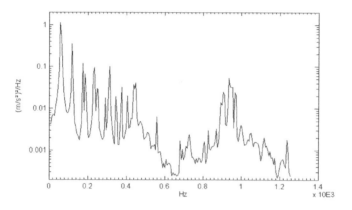

Figure 1.61. *Average PSD plus 3 standard deviations calculated from the 6 PSDs in Figure 1.60*

We will see in Volume 5 that the coefficient of variation is preferably calculated from the ERS and FDS determined with PSDs. It will be used in the calculation of an uncertainty coefficient during the development of a specification.

Chapter 2

Random Vibration Properties in the Frequency Domain

The frequency content of the random signal must produce useful information by comparison with the natural frequencies of the mechanical system which undergoes the vibration.

This chapter is concerned with power spectral density (PSD), its definition and properties. The following chapters will show that this spectrum provides a powerful tool to enable the description of random vibrations. It also provides basic data for many other analyses of signal properties.

2.1. Fourier transform

The Fourier transform of a non-periodic $\ell(t)$ signal, having a finite total energy, is given by the relationship:

$$L(\Omega) = \int_{-\infty}^{+\infty} \ell(t)\, e^{-i\Omega t}\, dt \qquad [2.1]$$

This expression is complex; it is therefore necessary in order to represent it graphically to plot:

– either the real and the imaginary part versus the angular frequency Ω;

– or the amplitude and the phase versus Ω. Very often, we limit ourselves to amplitude data. The curve thus obtained is called the *Fourier spectrum* [BEN 58].

80 Random Vibration

The random signals are not of finite energy. We can thus calculate only the Fourier transform of a sample of signal of duration T by assuming that this sample is representative of the whole phenomenon. It is in addition possible, starting from the expression of $L(\Omega)$, to return to the temporal signal by calculation of the inverse transform.

$$\ell(t) = \frac{1}{2\pi} \int_{-\infty}^{+\infty} L(\Omega)\, e^{i\Omega t}\, d\Omega \qquad [2.2]$$

We could envisage the comparison of two random vibrations (assumed to be ergodic) from their Fourier spectra calculated using samples of duration T. This work is difficult, as it assumes the comparison of four curves two by two, each transform being made up of a real part and an imaginary part (or amplitude and phase).

We could however limit ourselves to a comparison of the amplitudes of the transforms, by neglecting the phases. We will see in the following sections that, for reasons related to the randomness of the signal and the miscalculation which results from it, it is preferable to proceed with an average of the modules of Fourier transforms calculated for several signal samples (or, more exactly, an average of the squares of the amplitudes). This is the idea behind the PSD.

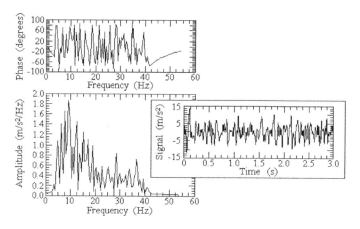

Figure 2.1. *Example of Fourier transform*

In an indirect way, the Fourier transform is thus used very often in the analysis of random vibrations.

2.2. Power spectral density

2.2.1. Need

The search for a criterion for estimating the severity of a vibration naturally results in the examination of the following characteristics:

– the maximum acceleration of the signal: this parameter neglects the smaller amplitudes which can excite the system for a prolonged length of time;

– the mean value of the signal: this parameter is not very significant as a criterion of severity, because negative accelerations are subtractive and the mean value is in general zero. If this is not the case, it does not produce information sufficient to characterize the severity of the vibration;

– the rms value: for a long time this was used to characterize the voltages in electrical circuits, the rms value being much more interesting data [MOR 55]:

- if the mean is zero, the rms value is in fact the standard deviation of instantaneous acceleration and is thus one of the characteristics of the statistical distribution,

- even if two or several signal samples have very different frequency contents, their rms values can be combined by using the square root of the sum of their squares.

This quantity is thus often used as a relative instantaneous severity criterion of the vibrations [MAR 58]. However, it has the disadvantage of being global data and of not revealing the distribution of levels according to frequency, which is nevertheless very important. For this purpose, a solution can be provided by [WIE 30]:

– filtering the signal $\ell(t)$ using a series of rectangular filters of central frequency f and bandwidth Δf (Figure 2.2);

– calculating the rms value L_{rms} of the signal collected at the output of each filter.

The curve which would give L_{rms} with respect to f would indeed be a description of the spectrum of signal $\ell(t)$, but the result would be different depending on the width Δf derived from the filters chosen for the analysis. So, for a stationary noise, we filter the supposed broadband signal using a rectangular filter of filter width Δf, centered around a central frequency f_c, the obtained response having the aspect of a stable, permanent signal. Its rms value is more or less constant over time. If, by preserving its central frequency, we reduce the filter width Δf, maintaining its gain, the output signal will seem unstable, fluctuating greatly over time (as well as its rms value), and more especially so if Δf is weaker.

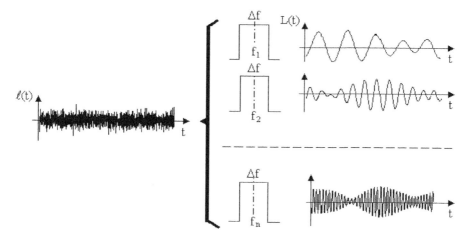

Figure 2.2. *Filtering of the random signal*

To obtain a characteristic value of the signal, it is thus necessary to calculate the mean over a much longer length of time, or to calculate the mean of several rms values for various samples of the signal. In addition, we note that the smaller Δf is, the more the signal response at the filter output has a low rms value [TIP 77]: a filter that is twice as wide leads to a greater rms value by a factor that is approximately equal to $\sqrt{2}$.

To be liberated from the width Δf, we instead consider the variations of $\dfrac{L_{rms}^2}{\Delta f}$ with f. The rms value is squared by analogy with electrical power. The resulting curve has for dimension the square of an acceleration divided by a frequency. It is thus expressed in $(m/s^2)^2/Hz$ or in g^2/Hz.

2.2.2. *Definition*

2.2.2.1. *Reminder: power dissipated in an electrical circuit*

If we consider a tension $u(t)$ applied to the terminals of a resistance $R = 1 \, \Omega$, passing current $i(t)$, the energy dissipated (Joule effect) in the resistance during time dt is equal to:

$$dE = R \, i^2(t) \, dt = i^2(t) \qquad [2.3]$$

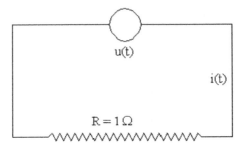

Figure 2.3. *Electrical circuit with source of tension and resistance*

The instantaneous power of the signal is thus:

$$P(t) = \frac{dE}{dt} = i^2(t) \qquad [2.4]$$

and the energy dissipated during time T, between t and t + T, is written:

$$E_T = \int_t^{t+T} i^2(t)\, dt \qquad [2.5]$$

P(t) depends on time t (if i varies with t). It is possible to calculate a mean power in the interval T using:

$$P_{m_T} = \frac{1}{T}\int_t^{t+T} p(t)\, dt = \frac{1}{T} E_T \qquad [2.6]$$

The total energy of the signal is therefore:

$$E = \int_{-\infty}^{+\infty} i^2(t)\, dt \qquad [2.7]$$

and the total mean power is:

$$P_m = \lim_{T\to\infty} \frac{1}{T}\int_{-T/2}^{+T/2} i^2(t)\, dt \qquad [2.8]$$

2.2.2.2. Mechanical vibrations

By analogy with these calculations, we define [BEN 58], [TUS 72] in vibration mechanics the *mean power* of an excitation $\ell(t)$ between $-T/2$ and $+T/2$ by:

$$P_m = \lim_{T \to \infty} \frac{1}{T} \int_{-T/2}^{+T/2} |\ell_T(t)|^2 \, dt \qquad [2.9]$$

where

$$\begin{cases} \ell_T = \ell(t) & \text{for } |t| \leq T/2 \\ \ell_T = 0 & \text{for } |t| > T/2 \end{cases}$$

Let us suppose that the function $\ell_T(t)$ has as a Fourier transform $L_T(f)$. According to Parseval's equality,

$$\int_{-\infty}^{+\infty} |\ell_T|^2 \, dt = \int_{-\infty}^{+\infty} |L_T(f)|^2 \, df \qquad [2.10]$$

yielding, since [JAM 47],

$$\int_{-T/2}^{+T/2} |\ell_T|^2 \, dt = \int_{-\infty}^{+\infty} |\ell_T(t)|^2 \, dt \qquad [2.11]$$

$$P_m = \lim_{T \to \infty} \frac{1}{T} \int_{-\infty}^{+\infty} |L_T(f)|^2 \, df = \lim_{T \to \infty} \frac{2}{T} \int_{0}^{\infty} |L_T(f)|^2 \, df \qquad [2.12]$$

This relation gives the mean *power* contained in $\ell(t)$ when all the frequencies are considered. Let us find the mean power contained in a frequency band Δf. In order to do this, let us assume that the excitation $\ell(t)$ is applied to a linear system with constant parameters whose weighting function is $h(t)$ and the transfer function is $H(f)$. The response $r_T(t)$ is given by the convolution integral

$$r_T(t) = \int_{0}^{\infty} h(\lambda) \, \ell_T(t - \lambda) \, d\lambda \qquad [2.13]$$

where λ is an integration constant. The mean power of the response is written:

$$P_{m\,response} = \lim_{T \to \infty} \frac{2}{T} \int_{0}^{T} r_T^2(t) \, dt \qquad [2.14]$$

i.e., according to Parseval's theorem:

$$P_{m\,response} = \lim_{T \to \infty} \frac{2}{T} \int_0^T |R_T(f)|^2 \, df \qquad [2.15]$$

If we take the Fourier transform of the two members of [2.13], we can show that:

$$R_T(f) = H(f) \, L_T(f) \qquad [2.16]$$

yielding

$$P_{m\,response} = \lim_{T \to \infty} \frac{2}{T} \int_0^\infty |H(f)|^2 \, |L_T(f)|^2 \, df \qquad [2.17]$$

Examples

1. If $H(f) = 1$ for any value of f,

$$P_{m\,response} = \lim_{T \to \infty} \int_0^\infty \frac{2|L_T(f)|^2}{T} \, df = P_{m\,input} \qquad [2.18]$$

a result which *a priori* is obvious.

2. If $H(f) = 1$ for $0 \le f - \dfrac{\Delta f}{2} \le f \le f + \dfrac{\Delta f}{2}$

$H(f) = 0$ elsewhere

$$P_{m\,response} = \lim_{T \to \infty} \int_{f-\Delta f/2}^{f+\Delta f/2} \frac{2|L_T(f)|^2}{T} \, df \qquad [2.19]$$

In this last case, let us set:

$$G_T(f) = \frac{2|L_T(f)|^2}{T} \qquad [2.20]$$

86 Random Vibration

The mean power corresponding to the record $\ell_T(t)$, finite length T, in the band Δf centered on f, is written:

$$P_T(f, \Delta f) = \int_{f-\Delta f/2}^{f+\Delta f/2} G_T(f)\,df \qquad [2.21]$$

and total mean power through out the record

$$P(f, \Delta f) = \lim_{T \to \infty} \int_{f-\Delta f/2}^{f+\Delta f/2} G_T(f)\,df \qquad [2.22]$$

We call PSD the quantity:

$$G(f) = \lim_{\Delta f \to \infty} \frac{P(f, \Delta f)}{\Delta f} \qquad [2.23]$$

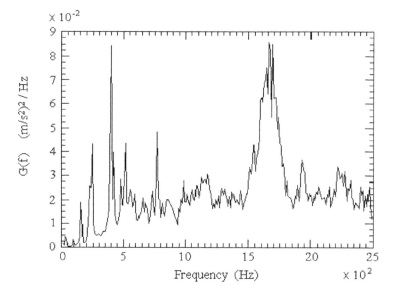

Figure 2.4. *Example of PSD (aircraft)*

NOTE.– *By using the angular frequency Ω, we would obtain*:

$$P(\Omega, \Delta\Omega) = \lim_{T \to \infty} \int_{\Omega-\Delta\Omega/\Omega}^{\Omega+\Delta\Omega/\Omega} G_T(\Omega)\,d\Omega \qquad [2.24]$$

with

$$G_T(\Omega) = \frac{2|L_T(\Omega)|^2}{2\pi T}$$ [2.25]

Taking into account the above relations, and [2.10] in particular, the PSD $G(f)$ can be written [BEA 72], [BEN 63], [BEN 80]:

$$G(f) = \lim_{\substack{\Delta f \to 0 \\ T \to \infty}} \frac{1}{T\,\Delta f} \int_0^T \ell_T^2(t, \Delta f)\, dt$$ [2.26]

where $\ell_T(t, \Delta f)$ is the part of the signal ranging between the frequencies $f - \Delta f/2$ and $f + \Delta f/2$. This relation shows that the PSD can be obtained by filtering the signal using a narrowband filter of given width, by squaring the response and by taking the mean of the results for a given time interval [BEA 72]. This method is used for analog calculations.

Expression [2.26] theoretically defines the PSD. In practice, this relation cannot be respected exactly since the calculation of $G(f)$ would require an infinite integration time and an infinitely narrow bandwidth.

NOTES.–

– *The function $G(f)$ is positive or zero whatever the value of f.*

– *The PSD was defined above for f ranging between 0 and infinity, which corresponds to the practical case. In a more general way, we could mathematically define $S(f)$ between $-\infty$ and $+\infty$, in such a way that*

$$S(-f) = S(f)$$ [2.27]

– *The pulsation $\Omega = 2\pi f$ is sometimes used as variable instead of f. If $G_\Omega(\Omega)$ is the corresponding PSD, we have*

$$G(f) = 2\pi G_\Omega(\Omega)$$ [2.28]

The relations between these various definitions of the PSD can be easily obtained starting from the expression of the rms value:

$$\ell_{rms}^2 = \int_{-\infty}^{+\infty} S(f)\, df = \int_0^\infty G(f)\, df = \int_0^\infty G_\Omega(\Omega)\, d\Omega = \int_{-\infty}^{+\infty} S(\Omega)\, d\Omega$$ [2.29]

We then deduce:

$$G(f) = 2\, S(f) \qquad [2.30]$$

$$G(f) = 2\,\pi\, G_\Omega(\Omega) \qquad [2.31]$$

$$G(f) = 4\,\pi\, S_\Omega(\Omega) \qquad [2.32]$$

NOTE.– *A sample of duration T of a stationary random signal can be represented by a Fourier series, the term a_i of the development in an exponential Fourier series being equal to:*

$$a_i = \frac{2}{T} \int_{-T/2}^{+T/2} \ell(t) \begin{cases} \sin\dfrac{2\pi k t}{T} \\ \cos\dfrac{2\pi k t}{T} \end{cases} dt \qquad [2.33]$$

The signal $\ell(t)$ can be written in complex form

$$\ell(t) = \sum_{k=-\infty}^{\infty} c_k\, e^{2\pi i k \frac{t}{T}} \qquad [2.34]$$

where $c_k = \dfrac{1}{2}(\alpha_i - \beta_i\, i)$.

The PSD can also be defined from this development in a Fourier series. It is shown that [PRE 54], [RAC 69], [SKO 59], [SVE 80]

$$G(f) = \lim_{T \to \infty} \frac{a_i^2}{2\,\Delta f} \qquad [2.35]$$

The PSD is a curve frequently used in the analysis of vibrations:

– either in a direct way, to compare the frequency contents of several vibrations, to calculate, in a given frequency band, the rms value of the signal, to transfer a vibration from one point in a structure to another, etc.;

– or as intermediate data, to evaluate certain statistical properties of the vibration (frequency expected, probability density of the peaks of the signal, number of peaks expected per unit time, etc.).

NOTE.– *The function $G(f)$, although called power, does not have its dimension. This term is often used because the square of the fluctuating quantity often appears in the expression for the power, but it is unsuitable here [LAL 95]. As such, it is often preferable to call it "acceleration spectral density" or "acceleration density" [BOO 56] or "PSD of acceleration" or "intensity spectrum" [MAR 58].*

2.3. Amplitude Spectral Density

The amplitude spectral density (ASD) is the spectrum defined by the square root of the PSD:

$$\text{ASD}(f) = \sqrt{\text{PSD}(f)} \qquad [2.36]$$

This is the image of the rms value in spectral density for a non-impulse signal. It is expressed in $m/s^2/Hz^{1/2}$ or in $g/Hz^{1/2}$

2.4. Cross-power spectral density

The PSD expression of a signal ℓ_1 can be written and, for T infinite, from the preceding

$$G_{\ell_1 \ell_1}(f) = \lim_{T \to \infty} \frac{2|L_1(f)|^2}{T} = \lim_{T \to \infty} \frac{2 L_1^*(f) L_1(f)}{T} \qquad [2.37]$$

The resulting curve $G_{\ell_1 \ell_1}(f)$ is also called the "autospectrum".

Similarly, the *cross-power spectral density* of two vibrations $\ell_1(t)$ and $\ell_2(t)$ is defined as the mean f products $L_1^*(f) L_2(f)$, L_1 and L_2 being respectively the Fourier transforms of $\ell_1(t)$ and $\ell_2(t)$ calculated between 0 and T over K blocks of points of the two signals.

From two samples of random signal records $\ell_1(t)$ and $\ell_2(t)$, the cross-power spectrum is thus defined for T infinite by

$$G_{\ell_1 \ell_2}(f) = \lim_{T \to \infty} \frac{2 L_1^*(f) L_2(f)}{T} \quad [2.38]$$

if the limit exists.

The cross-power spectral density is a complex quantity that is generally represented by its amplitude and phase.

2.5. Power spectral density of a random process

The PSD was defined above for only one function of time $\ell(t)$. Let us consider now the case where the function of time belongs to a random process, where each function will be noted $^i\ell(t)$. A sample of this signal of duration T will be denoted by $^i\ell_T(t)$, and its Fourier transform $^iL_T(f)$. Its PSD is

$$^iG_T(f) = \frac{2\left|^iL_T(f)\right|^2}{T} \quad [2.39]$$

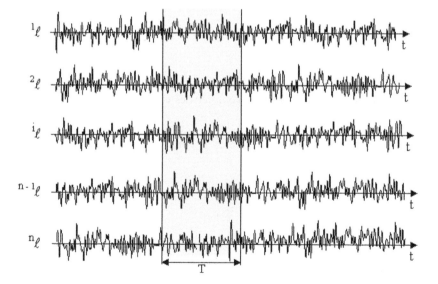

Figure 2.5. *Signal samples for the calculation of the PSD of a process*

By definition, the PSD of the random process is, over time T, equal to:

$$G_T(f) = \frac{\sum_{i=1}^{n} {}^i G_T(f)}{n} \qquad [2.40]$$

n being the number of functions ${}^i \ell(t)$ and, for T infinite,

$$G(f) = \lim_{T \to \infty} G_T(f) \qquad [2.41]$$

If the process is stationary and ergodic, the PSD of the process can be calculated starting from several samples of one recording only. The PSD is a real quantity.

2.6. Cross-power spectral density of two processes

As previously, we define the cross-power spectrum between two records of duration T, each one taken in one of the processes by:

$${}^i G_T(f) = \frac{2 \, {}^i L_1^* \, {}^i L_2}{T} \qquad [2.42]$$

The cross-power spectrum of the two processes is, over T,

$$G_T(f) = \frac{\sum_{i=1}^{n} {}^i G_T(f)}{n} \qquad [2.43]$$

and, for T infinite,

$$G(f) = \lim_{T \to \infty} G_T(f) \qquad [2.44]$$

Example 2.1.

Let us take a vibration sample lasting 50 s, with constant PSD and equal to 2 (m/s²)²/Hz between 1 Hz and 500 Hz, and then very small up to 650 Hz.

This vibration was applied to a linear one-degree-of-freedom system with a frequency of 200 Hz and an over-tension of 5.

Figure 2.6 shows autospectrums from the "excitation" vibration and response, Figure 2.7 shows the cross-power spectral density of the response and excitation superimposed to the excitation autospectrum.

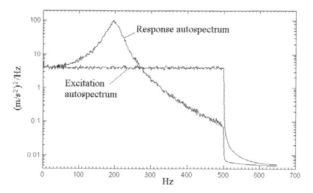

Figure 2.6. *Autospectrums of the vibration excitation and response of a one-degree-of-freedom 200 Hz system, Q = 5*

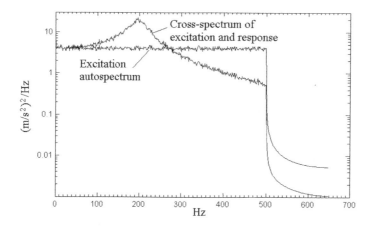

Figure 2.7. *Cross-power spectral density of the response and excitation, superimposed to the excitation autospectrum*

2.7. Relationship between the PSD and correlation function of a process

It is shown that, for a stationary process [BEN 58], [BEN 80], [JAM 47], [LEY 65], [NEW 75]:

$$G(f) = 2 \int_{-\infty}^{+\infty} R(\tau) e^{-2\pi i f \tau} d\tau \qquad [2.45]$$

$R(\tau)$ being an even function of τ, we obtain:

$$G(f) = 4 \int_{0}^{\infty} R(\tau) \cos(2\pi f \tau) d\tau \qquad [2.46]$$

If we take the inverse transform of $G(f)$ given in [2.44], it becomes:

$$R(\tau) = \frac{1}{2} \int_{-\infty}^{+\infty} G(f) e^{2\pi i f \tau} df \qquad [2.47]$$

i.e., since $G(f)$ is an even function of f [LEY 65]:

$$R(\tau) = \int_{0}^{\infty} G(f) \cos(2\pi f \tau) df \qquad [2.48]$$

and

$$R(0) = \overline{\ell^2(t)} = \int_{0}^{\infty} G(f) \, df = (\text{rms value})^2 \qquad [2.49]$$

NOTE.– *These relations, called "Wiener-Khinchine relations", can be expressed in terms of the angular frequency Ω in the form [BEN 58], [KOW 69], [MIX 69]:*

$$G(\Omega) = \frac{2}{\pi} \int_{0}^{\infty} R(\tau) \cos(\Omega \tau) d\tau \qquad [2.50]$$

$$R(\tau) = \int_{0}^{\infty} G(\Omega) \cos(\Omega \tau) d\tau \qquad [2.51]$$

2.8. Quadspectrum – cospectrum

The cross-power spectral density $G_{\ell u}(f)$ can be written in the form [BEN 80]:

$$G_{\ell u}(f) = 2 \int_{-\infty}^{+\infty} R_{\ell u}(\tau) e^{-2\pi i f \tau} d\tau = C_{\ell u}(f) - i\, Q_{\ell u}(f) \qquad [2.52]$$

where the function

$$C_{\ell u}(f) = 2\int_{-\infty}^{+\infty} R_{\ell u}(\tau)\cos(2\pi f \tau)\,d\tau \qquad [2.53]$$

is the *cospectrum* or *coincident spectral density*, and where

$$Q_{\ell u}(f) = 2\int_{-\infty}^{+\infty} R_{\ell u}(\tau)\sin(2\pi f \tau)\,d\tau \qquad [2.54]$$

is the *quadspectrum* or *quadrature spectral density function*.

We obtain:

$$R_{\ell u}(\tau) = \frac{1}{2}\int_0^\infty G_{\ell u}(f) e^{2\pi i f \tau}\,df + \frac{1}{2}\int_0^\infty G_{\ell u}^*(f) e^{-2\pi i f \tau}\,df \qquad [2.55]$$

$$R_{\ell u}(\tau) = \int_0^\infty \left[C_{\ell u}(f) \cos(2\pi f \tau) + Q_{\ell u}(f) \sin(2\pi f \tau) \right] df \qquad [2.56]$$

$$G_{\ell u}(f) = |G_{\ell u}(f)|\, e^{-i\theta_{\ell u}(f)} \qquad [2.57]$$

$$|G_{\ell u}(f)| = \sqrt{C_{\ell u}^2 + Q_{\ell u}^2(f)} \qquad [2.58]$$

$$\theta_{\ell u}(f) = \arctan\left(\frac{Q_{\ell u}(f)}{C_{\ell u}(f)}\right) \qquad [2.59]$$

2.9. Definitions

2.9.1. *Broadband process*

A broadband process is a random stationary process whose PSD $G(\Omega)$ has significant values in a frequency band or a frequency domain which is rigorously of the same order of magnitude as the central frequency of the band [PRE 56a].

Figure 2.8. *Wideband process*

Such processes appear in pressure fluctuations on the skin of a missile rocket (jet noise and turbulence of supersonic boundary layer).

2.9.2. White noise

When carrying out analytical studies, it is now usual to idealize the wideband process by considering a uniform spectral density $G(f) = G_0$.

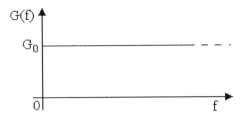

Figure 2.9. *White noise*

A process having such a spectrum is called *white noise* by analogy with white light which is composed from all the frequencies of the visible spectrum.

An ideal white noise, which is assumed to have a uniform density at all frequencies, is a theoretical concept, physically unrealizable, since the area under the curve would be infinite (and therefore so would the rms value). Nevertheless, the model ideal white noise is often used to simplify calculations and to obtain suitable orders of magnitude of the solution, in particular for the evaluation of the response of a one-degree-of-freedom system to wideband noise. This response is indeed primarily produced by the values of the PSD in the frequency band ranging between the half-power points. If the PSD does not vary too much in this interval, we can compare it at a first approximation with that of a white noise of the same amplitude. It should however be ensured that the results of this simplified analysis do indeed provide a correct approximation from what would be obtained with physically attainable excitation.

2.9.3. Band-limited white noise

We also use in the calculations the spectra of band-limited white noises, such as that in Figure 2.10, which are correct approximations to many realizable random processes on exciters.

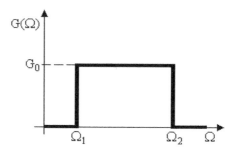

Figure 2.10. *Band-limited white noise*

2.9.4. Narrowband process

A narrowband process is a random stationary process whose PSD has significant values in one frequency band only or a frequency domain whose width is small compared to the value of the central frequency of the band [FUL 62].

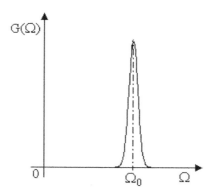

Figure 2.11. *PSD of narrowband noise*

The signal as a function of time $\ell(t)$ looks like a sinusoid of angular frequency Ω_0, with amplitude and phase varying randomly. There is only one peak between two zero crossings.

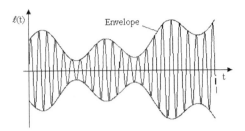

Figure 2.12. *Narrowband noise*

It is interesting to consider *individual cycles* and *envelopes,* whose significance we will note later on.

If the process is Gaussian, it is possible to calculate from $G(\Omega)$ the expected frequency of the cycles and the probability distribution of the points on the envelope.

These processes relate in particular to the response of low damped mechanical systems, when the excitation is a broadband noise.

2.9.5. *Colors of noise*

Many definitions assume a signal with components at all frequencies, with a power spectral density per unit of bandwidth proportional to $1/f^{\alpha}$ (power-law noise).

Color	α	PSD's slope
Brown noise Red noise Brownian noise Random walk noise Drunkard's walk noise	- 2	-6 db/octave
Pink noise Flicker noise	- 1	-3 db/octave
White noise	0	0
Blue noise Azure noise	+ 1	3 dB/octave
Violet noise Purple noise	+ 2	+ 6 dB/octave

Table 2.1. *PSD's slope according to noise color*

2.10. Autocorrelation function of white noise

Relation [2.48] can be also written, since $G(f) = 4\pi S(\Omega)$ [BEN 58], [CRA 63]:

$$R(\tau) = \int_{-\infty}^{+\infty} S(\Omega) \, e^{i\Omega\tau} \, d\Omega \qquad [2.60]$$

If $S(\Omega)$ is constant equal to S_0 when Ω varies, this expression becomes:

$$R(\tau) = 2\pi S_0 \int_{-\infty}^{+\infty} e^{2\pi i f \tau} \, df \qquad [2.61]$$

where the integral is the Dirac delta function $\delta(\tau)$, such that

$$\begin{cases} \delta(\tau) \to \infty & \text{when } \tau \to 0 \\ \delta(\tau) = 0 & \text{when } \tau = 0 \\ \int_{0^-}^{0^+} \delta(\tau) \, d\tau = 1 \end{cases} \qquad [2.62]$$

yielding

$$R(\tau) = 2\pi S_0 \, \delta(\tau) \qquad [2.63]$$

NOTE.– *If the PSD is defined by $G(\Omega)$ in $(0, \infty)$, this expression becomes*

$$R(\tau) = \pi \frac{G_0}{2} \delta(\tau) \qquad [2.64]$$

whilst, for $G(f) \in (0, \infty)$:

$$R(\tau) = \frac{1}{2} G_0 \, \delta(\tau) \qquad [2.65]$$

For $\tau = 0$, $R \to \infty$. Knowing that $R(0)$ is equal to the square of the rms value, the property of the white noise is verified (infinite rms value).

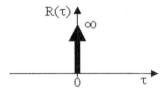

Figure 2.13. *Autocorrelation of a white noise*

It should be noted in addition that the correlation is zero between two arbitrary times.

An ideal white noise thus has an infinite intensity, but has no correlation whatsoever between past and present [CRA 63].

2.11. Autocorrelation function of band-limited white noise

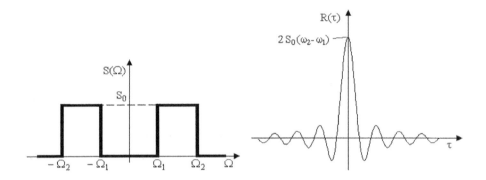

Figure 2.14. *Band-limited white noise*   **Figure 2.15.** *Autocorrelation of band-limited white noise*

From definition [2.61], we have, if $S(\Omega) = S_0$ [FUL 62],

$$R(\tau) = 2 S_0 \int_0^\infty e^{i \Omega \tau} \, d\Omega \qquad [2.66]$$

$$R(\tau) = 2 S_0 \int_{\Omega_1}^{\Omega_2} \cos(\Omega \tau) \, d\Omega \qquad [2.67]$$

$$R(\tau) = 2 S_0 \frac{\sin \Omega_2 \tau - \sin \Omega_1 \tau}{\tau} \qquad [2.68]$$

$R(\tau)$ can also be written

$$R(\tau) = \frac{4 S_0}{\tau} \cos\left(\frac{\Omega_1 + \Omega_2}{2} \tau\right) \sin\left(\frac{\Omega_2 - \Omega_1}{2} \tau\right) \qquad [2.69]$$

The rms value, given by [BEN 61a]

$$\ell_{rms} = \sqrt{R_\ell(0)} = \sqrt{2 S_0 (\Omega_2 - \Omega_1)} \qquad [2.70]$$

is finite. If τ tends towards zero, $R(\tau) \to 2 S_0 (\Omega_2 - \Omega_1)$ (square of the rms value). The correlation between the past and the present is non-zero, at least for small intervals. When the bandwidth is widened, the above case is obtained.

NOTES.–

1. *The result obtained for a white noise process is demonstrated in this particular case when* $\Omega_1 = 0$ *and* $\Omega_2 \to \infty$; *indeed, if* $\Omega_2 = 0$,

$$R(\tau) = \frac{4 S_0}{\tau} \cos\frac{\Omega_2 \tau}{2} \sin\frac{\Omega_2 \tau}{2} = \frac{2 S_0}{\tau} \sin \Omega_2 \tau$$

If $\Omega_2 \to \infty$ *[2.63]*,

$$R(\tau) \to 2 \pi S_0 \delta(\tau)$$

Conversely, if $R(\tau)$ *has this value,*

$$S(\Omega) = \frac{1}{2\pi} \int_{-\infty}^{+\infty} 2 \pi S_0 \delta(\tau) e^{-i \Omega \tau} d\tau = S_0$$

2. *If we set* $\Omega_1 = \Omega_0 - \frac{\Delta\Omega}{2}$ *and* $\Omega_2 = \Omega_0 + \frac{\Delta\Omega}{2}$, $R(\tau)$ *can be written* [COU 70]:

$$R(\tau) = \frac{4 S_0}{\tau} \cos(\Omega_0 \tau) \sin\left(\frac{\tau \Delta\Omega}{2}\right) \qquad [2.71]$$

If $\tau \to 0$,

$$R(0) \to 2 S_0 \Delta\Omega$$

yielding

$$\rho = \frac{R(\tau)}{R(0)} = \frac{2}{\tau \Delta\Omega} \cos \Omega_0 \tau \; \sin \frac{\tau \Delta\Omega}{2} \qquad [2.72]$$

3. *If, in practice, the noise is defined only for the positive frequencies, expressions [2.68] and [2.70] become*

$$R(\tau) = G_0 \frac{\sin \Omega_2 \tau - \sin \Omega_1 \tau}{\tau} \qquad [2.73]$$

$$\ell_{rms} = \sqrt{R_\ell(0)} = \sqrt{G_0(\Omega)(\Omega_2 - \Omega_1)} = \sqrt{G_0(f)(f_2 - f_1)} \qquad [2.74]$$

2.12. Peak factor

The *peak factor, peak ratio* or *crest factor* F_p of a signal can be defined as the ratio of its maximum value (positive or negative) to its standard deviation (or to its rms value). For a sinusoidal signal, this ratio is equal to $\sqrt{2}$ (≈ 1.414). For a signal made up of periodic rectangular waveforms, it equals 1, while for saw tooth waveforms, it is approximately equal to 1.73.

In the case of a random signal, the probability of finding a peak of given amplitude is an increasing function of the duration of the signal. The peak factor is thus undefined and extremely large. Such a signal will thus necessarily have peaks which will be truncated because of the limitation of the dynamics of the analyzer. An error in the PSD calculation will result from this.

2.13. Effects of truncation of peaks of acceleration signal on the PSD

Let us consider a random signal $\ell(t)$ of rms value ℓ_{rms}. If the signal filtered in a filter of width Δf has its values truncated higher than ℓ_0 (or if the signal was truncated during measurement), the calculated PSD is equal to

$$G'(f) = \frac{\overline{\ell'^2_{\Delta f}(f)}}{\Delta f} \text{ instead of } G(f) = \frac{\overline{\ell^2_{\Delta f}(f)}}{\Delta f}$$

Let us set:

$$F_p = \frac{\ell_0}{\ell_{rms\Delta f}} \qquad [2.75]$$

The error will thus be, at frequency f,

$$e = 100\left[1 - \frac{G'(f)}{G(f)}\right] \qquad [2.76]$$

with

$$\frac{G'(f)}{G(f)} = \frac{\overline{\ell'^2_{\Delta f}(f)} / \Delta f}{\overline{\ell^2_0(f)} / \Delta f} = \frac{\overline{\ell'^2_{\Delta f}(f)}}{\overline{\ell^2_0(f)}} = \rho \qquad [2.77]$$

It is shown that [PIE 64], for a Gaussian signal, the error varies according to the peak factor F_p according to the law

$$e = 100\,(1 - \rho) \qquad [2.78]$$

where

$$\rho = 2\,F_p^2 \int_{F_p}^{\infty} p(x)\,dx + 2\int_0^{F_p} p(x)\,dx - 2\,F_p\,p(x) \qquad [2.79]$$

and

$$p(x) = \frac{1}{\sqrt{2\pi}} \exp\left(-\frac{x^2}{2}\right) \qquad [2.80]$$

The variations of the error e according to F_p are represented in Figure 2.16.

Random Vibration Properties in the Frequency Domain   103

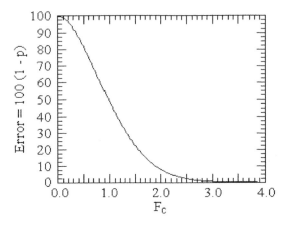

Figure 2.16. *Error versus peak factor (from [PIE 64])*

The calculation of ρ can be simplified if it is noted that:

$$\frac{1}{\sqrt{2\pi}} \int_{-\infty}^{-F_p} e^{-\frac{x^2}{2}} dx + \frac{1}{\sqrt{2\pi}} \int_{-F_p}^{F_p} e^{-\frac{x^2}{2}} dx + \frac{1}{\sqrt{2\pi}} \int_{F_p}^{\infty} e^{-\frac{x^2}{2}} dx = 1$$

and that the probability density is symmetric about the y-axis:

$$\int_{F_p}^{\infty} p(x) \, dx = \frac{1}{2} - \int_{0}^{F_p} p(x) \, dx$$

ρ is then written:

$$\rho = 2 F_p^2 \left[\frac{1}{2} - \int_{0}^{F_p} p(x) \, dx \right] + 2 \int_{0}^{F_p} p(x) \, dx - 2 F_p \, p(F_p)$$

$$\rho = 2 \left(1 - F_p^2\right) \int_{0}^{F_p} p(x) \, dx + F_p^2 - 2 F_p \, p(F_p) \qquad [2.81]$$

The integral $\int_{0}^{F_p} p(x) \, dx$ can be calculated using the error function, knowing that this function can be defined in the form [LAL 94] (Appendix A4.1.2):

104   Random Vibration

$$\text{erf}_2\left(F_p\right) = \frac{1}{\sqrt{2\pi}} \int_0^{F_p} e^{-\frac{x^2}{2}} \, dx \qquad [2.82]$$

The influence of a truncation of the peaks of a random acceleration signal on its PSD is shown in the following example.

Acceleration signal selected for study

The signal considered is a sample of duration 20 seconds of a white noise over a bandwidth of 10–2,000 Hz, of rms value $\ddot{x}_{rms} = 44.61$ m/s^2 (PSD amplitude: 1 (m/s^2)^2/Hz). The sampling frequency is chosen equal to 30 kHz, much larger than that imposed by Shannon's theorem, in order to allow later on a correct calculation of the extreme response spectra and of the fatigue damage spectra. This signal was then truncated with various acceleration values: $\pm 5\,\ddot{x}_{rms}$, $\pm 4.5\,\ddot{x}_{rms}$ ..., until $\pm 0.5\,\ddot{x}_{rms}$.

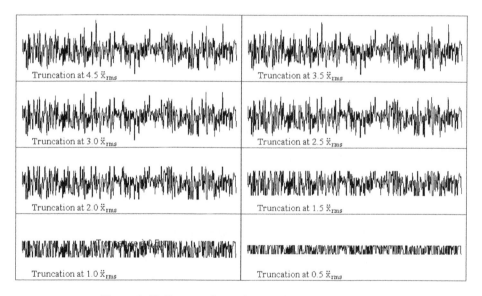

Figure 2.17. *Truncated signals (samples of duration 0.1 s)*

Power spectral densities obtained

The spectral densities of these signals were calculated between 10 Hz and 3,000 Hz. We observe from the PSD (Figure 2.18) that:

– truncation causes the amplitude of the PSD to decrease uniformly in the defined bandwidth (between 10 Hz and 2,000 Hz);

– this reduction is only sensitive if we clip the peaks below $2\,\ddot{x}_{rms}$ approximately;

– truncation increases the amplitude of the PSD beyond its specified bandwidth (2,000 Hz). This effect is related to the mode of truncation selected (clean cut-off at the peaks and no non-linear attenuation, which would smooth out the signal in the zone concerned).

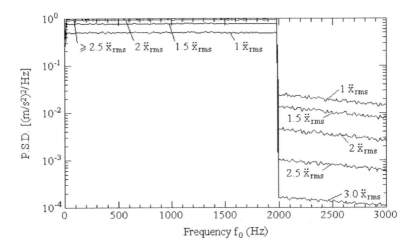

Figure 2.18. *PSD of the truncated signals*

2.14. Standardized PSD/density of probability analogy

Standardized PSD is the term given to the quantity [WAN 45]:

$$G_N = \frac{G(\Omega)}{\ell_{rms}^2} \qquad [2.83]$$

It should be noted that the standardized PSD and the probability density function have common properties:

– they are non-negative functions;

– they have a unit area under the curve;

– if we set $R(\Omega) = \int_0^\Omega G_N(u)\,du$, $R(\Omega)$ increases in a monotonous way from zero ($\Omega = 0$) to 1 (for Ω infinite). $R(\Omega)$ can thus be regarded as the analog of the distribution function of $G(\Omega)$.

2.15. Spectral density as a function of time

In practice, the majority of the physical processes are, to a certain degree, non-stationary, i.e. their statistical properties vary with time. Very often, however, the excitation is clearly non-stationary over a long period of time, but, for small intervals, which are still long with respect to the time of response of the dynamic system, the excitation can be regarded as stationary. Such a process is known as *quasi-stationary*. It can be analyzed for two aspects [CRA 83]:

– study of the stationary parts by calculation of the PSD whose parameters are functions slowly variable with time,

– study of the long-term behavior, described for example by a cross-probability distribution for the parameters slowly variable with the PSD.

The non-stationary process can also be of short duration. This is particularly the case for a mechanical oscillator at rest suddenly exposed to a stationary random excitation; there is a phase of transitory response, therefore it is non-stationary. Many studies have been conducted on this subject [CHA 72], [HAM 68], [PRI 67], [ROB 71], [SHI 70b]. Various solutions were obtained, among those of T.K. Caughey and H.J. Stumpf [CAU 61] (Chapter 8), R.L. Barnoski and J.R. Maurer [BAR 69] and Y.K. Lin [LIN 67]. Other definitions were also proposed for the PSD of non-stationary phenomena [MAR 70].

2.16. Sum of two random processes

Define the random process z according to

$$z = x \pm y \qquad [2.84]$$

in the interval [0, T]. The power spectral density of the random process $z = x \pm y$ is given by [HOW 02]

$$G_z(f) = G_x(f) + G_y(f) \pm 2\,\mathrm{Re}\!\left[G_{xy}(f)\right] \qquad [2.85]$$

where G_x and G_y, respectively, are the power spectral densities of x and y, and G_{xy} is the cross power spectral density between x and y.

If x is independent of y, the power spectral density of z is

$$G_z(f) = G_x(f) + G_y(f) \pm \frac{2}{T} \operatorname{Re}\left[\overline{X}(f)\overline{Y}^*(f)\right] \quad [2.86]$$

where $\overline{X}$ and $\overline{Y}^*$ are mean Fourier transforms.

For the case where x and y are independent and stationary, such that their respective means, denoted μ_x and μ_y are constant on [0, T], the power spectral density is

$$G_z(f) = G_x(f) + G_y(f) \pm 2 \operatorname{Re}\left[\mu_x \mu_y^*\right] T \operatorname{sinc}^2(f\,T) \quad [2.87]$$

If the mean of processes x and y is zero, the PSD of z(t) is equal to [ANT 06], [HOW 02], [PRI 04]:

$$G_z(f) = G_x(f) + G_y(f) \quad [2.88]$$

More generally, if

$$z(t) = a\,x(t) + b\,y(t) \quad [2.89]$$

we have

$$G_z(f) = a^2\,G_x(f) + b^2\,G_y(f) \quad [2.90]$$

This property can be used to remove noise from a signal if we have a recording of the single noise.

In the same way, the variance of the sum of non-correlated noises is such that:

$$\sigma_z^2 = \sigma_x^2 + \sigma_y^2 \quad [2.91]$$

The notion of non-correlation is fundamental here since the sum of correlated noises can differ radically from the result above. For example, if a noise is added to another noise identical to itself:

$$n(t) + n(t) = 2\, n(t)$$

the PSD is equal to 2 Gn(f), whereas for non-correlated noises, we would have 4 Gn(f).

We recall that the cross-correlation between 2 stochastic processes is defined by

$$R_{x,y}(\tau) = E\{x(t)\, y(t+\tau)\}$$

2.17. Relationship between the PSD of the excitation and the response of a linear system

We can easily show that [BEN 58], [BEN 62], [BEN 63], [BEN 80], [CRA 63]:

– if the excitation is a random stationary process, the response of a linear system is itself stationary,

– if the excitation is ergodic, the response is also ergodic.

Let us consider one of the functions $^i\ell(t)$ of a process (whether stationary or not); the response of a linear system can be written:

$$^i u(t) = \int_0^\infty h(\lambda)\; ^i\ell(t-\lambda)\, d\lambda \qquad [2.92]$$

yielding:

$$^i u(t_1)\; ^i u(t_2) = \int_0^\infty h(\lambda)\; ^i\ell(t_1-\lambda)\, d\lambda \int_0^\infty h(\mu)\; ^i\ell(t_2-\mu)\, d\mu$$

$$^i u(t_1)\; ^i u(t_2) = \int_0^\infty \int_0^\infty h(\lambda)\, h(\mu)\; ^i\ell(t_1-\lambda)\; ^i\ell(t_2-\mu)\, d\lambda\, d\mu \qquad [2.93]$$

Ensemble average

$$R_u(t_1, t_2) = E\big[u(t_1)\, u(t_2)\big] \qquad [2.94]$$

$$R_u(t_1, t_2) = \int_0^\infty \int_0^\infty h(\lambda)\, h(\mu)\, R_\ell(t_1-\lambda, t_2-\mu)\, d\lambda\, d\mu \qquad [2.95]$$

where

$$R_\ell(t_1 - \lambda, t_2 - \mu) = E[\ell(t_1 - \lambda) \ \ell(t_2 - \mu)] \quad [2.96]$$

Example of a stationary process

In this case,

$$R_u(t_1, t_2) = R_u(0, t_2 - t_1) = R_u(t_2 - t_1) = R_u(\tau)$$

and

$$R_u(\tau) = \int_0^\infty \int_0^\infty h(\lambda) \ h(\mu) \ R_\ell(\tau, \lambda - \mu) \ d\lambda \ d\mu \quad [2.97]$$

In addition, we have seen that [2.45]:

$$G(f) = 2 \int_{-\infty}^{+\infty} R(\tau) e^{-2\pi i f \tau} \ d\tau$$

The PSD of the response can be calculated from this expression [CRA 63], [JEN 68]:

$$G_u(f) = 2 \int_{-\infty}^{+\infty} R_u(\tau) e^{-2\pi i f \tau} \ d\tau$$

$$G_u(f) = 2 \int_{-\infty}^{+\infty} e^{-2\pi i f \tau} \left[\int_0^\infty \int_0^\infty h(\lambda) h(\mu) R_\ell(\tau + \lambda - \mu) d\lambda \ d\mu \right] d\tau$$

$$G_u(f) = 2 \int_0^\infty h(\lambda) e^{2\pi i f \lambda} \ d\lambda \int_0^\infty h(\mu) e^{-2\pi i f \mu} \ d\mu$$

$$\int_{-\infty}^{+\infty} R_\ell(\tau + \lambda - \mu) e^{-2\pi i f (\tau + \lambda - \mu)} \ d\tau$$

$$G_u(f) = H^*(f) \ H(f) \ G_\ell(f)$$

$$G_u(f) = |H(f)|^2 \ G_\ell(f) \quad [2.98]$$

Depending on the angular frequency, this expression becomes:

$$G_u(\Omega) = |H(j\Omega)|^2 \ G_\ell(\Omega) \quad [2.99]$$

NOTE.– *This result can be found starting with a Fourier series development of the excitation $\ell(t)$. Let us set $u(t)$ as the response at a point of the system. With each frequency f_j, the response is equal to H_j times the input (H_j = a real number). Thus, $u(t)$ can also be expressed in the form of a Fourier series, each term of $u(t)$ being equal to the corresponding term of $\ell(t)$ modified by the factor H_j and phase φ_j:*

$$u(t) = \sum_j u_j \, H_j \, \sin\left(\frac{2\pi j}{T} t + \varphi_j\right) \qquad [2.100]$$

i.e.

$$u(t) = \sum_j u_j \, H_j \left(\cos \varphi_j \, \sin\frac{2\pi j t}{T} + \sin \varphi_j \, \cos\frac{2\pi j t}{T} \right) \qquad [2.101]$$

The rms value of $u(t)$ is equal to

$$u^2_{rms} = \frac{1}{T} \sum_{j=1}^{\infty} u_j^2 \, H_j^2 \left(\frac{T}{2} \cos^2 \varphi_j + \frac{T}{2} \sin^2 \varphi_j \right) \qquad [2.102]$$

When $T \to \infty$,

$$u(t) \to \int_0^\infty G_\ell(f) \, H^2(f) \, df \qquad [2.103]$$

Knowing that, if $G_u(f)$ is the PSD of the response, $u^2_{rms} = \int_0^\infty G_u(f) \, df$, it becomes

$$G_u(f) = H^2(f) \, G_\ell(f) \qquad [2.104]$$

This method can be used for the measurement of the transfer function of a structure undergoing a pseudo-random vibration (random vibration of finite duration, possibly repeated several times). The method consists of applying white noise of duration T to the material, in measuring the response at a point and in determining the transfer function by term to term calculation of the ratio of the input and output coefficients of the Fourier series development.

2.18. Relationship between the PSD of the excitation and the cross-power spectral density of the response of a linear system

$$^{i}u(t) = \int_{0}^{\infty} h(\lambda) \, ^{i}\ell(t-\lambda) \, d\lambda$$

$$^{i}u(t) \, ^{i}u(t+\tau) = \int_{0}^{\infty} h(\lambda) \, ^{i}\ell(t) \, ^{i}\ell(t+\tau-\lambda) \, d\lambda$$

If the process is stationary, the ensemble average is:

$$R_{\ell u}(\tau) = \int_{0}^{\infty} h(\lambda) \, R_{\ell}(\tau-\lambda) \, d\lambda$$

and the cross-spectrum:

$$G_{u}(f) = 2 \int_{-\infty}^{+\infty} e^{-2\pi i f \tau} \left[\int_{0}^{\infty} h(\lambda) R_{\ell}(\tau-\lambda) d\lambda \right] d\tau$$

$$G_{\ell u}(f) = \int_{-\infty}^{+\infty} h(\lambda) e^{-2\pi i f \lambda} \, d\lambda \left[2 \int_{-\infty}^{\infty} R_{\ell}(\tau-\lambda) e^{-2\pi i f (\tau-\lambda)} d\tau \right]$$

$$G_{\ell u}(f) = H(f) \, G_{\ell}(f) \qquad [2.105]$$

NOTE.– *If we set:*

$$G_{\ell u}(f) = |A(f)| \, e^{i\varphi(f)}$$

the transfer function $H(f)$ can be written, knowing that the PSD $G_{\ell}(f)$ is a real function

$$\begin{cases} |H(f)| = \dfrac{|A(f)|}{G_{\ell}(f)} \\ \psi(f) = \varphi(f) \end{cases} \qquad [2.106]$$

2.19. Coherence function

The coherence function between two signals $\ell(t)$ and $u(t)$ is defined by [BEN 63], [BEN 78], [BEN 80], [ROT 70]:

$$\gamma_{\ell u}^2(f) = \frac{|G_{\ell u}(f)|^2}{G_{\ell\ell}(f)\, G_{uu}(f)} \qquad [2.107]$$

This function is a measure of the effect of input on response of a system.

In an ideal case,

$$G_{\ell u} = H(f)\, G_{\ell\ell}$$

and

$$\gamma_{\ell u}^2 = 1$$

Example 2.2.

Let us take a wideband random vibration and the response of a one-degree-of-freedom linear mechanical system with a frequency of 300 Hz and Q factor 10 for this noise.

The coherence function of this vibration and response is shown in Figure 2.19. We can observe that this function is still very close to 1 where the PSD has significant amplitude, which makes sense since the response signal is the result of a calculation (absence of unwanted noise).

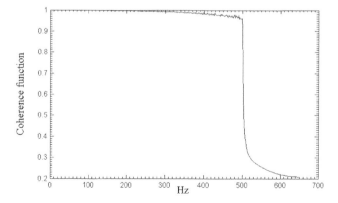

Figure 2.19. *Coherence function of "input" vibrations and response of a one-degree-of-freedom system*

$\gamma_{\ell u}$ is in addition zero if the signals $\ell(t)$ and $u(t)$ are completely uncorrelated. In general, $\gamma_{\ell u}$ lies between 0 and 1 for the following reasons:

- the presence of noise in measurements;
- the non-linear relationship between $\ell(t)$ and $u(t)$;
- the response $u(t)$ is due to other inputs than $\ell(t)$.

Presence of noise on the input

If the noise is present only on the input, we set $\ell(t) = s(t) + b(t)$ where s(t) is the pure signal and b(t) the noise. We have in the same way, for the PSD,

$$G_{\ell\ell}(f) = G_{ss}(f) + G_{bb}(f)$$

$$G_{uu}(f) = |H(f)|^2 G_{ss}$$

$$G_{\ell u}(f) = H(f) G_{ss}$$

$$\gamma_\ell^2 G_{\ell\ell} = G_{ss} \qquad [2.108]$$

$$\gamma_{\ell u}^2 = \frac{G_{ss}}{G_{\ell\ell}} = 1 - \frac{G_{bb}}{G_{\ell\ell}} \qquad [2.109]$$

Presence of noise on the response

Let us consider the case where noise exists only with the response. Setting G_{vv} as the PSD of the response without noise and G_{nn} that of the noise alone, it becomes:

$$G_{uu}(f) = G_{vv} + G_{nn}$$

where

$$G_{vv}(f) = |H(f)|^2 G_{\ell\ell}(f)$$

$$G_{vv} = \frac{|G_{\ell u}|^2}{|G_{\ell\ell}|} \quad G_{\ell\ell} = \gamma_{\ell u}^2 G_{uu} \qquad [2.110]$$

yielding

$$\gamma_{\ell u}^2 = \frac{G_{vv}}{G_{uu}} = \frac{G_{uu} - G_{nn}}{G_{uu}} = 1 - \frac{G_{nn}}{G_{uu}} \qquad [2.111]$$

The quantity $\gamma_{\ell u}^2 G_{uu}$ is called the *coherent ouput power spectrum*.

2.20. Transfer function calculation from random vibration measurements

2.20.1. *Theoretical relations*

Consider a stationary random vibration s(t) applied to a linear mechanical system and its response v(t). In the absence of noise on the input and on the response, measures $\ell(t)$ and u(t) of these signals are identical to s(t) and v(t) respectively. Their Fourier transforms are linked by

$$U(f) = H(f) L(f) \qquad [2.112]$$

where H(f) of the transfer function of the system.

We will square each member of this equation:

$$U^*(f) U(f) = H^*(f) H(f) L^*(f) L(f) \qquad [2.113]$$

or, according to the autospectrum definition,

$$G_{uu}(f) = |H(f)|^2 G_{\ell\ell}(f) \qquad [2.114]$$

which is simply expression [2.98] already established.

Yielding a first expression of the transfer function, which we will call $H_1(f)$:

$$|H_1(f)| = \sqrt{\frac{G_{uu}(f)}{G_{\ell\ell}(f)}} \qquad [2.115]$$

where $H_1(f)$ contains no information on the phase.

Going back to equation [2.112] and multiplying its two members by the conjugate Fourier transform of L(f):

$$L^*(f)\,U(f) = H(f)\,L^*(f)\,L(f) \qquad [2.116]$$

or, according to the PSD definition (autospectrum and cross-power spectral density):

$$G_{\ell u}(f) = H(f)\,G_{\ell\ell}(f) \qquad [2.117]$$

yielding a second expression of the transfer function:

$$H_2(f) = \frac{G_{\ell u}(f)}{G_{\ell\ell}(f)} \qquad [2.118]$$

Finally, multiply both sides of relation [2.112] by the conjugate $U^*(f)$ of $U(f)$. It becomes:

$$U^*(f)\,U(f) = H(f)\,U^*(f)\,L(f) \qquad [2.119]$$

$$G_{uu}(f) = H(f)\,G_{u\ell}(f) \qquad [2.120]$$

Yielding the last expression of $H(f)$:

$$H_3(f) = \frac{G_{uu}(f)}{G_{u\ell}(f)} \qquad [2.121]$$

These three conditions are theoretically similar as long as:

– signals $\ell(t)$ and $u(t)$ have no noise;

– there is no other vibration source contributing to the response $u(t)$;

– PSDs are calculated with a very low statistical error (see Chapter 4).

From definition [2.107], we can say

$$\gamma^2(f) = \frac{\left|G_{\ell u}(f)^2\right|}{G_{\ell\ell}(f)\,G_{uu}(f)} = \frac{G_{\ell u}(f)}{G_{\ell\ell}(f)}\,\frac{G_{\ell u}(f)}{G_{uu}(f)}$$

Transfer functions are therefore linked to the coherence function by:

$$\gamma^2(f) = \frac{H_2(f)}{H_3(f)} \qquad [2.122]$$

2.20.2. Presence of noise on the input

This situation can occur in practice during a test on an exciter. With stationary random vibration, the mechanical response of the specimen shows peaks at resonance frequencies. At these frequencies, the power of the exciter may not be sufficient for maintaining the excitation at the desired value. In this case, the level of excitation decreases and can be close to the background noise.

Consider an "input" vibration s(t) producing a response u(t) in the output of a mechanical system. We presume here that the input has a noise b(t) that is not correlated with s(t).

$$\ell(t) = s(t) + b(t) \qquad [2.123]$$

The Fourier transform of $\ell(t)$ is equal to

$$L(f) = S(f) + B(f) \qquad [2.124]$$

Since the response is noiseless, the measured value u(t) is equal to v(t). Knowing that

$$U(f) = H(f) L(f) \qquad [2.125]$$

it comes to

$$G_{\ell\ell}(f) = G_{ss}(f) + G_{bb}(f) \qquad [2.126]$$

First definition

The transfer function defined by $H_1(f)$ is equal to

$$H_1(f) = \frac{G_{uu}(f)}{G_{\ell\ell}(f)} = |H^2(f)| \frac{G_{ss}}{G_{ss} + G_{bb}} \qquad [2.127]$$

$$H_1(f) = |H^2(f)| \frac{1}{1 + \frac{G_{nn}}{G_{ss}}} \qquad [2.128]$$

$H_1(f)$ moves away from the theoretical H(f) value of the transfer function as the signal-to-noise ratio becomes greater. The calculated value is lower than the real value.

Second definition

The response and excitation autospectrums are linked by

$$G_{uu}(f) = H^2 G_{ss}(f) \qquad [2.129]$$

However,

$$G_{\ell u}(f) = G_{su}(f) + G_{bu}(f) = G_{su}(f) \qquad [2.130]$$

since $G_{bu} = 0$. Hence

$$G_{\ell u}(f) = G_{su}(f) = H^2(f) G_{ss}(f) \qquad [2.131]$$

and

$$H_2(f) = \frac{G_{\ell u}(f)}{G_{\ell \ell}(f)} = H(f) \frac{G_{ss}}{G_{ss} + G_{bb}} \qquad [2.132]$$

$$H_2(f) = H(f) \frac{1}{1 + \frac{G_{nn}}{G_{ss}}} \qquad [2.133]$$

Third definition

The definition of transfer function $H_3(f)$ leads to

$$H_3(f) = \frac{G_{uu}(f)}{G_{u\ell}(f)} = \frac{H^2(f) G_{ss}}{H^*(f) G_{ss}} = H(f) \qquad [2.134]$$

The H_3 function is then impervious to the noise present in the input measure. We can show that it is the best estimation of function $H(f)$ [BEN 80].

Coherence function

$$\gamma^2(f) = \frac{H_2(f)}{H_3(f)} = \frac{1}{1 + \frac{G_{bb}(f)}{G_{ss}(f)}} \qquad [2.135]$$

Signal over noise ratio

$$R(f) = \frac{G_{ss}(f)}{G_{bb}(f)} = \frac{\gamma^2(f)}{1-\gamma^2(f)} \quad [2.136]$$

2.20.3. *Presence of noise on the response*

We now consider a mechanical system experiencing a random vibration s(t) without noise. The measured signal $\ell(t)$ is therefore directly s(t).

Let v(t) be the response of the system; its measure is carried out in the presence of a noise n(t). We have:

$$u(t) = v(t) + n(t) \quad [2.137]$$

$$U(f) = V(f) + N(f) \quad [2.138]$$

First definition

$$G_{uu}(f) = G_{vv}(f) + G_{nn}(f) = |H(f)|^2 G_{ss}(f) + G_{nn}(f) \quad [2.139]$$

$$|H_1(f)|^2 = \frac{G_{uu}(f)}{G_{\ell\ell}(f)} = \frac{G_{uu}(f)}{G_{ss}(f)} = \frac{G_{vv}(f) + G_{nn}(f)}{G_{ss}(f)}$$

$$|H_1(f)|^2 = \frac{|H(f)|^2 G_{ss}(f) + G_{nn}(f)}{G_{ss}(f)}$$

$$|H_1(f)|^2 = |H(f)|^2 \frac{1}{1 + \frac{G_{nn}(f)}{G_{ss}(f)}} \quad [2.140]$$

Second definition

$$G_{\ell u}(f) = G_{\ell v}(f) + G_{\ell n}(f) = G_{\ell v}(f) = H(f) G_{\ell\ell}(f) = H(f) G_{ss}(f) \quad [2.141]$$

since $G_{\ell n}(f) = 0$.

Hence

$$H_2(f) = \frac{G_{\ell u}(f)}{G_{\ell \ell}(f)} = \frac{G_{su}(f)}{G_{ss}(f)} = H(f) \qquad [2.142]$$

This is the best estimation of H(f) in these conditions [HER 84a].

Third definition

$$H_3(f) = \frac{G_{uu}(f)}{G_{y\ell}(f)} = \frac{|H(f)|^2 G_{\ell\ell}(f) + G_{nn}(f)}{H^*(f) G_{\ell\ell}(f)} \qquad [2.143]$$

$$H_3(f) = H(f)\left(1 + \frac{G_{nn}(f)}{G_{vv}(f)}\right) \qquad [2.144]$$

Function $H_3(f)$ overestimates the transfer function because of noise. The phase that is determined from the cross-power spectral density is correct.

Coherence function

$$\gamma^2(f) = \frac{H_2(f)}{H_3(f)} = \frac{1}{1 + \frac{G_{nn}(f)}{G_{vv}(f)}} \qquad [2.145]$$

Knowing that

$$\gamma^2(f) = \frac{|G_{\ell u}(f)^2|}{G_{\ell\ell}(f) G_{uu}(f)} = \frac{G_{\ell\ell}(f) G_{vv}(f)}{G_{\ell\ell}(f) G_{uu}(f)} = \frac{G_{vv}(f)}{G_{uu}(f)} \qquad [2.146]$$

Relation [2.146] shows that the response can be extracted from the noisy signal with the help of

$$G_{vv}(f) = \gamma^2(f) G_{uu}(f) \qquad [2.147]$$

whereas the noise is given by

$$G_{nn}(f) = \left(1 - \gamma^2(f)\right) G_{uu}(f) \qquad [2.148]$$

Signal-to-noise ratio

$$R(f) = \frac{G_{vv}(f)}{G_{nn}(f)} = \frac{\gamma^2(f)}{1-\gamma^2(f)} \qquad [2.149]$$

2.20.4. *Presence of noise on the input and response*

Finally, we consider the more general case where the input and response are both affected by noises that are not correlated between each other:

$$\ell(t) = s(t) + b(t) \qquad [2.150]$$

$$u(t) = v(t) + n(t) \qquad [2.151]$$

$$G_{\ell\ell}(f) = G_{ss}(f) + G_{bb}(f) \qquad [2.152]$$

$$G_{uu}(f) = G_{vv}(f) + G_{nn}(f) = |H(f)|^2 G_{ss}(f) + G_{nn}(f) \qquad [2.153]$$

First definition

$$|H_1(f)|^2 = |H(f)|^2 \frac{1+\dfrac{G_{nn}(f)}{G_{vv}(f)}}{1+\dfrac{G_{bb}(f)}{G_{ss}(f)}} \qquad [2.154]$$

Second definition

$$H_2(f) = H(f) \frac{1}{1+\dfrac{G_{bb}(f)}{G_{ss}(f)}} \qquad [2.155]$$

Third definition

$$H_3(f) = H(f)\left(1+\frac{G_{nn}(f)}{G_{vv}(f)}\right) \qquad [2.156]$$

$H_2(f)$ underestimates the transfer function, whereas $H_3(f)$ overestimates it. Depending on the values of the signal-to-noise ratios, $H_1(f)$ can overestimate or underestimate the transfer function. The phases, calculated from the cross-spectrum, are correct.

Coherence function

$$\gamma^2(f) = \frac{H_2(f)}{H_3(f)} = \frac{1}{\left(1 + \frac{G_{bb}(f)}{G_{ss}(f)}\right)\left(1 + \frac{G_{nn}(f)}{G_{vv}(f)}\right)} \qquad [2.157]$$

2.20.5. *Choice of transfer function*

In the ideal case of a linear system where there is no noise on the excitation and response, the coherence function is equal to 1 and the transfer function $H(f)$ is equally given by $|H_1(f)|$, $H_2(f)$ or $H_3(f)$ [BEN 80], [HER 84a].

	Input without noise/ Output with noise	Input with noise/ Output without noise	Input with noise/ Output with noise
$|H_1(f)|$	Overestimates H(f)	Underestimates H(f)	Over or underestimates H(f) according to signal-to-noise ratios at input and response
$H_2(f)$	Best representation of H(f)	Underestimates H(f)	Underestimates H(f)
$H_3(f)$	Overestimates H(f)	Impervious to noise on input	Overestimates H(f)

Table 2.2. *Comparison of the different transfer functions*

In practice, it is preferable to use [HER 84a]:

– $H_2(f)$ when the response has noise or when there are many independent inputs;

– $H_3(f)$ when there is noise on excitation or in the presence of "leakage" at resonance frequency (resolution bias).

Reality is generally more complex:

– using $H_2(f)$ and $H_3(f)$ is often useful according to the frequency range, for example $H_2(f)$ for anti-resonances, where the noise on the response signal tends to dominate, and $H_3(f)$ for the peaks, where the noise linked to "leakage" tends to occur,

– since the noise is present on input and output, $H_2(f)$ and $H_3(f)$ can provide the lower and higher limits of the transfer function.

Example 2.3.

Let us take a random vibration lasting 50 s, defined by a PSD close to 0.1 $(m/s^2)^2$/Hz between 10 Hz and 500 Hz (rms value: 7 m/s^2) and the response of a one-degree-of-freedom linear system at a frequency of $f_0 = 250$ Hz with an over-tension $Q = 10$ (rms value: approximately 20 m/s^2).

Figure 2.20 shows the transfer functions H_1, H_2 and H_3 of the system calculated from relations [2.115], [2.118] and [2.121] respectively. We can verify that they are identical.

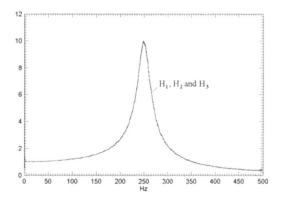

Figure 2.20. *Transfer functions H_1, H_2 and H_3 without noise*

A constant PSD noise between 10 Hz and 500 Hz with the same rms value as the excitation was added to the response before calculation of transfer functions. Resulting transfer functions are superimposed to Figure 2.21 with the theoretical transfer function (function H_1 in Figure 2.20). H_1 and H_3 are higher than H_2, which is very close to the reference.

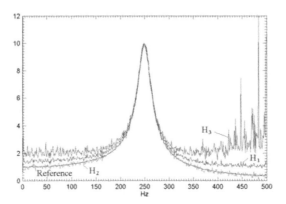

Figure 2.21. *Transfer functions H_1, H_2 and H_3 with noise on the response*

This same noise was added to the "input" vibration, since the response is presumed to be noiseless here. Function H_3 is then the closest to the theoretical function, as H_1 and H_2 are lower (Figure 2.22).

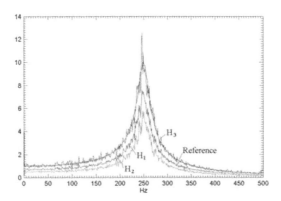

Figure 2.22. *Transfer functions H_1, H_2 and H_3 with noise on input*

Finally, the same noise was added to random input and response vibrations. The H_1 curve, the closest to the reference, is included between H_2 (the shortest) and H_3 (the tallest) curves (Figure 2.23).

124   Random Vibration

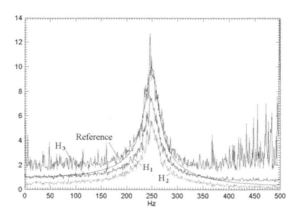

Figure 2.23. *Transfer functions H_1, H_2 and H_3 with noise on input and response*

Example 2.4.

Let us take a random vibration lasting 50 s, with a constant PSD equal to 2 $(m/s^2)^2/Hz$ between 1 Hz and 500 Hz, and very small up to 650 Hz. This vibration (signal according to time) was applied to a one-degree-of-freedom linear system with a frequency of 200 Hz and a Q factor of 5.

Figure 2.24 shows the transfer function determined by two methods: from the ratio between excitation and response autospectrums, or from the cross-spectrum. This example being based on a numerical calculation of the response, the results are very similar.

However, we observe that, in the frequency band greater than 500 Hz where the input PSD is very low, the autospectrum ratio leads to a faulty transfer function, which is not the case with the cross-spectrum.

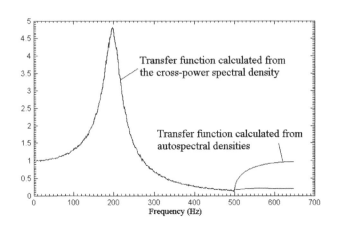

Figure 2.24. *Transfer function calculated from autospectrum and cross-spectrum*

Chapter 3

Rms Value of Random Vibration

3.1. Rms value of a signal as a function of its PSD

We saw that [2.26]:

$$G(f) = \lim_{\substack{\Delta f \to 0 \\ T \to \infty}} \frac{1}{T\,\Delta f} \int_0^T \ell_T^2(t, \Delta f)\, dt$$

$G(f)$ is the square of the rms value of the signal filtered by a filter Δf whose width tends towards zero, centered around f. To obtain the total rms value ℓ_{rms} of the signal, taking into account all the frequencies, it is thus necessary to calculate

$$\ell_{rms}^2 = \int_{0^-}^{\infty} G(f)\, df \qquad [3.1]$$

The notation 0^- means that integration is carried out in a frequency interval covering $f = 0$, while 0^+ indicates that the interval does not include the limiting case $f = 0$. In a given frequency band f_1, f_2 $(f_2 > f_1)$,

$$\ell_{rms}^2 = \int_{f_1}^{f_2} G(f)\, df \qquad [3.2]$$

128 Random Vibration

The square of the rms value of the signal in a limited frequency interval f_1, f_2 is equal to the area under the curve $G(f)$ in this interval. In addition:

$$\int_{0^+}^{\infty} G(f)\,df = s_\ell^2 \qquad [3.3]$$

where

– s_ℓ^2 is the variance of the signal without its continuous component; and

– s_ℓ is the standard deviation of the signal.

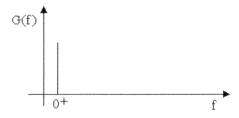

Figure 3.1. *Non-zero lower limit of the PSD*

In addition:

$$\int_{0^-}^{0^+} G(f)\,df = (\bar{\ell})^2 \qquad [3.4]$$

$\bar{\ell}$ is the mean value of the signal. We thus have:

$$\ell_{rms}^2 = s_\ell^2 + (\bar{\ell})^2 \qquad [3.5]$$

Lastly, for $f \neq 0$, we have:

$$\int_{f^-}^{f^+} G(f)\,df = 0 \qquad [3.6]$$

A purely random signal does not have a discrete frequency component.

NOTE.– *The mean value $\overline{\ell}$ corresponding to the continuous component of the signal can originate in:*

– the shift due to the measuring equipment, the mean value of the signal being actually zero. This component can be eliminated, either by centering the signal $\overline{\ell}$ before the calculation of its PSD or by calculating the PSD between $f_1 = 0 + \varepsilon$ and f_2 (ε being a positive constant different from zero, arbitrarily small);

– the permanent acceleration, constant or slowly variable, corresponding to a rigid body movement of the vehicle (for example, static acceleration in phase propulsion of a launcher using propellant). We often dissociate this by filtering such static acceleration of vibrations which are superimposed on it, the consideration of static and dynamic phenomena being carried out separately. It is however important to be able to identify and measure these two parameters, in order to be able to study the combined effects of them, for example during calculations of fatigue strength, if necessary (using the Goodman or Gerber rule, etc. (see Volume 4) or of reaction to extreme stress.

Static and dynamic accelerations are often measured separately by different sensors, the vibration measuring equipment not always covering the DC component. Except for particular cases, we will always consider in what follows the case of zero mean signals. We know that, in this case, the rms value of the signal is equal to its standard deviation.

Obtained by calculation of a mean square value, the PSD is an incomplete description of the signal $\ell(t)$. There is loss of information on phase. Two signals of comparable nature and of different phases will have the same PSD.

Example 3.1.

Let us consider a stationary random acceleration $\ddot{x}(t)$ having a uniform power spectral density given by:

$G_{\ddot{x}}(f) = 0.0025$ $(m/s^2)^2/Hz$

in the frequency domain ranging between $f_1 = 10$ Hz and $f_2 = 1{,}000$ Hz, and zero elsewhere.

130   Random Vibration

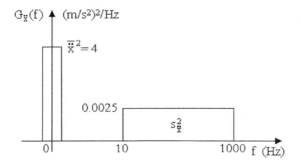

Figure 3.2. *PSD of a signal having a continuous component*

Let us suppose in addition that the continuous component of the signal is equal to $\bar{\ddot{x}} = 2$ m/s². Let us calculate the rms value and the standard deviation of the signal. The mean square value of the signal is given by relation [3.5]:

$$\overline{\ddot{x}^2} = \bar{\ddot{x}}^2 + s_{\ddot{x}}^2 = \int_{0^-}^{\infty} G_{\ddot{x}}(f)\, df$$

$$s_{\ddot{x}}^2 = \int_{0^+}^{\infty} G_{\ddot{x}}(f)\, df$$

$$s_{\ddot{x}}^2 = \int_{10}^{1000} 0.0025\, df$$

$$s_{\ddot{x}}^2 = (1000 - 10)\, 0.0025 = 2.475\ (\text{m/s}^2)^2$$

$$\bar{\ddot{x}}^2 = 4\ (\text{m/s}^2)^2$$

yielding the mean square value

$$\overline{\ddot{x}^2} = 4 + 2.475 = 6.475\ (\text{m/s}^2)^2$$

and the rms value

$$\sqrt{\overline{\ddot{x}^2}} \approx 2.545\ \text{m/s}^2$$

while the standard deviation is equal to $s_{\ddot{x}} = 1.573$ m/s².

The random signals are in general centered before the calculation of the spectral density, so that $\overline{\ddot{x}^2} = s_{\ddot{x}}^2$.

3.2. Relationships between the PSD of acceleration, velocity and displacement

Let us set as $\ell(t)$ a random signal with Fourier transform $L(f)$; by definition, we have:

$$L(f) = \int_{-\infty}^{+\infty} \ell(t) \, e^{-2\pi i f t} \, dt \qquad [3.7]$$

and

$$\ell(t) = \int_{-\infty}^{+\infty} L(f) \, e^{2\pi i f t} \, df \qquad [3.8]$$

yielding

$$\dot{\ell}(t) = \frac{d\ell}{dt} = \int_{-\infty}^{+\infty} 2\pi i f \, L(f) \, e^{2\pi i f t} \, dt$$

$$\dot{\ell}(t) = \int_{-\infty}^{+\infty} \dot{L}(f) \, e^{2\pi i f t} \, dt \qquad [3.9]$$

By identification, it becomes:

$$\dot{L}(f) = 2\pi i f \, L(f) \qquad [3.10]$$

The conjugate expressions of $L(f)$ and of $\dot{L}(f)$ are obtained by replacing i with $-i$. If $G_\ell(f)$ and $G_{\dot{\ell}}(f)$ are respectively the PSD of $\ell(t)$ and of $\dot{\ell}(t)$, we thus obtain, since these quantities are functions of the products $L^*(f)\,L(f)$ and $\dot{L}^*(f)\,\dot{L}(f)$ [HUS 56], [LEY 65], [LIN 67], [THO 08],

$$G_{\dot{\ell}}(f) = 4\pi^2 f^2 \, G_\ell(f) \qquad [3.11]$$

yielding

$$\dot{\ell}_{rms}^2 = \int_0^{+\infty} G_{\dot{\ell}}(f) \, df \qquad [3.12]$$

$$\dot{\ell}_{rms}^2 = \int_0^{+\infty} (2\pi f)^2 \, G_\ell(f) \, df \qquad [3.13]$$

132   Random Vibration

and, in the same way,

$$\ddot{\ell}^2_{rms} = \int_0^{+\infty} (2\pi f)^4 \, G_\ell(f) \, df \qquad [3.14]$$

NOTES.–

1. *These relations use an integral with respect to the frequency between 0 and $+\infty$. In practice, the PSD is calculated only for one frequency interval (f_1, f_2). The initial frequency f_1 is a function of the duration of the sample selected; this duration being necessarily limited, f_1 cannot always be taken as low as would be desirable.*

The limit f_2 is if possible selected to be sufficiently large so that all the frequency content is described. It is not always possible for certain phenomena, if only because of the measuring equipment. A value often used is, for example, 2,000 Hz. However, the integral necessary for the evaluation of $\ddot{\ell}^2_{rms}$ includes a term in f^4 which makes it very sensitive to the high frequencies.

In the calculation of all the expressions utilizing $\ddot{\ell}^2_{rms}$, as will be the case for the irregularity parameter r which we will define later, the result could be spoilt, having considerable error in the event of an inappropriate choice of the limits f_1 and f_2.

J. Schijve [SCH 63] considers that the high frequency/small amplitude peaks have little influence on the fatigue suffered by the materials and propose limiting integration to approximately 1,000 Hz (for vibratory aircraft environments).

2. *It is known that the rms value of a sinusoidal acceleration signal is related to the corresponding velocity and the displacement by*

$$\ddot{\ell} = 2\pi f \, \dot{\ell} = (2\pi f)^2 \, \ell \qquad [3.15]$$

These relationships apply at first approximation to the rms values of a very narrowband random signal of central frequency f.

This makes it possible to differently demonstrate relations [3.13] and [3.14]. The PSD of a signal $\ddot{\ell}(t)$ is indeed calculated while filtering $\ddot{\ell}(t)$ using a filter of width Δf whose central frequency varies in the definition interval of the PSD, the result being squared and divided by Δf for each point of the PSD. We thus obtain [CUR 64], [DEE 71], [HIM 59]:

$$G_{\dot{\ell}} = (2\pi f)^2 G_{\ell} \qquad [3.16]$$

$$G_{\ddot{\ell}} = (2\pi f)^2 G_{\dot{\ell}} = (2\pi f)^4 G_{\ell} \qquad [3.17]$$

yielding [OSG 69], [OSG 82]:

$$\dot{\ell}^2_{rms} = \int_0^{\infty} \frac{G_{\ddot{\ell}}(f)}{(2\pi f)^2}\, df \qquad [3.18]$$

and

$$\ell^2_{rms} = \int_0^{\infty} \frac{G_{\ddot{\ell}}(f)}{(2\pi f)^4}\, df \qquad [3.19]$$

We can deduce from these relations the rms value of the displacement of a very narrowband noise [BAN 78], [OSG 69]:

$$\ell_{rms} = \frac{\ddot{\ell}_{rms}}{4\pi^2 f^2} \qquad [3.20]$$

3.3. Graphical representation of the PSD

We will consider here the most frequent case where the vibratory signal to analyze is an acceleration. The PSD is the subject of four general presentations:

– the first with the frequency on the x axis (Hz), the amplitude of the PSD on the y axis $[(m/s^2)^2/Hz]$, the points being regularly distributed by frequency (constant filter width Δf throughout the whole range of analysis);

– the second, encountered primarily in acoustics problems, uses an analysis in the $\frac{1}{n}$th octave, the filter width being thus variable with the frequency; we find more often in this case the ordinates expressed in decibels (dB). The number of decibels is then given by:

$$n_{dB} = 10 \log \frac{G}{G_0} \qquad [3.21]$$

where

134   Random Vibration

   – G is the amplitude of the measured PSD,

   – G_0 is a reference value, selected equal to 10^{-12} $(m/s^2)^2/Hz$ in general,

or, if we consider the rms value in each band of analysis, by

$$n_{dB} = 20 \log \frac{a}{a_0} \qquad [3.22]$$

where

   a = rms value of the signal in the selected band of analysis,

   a_0 = reference value of (10^{-6} m/s^2),

   – sometimes, the analysis in the $\frac{1}{n}$-th octave is carried out by indicating in ordinates the rms value obtained in each filter. For a noise whose PSD varies little with the frequency (close to white noise), the rms value obtained varies with the bandwidth of the filter;

   – relationships [3.17] show that the PSD can also be plotted on a four-coordinate nomographic grid on which the PSD value can be directly read for acceleration, velocity and displacement [HIM 59].

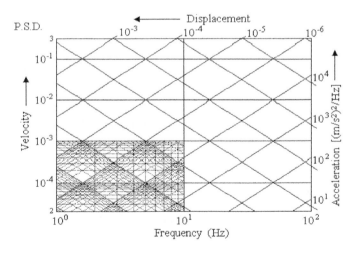

Figure 3.3. *Four-coordinate representation [HIM 59]*

3.4. Practical calculation of acceleration, velocity and displacement rms values

3.4.1. *General expressions*

The rms values of acceleration, velocity and displacement are more particularly useful for evaluation of feasibility of a specified random vibration on a test facility (electrodynamic shaker or hydraulic vibration machine). Control in a general way being carried out from a PSD of acceleration, we will in this case temporarily abandon the generalized coordinates. We saw that the rms value $\ddot{x}_{rms}$ of a random vibration $\ddot{x}(t)$ of PSD $G(f)$ is equal to:

$$\ddot{x}_{rms}^2 = \int_0^\infty G(f) \, df$$

The rms velocity and displacement corresponding to this signal of acceleration are respectively given by:

$$v_{rms}^2 = \int_0^\infty \frac{G(f)}{(2\pi f)^2} \, df \qquad [3.23]$$

and

$$x_{rms}^2 = \int_0^\infty \frac{G(f)}{(2\pi f)^4} \, df \qquad [3.24]$$

In the general case where the PSD $G(f)$ is not constant, the calculation of these three parameters is made by numerical integration between the two limits f_1 and f_2 of the definition interval of $G(f)$. When $G(f)$ can be represented by a succession of horizontal or arbitrary slope straight line segments, it is possible to obtain analytical expressions.

3.4.2. *Constant PSD in frequency interval*

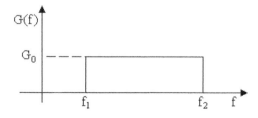

Figure 3.4. *Constant PSD between two frequencies*

In this very simple case where the PSD is constant between f_1 and f_2, $G(f) = G_0$, yielding:

$$\ddot{x}_{rms} = \sqrt{G_0(f_2 - f_1)} \qquad [3.25]$$

$$v_{rms}^2 = \frac{G_0}{4\pi^2} \int_{f_1}^{f_2} \frac{df}{f^2}$$

$$v_{rms} = \frac{1}{2\pi}\sqrt{G_0\left(\frac{1}{f_1} - \frac{1}{f_2}\right)} \qquad [3.26]$$

$$x_{rms} = \frac{1}{4\pi^2}\sqrt{\frac{G_0}{3}\left(\frac{1}{f_1^3} - \frac{1}{f_2^3}\right)} \qquad [3.27]$$

NOTE.− *The rms displacement x_{rms} can also be written as a function of rms acceleration:*

$$x_{rms} = \frac{1}{4\pi^2}\sqrt{\frac{1}{3}\frac{\ddot{x}_{rms}^2}{f_2 - f_1}\left(\frac{1}{f_1^3} - \frac{1}{f_2^3}\right)}$$

$$\frac{x_{rms}}{\ddot{x}_{rms}} = \frac{1}{4\pi^2\sqrt{3}}\left[\frac{f_1^2 + f_1 f_2 + f_2^2}{f_1^3 f_2^3}\right]^{\frac{1}{2}} \qquad [3.28]$$

If $f_1 \ll f_2$ [CRE 56]

$$x_{rms} = \frac{\ddot{x}_{rms}}{4\pi^2 f_1 \sqrt{3 f_1 f_2}} \qquad [3.29]$$

3.4.3. PSD comprising several horizontal straight line segments

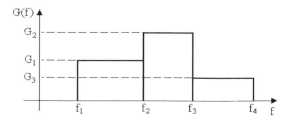

Figure 3.5. *PSD comprising horizontal segments*

We then have [SAN 63]:

$$\ddot{x}_{rms} = \sqrt{\sum_i G_i (f_{i+1} - f_i)} \qquad [3.30]$$

$$v_{rms} = \frac{1}{2\pi} \sqrt{\sum_i G_i \left(\frac{1}{f_i} - \frac{1}{f_{i+1}} \right)} \qquad [3.31]$$

$$x_{rms} = \frac{1}{4\pi^2} \sqrt{\sum_i \frac{G_i}{3} \left(\frac{1}{f_i^3} - \frac{1}{f_{i+1}^3} \right)} \qquad [3.32]$$

3.4.4. PSD defined by a linear segment of arbitrary slope

It is essential in this case to specify in which scales the segment of straight line is plotted.

Linear-linear scales

Between the frequencies f_1 and f_2, the PSD $G(f)$ obeys $G(f) = a\,f + b$, where a and b are constants such that, for $f = f_1$, $G = G_1$ and for $f = f_2$, $G = G_2$, yielding

$$a = \frac{G_2 - G_1}{f_2 - f_1} \text{ and } b = \frac{f_1 G_2 - f_2 G_1}{f_1 - f_2}.$$

$$\ddot{x}_{rms}^2 = \int_{f_1}^{f_2} G(f)\,df = \int_{f_1}^{f_2} (a\,f + b)\,df$$

$$\ddot{x}_{rms} = \sqrt{\frac{(f_2 - f_1)(G_2 + G_1)}{2}} \quad [3.33]$$

$$v_{rms}^2 = \frac{1}{4\pi^2}\left[a \ln \frac{f_2}{f_1} - b\left(\frac{1}{f_2} - \frac{1}{f_1}\right)\right]$$

$$v_{rms} = \frac{1}{2\pi}\sqrt{\frac{G_2 - G_1}{f_2 - f_1} \ln\left(\frac{f_2}{f_1}\right) + \frac{G_1}{f_1} - \frac{G_2}{f_2}} \quad [3.34]$$

$$x_{rms}^2 = \frac{1}{16\pi^4}\int_{f_1}^{f_2} \frac{a f + b}{f^4} df$$

$$x_{rms} = \frac{1}{4\pi^2 f_1 f_2}\sqrt{\frac{G_2 - G_1}{2}(f_1 + f_2) + \frac{1}{3}\left(\frac{G_1}{f_1} - \frac{G_2}{f_2}\right)\left(f_1^2 + f_1 f_2 + f_2^2\right)} \quad [3.35]$$

Linear-logarithmic scales

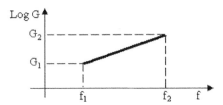

Figure 3.6. *Segment of straight line in lin-log scales*

The PSD can be expressed analytically in the form:

$$\ln G = a f + b$$

where $a = \dfrac{\ln G_2 - \ln G_1}{f_2 - f_1}$ and $b = \dfrac{f_1 \ln G_2 - f_2 \ln G_1}{f_1 - f_2}$.

$$\ddot{x}_{rms}^2 = \int_{f_1}^{f_2} e^{a f + b} df = \frac{e^b}{a}\left(e^{a f_2} - e^{a f_1}\right)$$

Rms Value of Random Vibration 139

$$\ddot{x}_{rms} = \sqrt{\frac{G_1}{a} \left[e^{a(f_2 - f_1)} - 1 \right]}$$ [3.36]

(if $a \neq 0$, i.e. if $G_2 \neq G_1$)

$$v_{rms}^2 = \frac{1}{4\pi^2} \int_{f_1}^{f_2} \frac{e^{af}}{f^2} \, df + \frac{e^b}{4\pi^2} \int_{f_1}^{f_2} \frac{df}{f^2}$$ [3.37]

Knowing that $\int \frac{e^{af}}{f^2} \, df = -\frac{e^{af}}{f} + a \int \frac{e^{af}}{f} \, df$, this integral can be calculated by a development in series (Appendix A4.2):

$$\int \frac{e^{af}}{f} \, df = \ln |f| + \frac{af}{1!} + \frac{a^2 f^2}{2 \cdot 2!} + \cdots + \frac{a^n f^n}{n \cdot n!} + \cdots$$

In the same way:

$$x_{rms}^2 = \frac{e^b}{16\pi^4} \int_{f_1}^{f_2} \frac{e^{af}}{f^4} \, df$$ [3.38]

The integral is calculated as above, from (Appendix A4.2):

$$\int \frac{e^{af}}{f^4} \, df = -\frac{e^{af}}{3f^3} + \frac{a}{3} \int \frac{e^{af}}{f^3} \, df$$

$$\int \frac{e^{af}}{f^4} \, df = -\frac{e^{af}}{3f^3} - \frac{a \, e^{af}}{6 f^2} - \frac{a^2 \, e^{af}}{6 f} + \frac{a^3}{6} \int \frac{e^{af}}{f} \, df$$

Particular case where $G_2 = G_1$

In this case:

$$\ddot{x}_{rms} = \sqrt{e^b (f_2 - f_1)} = \sqrt{G_1 (f_2 - f_1)}$$ [3.39]

140 Random Vibration

$$v_{rms} = \frac{1}{2\pi}\sqrt{G_1\left(\frac{1}{f_1} - \frac{1}{f_2}\right)} \qquad [3.40]$$

$$x_{rms} = \frac{1}{4\pi^2}\sqrt{\frac{G_1}{3}\left(\frac{1}{f_1^3} - \frac{1}{f_2^3}\right)} \qquad [3.41]$$

Logarithmic-linear scales

In these scales, the segment of straight line has as an analytical expression:

$$G = a \ln f + b$$

with $a = \dfrac{G_2 - G_1}{\ln f_2 - \ln f_1}$ and $b = \dfrac{G_2 \ln f_1 - G_1 \ln f_2}{\ln f_1 - \ln f_2}$.

$$\ddot{x}_{rms}^2 = a\left(f_2 \ln f_2 - f_1 \ln f_1\right) + \left(f_2 - f_1\right)(b - a) \qquad [3.42]$$

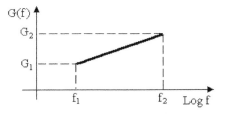

Figure 3.7. *Segment of straight line in log-lin scales*

$$v_{rms}^2 = \frac{a}{4\pi^2}\left(\frac{\ln f_1}{f_1} - \frac{\ln f_2}{f_2}\right) + \left(\frac{1}{f_1} - \frac{1}{f_2}\right)\left(\frac{a+b}{4\pi^2}\right) \qquad [3.43]$$

$$x_{rms}^2 = \frac{a}{48\pi^4}\left(\frac{\ln f_1}{f_1^3} - \frac{\ln f_2}{f_2^3}\right) - \frac{a+3b}{144\pi^4}\left(\frac{1}{f_2^3} - \frac{1}{f_1^3}\right) \qquad [3.44]$$

Logarithmic-logarithmic scales

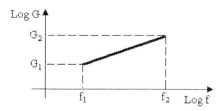

Figure 3.8. *Segment of a straight line in logarithmic scales*

The PSD is such that:

$$\ln G(f) = \ln G_1 + b \left(\ln f - \ln f_1\right)$$

hence

$$G(f) = G_1 \left(\frac{f}{f_1}\right)^b$$

The constant b is calculated from the coordinates of the point f_2, G_2:

$$b = \frac{\ln \dfrac{G_2}{G_1}}{\ln \dfrac{f_2}{f_1}}$$

Rms acceleration [PRA 70]:

$$\ddot{x}_{rms}^2 = \int_{f_1}^{f_2} G_1 \left(\frac{f}{f_1}\right)^b df$$

If $b \neq -1$:

$$\ddot{x}_{rms} = \sqrt{\frac{f_2 G_2 - f_1 G_1}{b+1}} \qquad [3.45]$$

If b = − 1:

$$\ddot{x}^2_{rms} = f_1 G_1 \int_{f_1}^{f_2} \frac{df}{f}$$

$$\ddot{x}_{rms} = \sqrt{f_1 G_1 \ln\frac{f_2}{f_1}} \qquad [3.46]$$

$$v^2_{rms} = \frac{G_1}{4\pi^2 f_1^b} \int_{f_1}^{f_2} f^{b-2} \, df$$

If b ≠ 1:

$$v_{rms} = \frac{1}{2\pi} \sqrt{\frac{G_1}{(b-1)f_1}\left[\left(\frac{f_2}{f_1}\right)^{b-1} - 1\right]} = \frac{1}{2\pi}\sqrt{\frac{1}{b-1}\left(\frac{G_2}{f_2} - \frac{G_1}{f_1}\right)} \qquad [3.47]$$

If b = 1:

The parameter b can be equal to 1 only if $\frac{G_2}{G_1} = \frac{f_2}{f_1}$, the commonplace case $G_1 = G_2$ and $f_1 = f_2$ being excluded. On this assumption, $G = G_1 \frac{f}{f_1}$ and

$$v_{rms} = \frac{1}{2\pi}\sqrt{\frac{G_1}{f_1}\ln\frac{f_2}{f_1}} \qquad [3.48]$$

$$x^2_{rms} = \int_{f_1}^{f_2} \frac{1}{16\pi^4 f^4} \frac{G_1}{f_1^b} f^b \, df$$

If b ≠ 3:

$$x_{rms} = \frac{1}{4\pi^2}\sqrt{\frac{1}{b-3}\frac{G_1}{f_1^3}\left[\left(\frac{f_2}{f_1}\right)^{b-3} - 1\right]} = \frac{1}{4\pi^2}\sqrt{\frac{1}{b-3}\left(\frac{G_2}{f_2^3} - \frac{G_1}{f_1^3}\right)} \qquad [3.49]$$

If b = 3:

$$x_{rms} = \frac{1}{4\pi^2} \sqrt{\frac{G_1}{f_1^3} \ln\frac{f_2}{f_1}} \qquad [3.50]$$

In logarithmic scales, a straight line segment is sometimes defined by three of the four values corresponding to the coordinates of the first and the last point, supplemented by the slope of the segment. The slope R, expressed in dB/octave, can be calculated as follows:

– the number N of dB is given by

$$N = 10 \log_{10} \frac{G_2}{G_1} \qquad [3.51]$$

– the number of octaves n between f_1 and f_2 is, by definition, such that $\frac{f_2}{f_1} = 2^n$, yielding:

$$n = \frac{\log_{10} \frac{f_2}{f_1}}{\log_{10} 2}$$

and

$$R = 10 \log_{10}(2) \frac{\log_{10} G_2/G_1}{\log_{10} f_2/f_1} \qquad [3.52]$$

$$R = 10 \log_{10}(2) b \approx 3.01\, b \qquad [3.53]$$

Let us set $\alpha = 10 \log_{10}(2)$. It becomes, by replacing b with $\frac{R}{\alpha}$ in the preceding expressions [CUR 71]:

$$\ddot{x}_{rms}^2 = \frac{\alpha}{R + \alpha} (f_2 G_2 - f_1 G_1)$$

144   Random Vibration

If $R \neq -\alpha$:

$$\ddot{x}^2_{rms} = \frac{\alpha f_1 G_1}{R+\alpha}\left[\left(\frac{f_2}{f_1}\right)^{\frac{R}{\alpha}+1} - 1\right] \qquad [3.54]$$

This can also be written [SAN 66]:

$$\ddot{x}^2_{rms} = \frac{f_2 G_2}{\frac{R}{\alpha}+1}\left(1 - \frac{f_1 G_1}{f_2 G_2}\right) = \frac{f_2 G_2}{\frac{R}{\alpha}+1}\left[1 - \left(\frac{f_1}{f_2}\right)^{1+\frac{R}{\alpha}}\right] \qquad [3.55]$$

or [OSG 82]:

$$\ddot{x}^2_{rms} = \frac{\alpha G_2}{\alpha + R}\left[f_2 - f_1\left(\frac{f_1}{f_2}\right)^{\frac{R}{\alpha}}\right] \qquad [3.56]$$

Reference [SAN 66] gives this expression for an increasing slope and, for a decreasing slope,

$$\ddot{x}^2_{rms} = \frac{f_1 G_1}{\frac{R}{\alpha}-1}\left[1 - \left(\frac{f_2}{f_1}\right)^{1-\frac{R}{\alpha}}\right] \qquad [3.57]$$

if $R \neq -\alpha$, or

$$\ddot{x}^2_{rms} = \frac{\alpha f_2 G_2}{R+\alpha}\left[1 - \left(\frac{f_1}{f_2}\right)^{\frac{R}{\alpha}+1}\right] \qquad [3.58]$$

For $R = -\alpha$:

$$\ddot{x}^2_{rms} = f_1 G_1 \ln\frac{f_2}{f_1} = f_2 G_2 \ln\frac{f_2}{f_1} \qquad [3.59]$$

If $R \neq \alpha$:

$$v^2_{rms} = \frac{\alpha}{4\pi^2(R-\alpha)}\frac{G_1}{f_1}\left[\left(\frac{f_2}{f_1}\right)^{\frac{R-\alpha}{\alpha}}-1\right] = \frac{\alpha}{4\pi^2(R-\alpha)}\frac{G_2}{f_2}\left[1-\left(\frac{f_1}{f_2}\right)^{\frac{R-\alpha}{\alpha}}\right]$$

$$[3.60]$$

For $R = \alpha$:

$$v^2_{rms} = \frac{G_1}{4\pi^2 f_1}\ln\frac{f_2}{f_1} = \frac{G_2}{4\pi^2 f_2}\ln\frac{f_2}{f_1} \qquad [3.61]$$

If $R \neq 3\alpha$:

$$x^2_{rms} = \frac{\alpha G_1}{16\pi^4 f_1^3 (R-3\alpha)}\left[\left(\frac{f_2}{f_1}\right)^{\frac{R-3\alpha}{\alpha}}-1\right] = \frac{\alpha G_2}{16\pi^4 f_2^3 (R-3\alpha)}\left[1-\left(\frac{f_2}{f_1}\right)^{\frac{R-3\alpha}{\alpha}}\right]$$

$$[3.62]$$

For $R = 3\alpha$:

$$x^2_{rms} = \frac{1}{16\pi^4}\frac{G_1}{f_1^3}\ln\frac{f_2}{f_1} = \frac{1}{16\pi^4}\frac{G_2}{f_2^3}\ln\frac{f_2}{f_1} \qquad [3.63]$$

Figures 3.9, 3.10 and 3.11 respectively show $\dfrac{\ddot{x}_{rms}^2}{f_1 G_1}$, $\dfrac{f_1 v_{rms}^2}{G_1}$ and $\dfrac{f_1^3 x_{rms}^2}{G_1}$ versus $\dfrac{f_2}{f_1}$, for different values of R.

Abacuses of this type can be used to calculate the rms value of $\ddot{x}$, v or x from a spectrum made up of straight line segments on logarithmic scales [HIM 64].

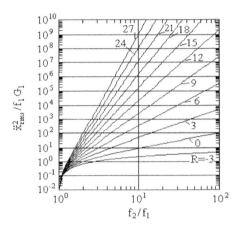

Figure 3.9. *Reduced rms acceleration versus* f_2/f_1 *and R*

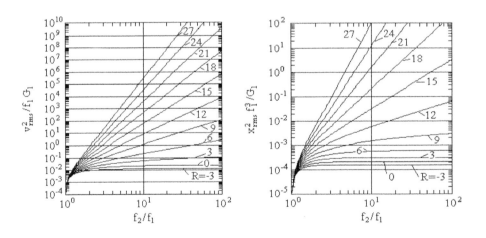

Figure 3.10. *Reduced rms velocity versus* f_2/f_1 *and R*

Figure 3.11. *Reduced rms displacement versus* f_2/f_1 *and R*

3.4.5. *PSD comprising several segments of arbitrary slopes*

Whatever the scales chosen, the rms value of a PSD made up of several straight line segments of arbitrary slope will be such that in [OSG 82], [SAN 63]:

$$\ddot{x}_{rms} = \sqrt{\sum_i \ddot{x}_{i\,rms}^2} \qquad [3.64]$$

$\ddot{x}_{i\,rms}^2$ being calculated starting from the relations above. In the same way:

$$v_{rms} = \sqrt{\sum_i v_{i\,rms}^2} \qquad [3.65]$$

and

$$x_{rms} = \sqrt{\sum_i x_{i\,rms}^2} \qquad [3.66]$$

3.5. Rms value according to the frequency

In order to determine what frequencies contribute the most to the global rms value of a random vibration, we can draw a curve that gives the rms value according to the frequency with the following relation:

$$\ddot{x}_{rms}(f) = \sqrt{\int_{f_1}^{f} G(f)\,df} \qquad [3.67]$$

where f_1 is the initial frequency of the PSD $G(f)$.

Example 3.2.

Let us take the "airplane" vibration defined by the PSD in Figure 3.12.

Figure 3.13 shows the rms value according to the frequency calculated from relation [3.67].

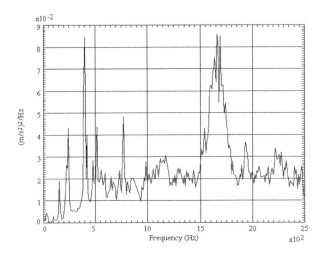

Figure 3.12. *PSD of an airplane vibration*

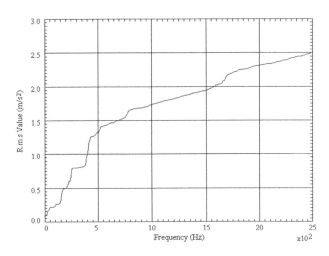

Figure 3.13. *Rms value versus frequency*

Rms Value of Random Vibration   149

This last curve could be drawn in a reduced form by dividing the rms values by the global rms value calculated on the complete frequency domain (ratio

$$\frac{\sqrt{\int_{f_1}^{f} G(f)\,df}}{\sqrt{\int_{f_1}^{f_2} G(f)\,df}}$$, possibly expressed in percent).

3.6. Case of periodic signals

It is known that any periodic signal can be represented by a Fourier series in accordance with:

$$\ell(t) = L_0 + \sum_{i=1}^{n} L_n \sin(2\pi n f_1 t + \phi_n) \qquad [3.68]$$

The PSD is equal to [2.26]:

$$G(f) = \lim_{\substack{T \to \infty \\ \Delta f \to 0}} \frac{1}{T\,\Delta f} \int_0^T \ell_T^2(t, \Delta f)\,dt$$

is zero for $f \neq f_n$ (with $f_n = n\,f_1$) and infinite for $f = f_n$ since the spectrum of $\ell(t)$ is a discrete spectrum, in which each component L_n has zero width Δf.

If we wish to standardize the representations and to be able to define the PSD of a periodic function, so that the integral $\int_0^\infty G(f)\,df$ is equal to the mean square value of $\ell(t)$, we must consider that each component is related to the Dirac delta function, the area under the curve of this function being equal to the mean square value of the component. With this definition,

$$G(f) = \sum_{n=0}^{\infty} \ell_{\text{rms}_n}^2\,\delta(f - nf_1) \qquad [3.69]$$

where $\ell_{\text{rms}_n}^2$ is the mean square value of the nth harmonic $\ell_n(t)$ defined by

$$\ell_{\text{rms}_n}^2 = \frac{1}{T_n} \int_0^{T_n} \ell_n^2(t)\,dt \qquad [3.70]$$

$$T_n = \frac{1}{n f_1} \qquad [3.71]$$

(n = 1, 2, 3, ...). $\ell_n(t)$ is the value of the nth component and

$$\ell_{rms0}^2 = \frac{1}{T} \int_0^T \ell_0^2(t)\, dt = \left(\bar{\ell}\right)^2$$

where T is arbitrary and $\bar{\ell}$ is the mean value of the signal $\ell(t)$. The Dirac delta function $\delta(f - n\, f_1)$ at the frequency f_n is such that:

$$\int_{f_n - \varepsilon}^{f_n + \varepsilon} \delta(f - f_n)\, df = 1 \qquad [3.72]$$

and

$$\delta(f - f_n) = 0 \qquad [3.73]$$

for $f \neq f_n$ (ε = positive constant different from zero, arbitrarily small). The definition of the PSD in this particular case of a periodic signal does not require taking the limit for infinite T, since the mean square value of a periodic signal can be calculated over only one period or a whole number of periods.

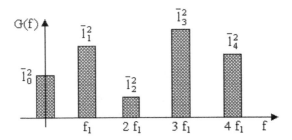

Figure 3.14. *PSD of a periodic signal*

The chart of the PSD of a periodic signal is that of a discrete spectrum, the amplitude of each component being proportional to the area representing its mean square value (and not its amplitude).

We have, with the preceding notations, relationships of the same form as those obtained for a random signal:

$$\int_{0^-}^{\infty} G(f)\, df = \sum_{n=0}^{\infty} \overline{\ell_n^2} = \overline{\ell^2} = s_\ell^2 + \left(\overline{\ell}\right)^2 \qquad [3.74]$$

$$\int_{0^+}^{\infty} G(f)\, df = \sum_{n=1}^{\infty} \overline{\ell_n^2} = s_\ell^2 \qquad [3.75]$$

$$\int_{0^-}^{0^+} G(f)\, df = \left(\overline{\ell}\right)^2 \qquad [3.76]$$

and, between two frequencies f_i and f_j ($f_i = i\,f_1 - \varepsilon$, $f_j = j\,f_1 + \varepsilon$, i and j integers, $j > i$):

$$\int_{i\,f_1 - \varepsilon}^{j\,f_1 + \varepsilon} G(f)\, df = \sum_{n=i}^{j} \overline{\ell_n^2} = \overline{\ell^2} \qquad [3.77]$$

Lastly, if for a random signal, we had:

$$\int_{f^-}^{f^+} G(f)\, df = 0 \qquad [3.78]$$

we have here:

$$\int_{f^-}^{f^+} G(f)\, df = \begin{cases} \overline{\ell_n^2} & \text{for } f = n\, f_1 \\ 0 & \text{for } f \neq n\, f_1 \text{ and } f \neq 0 \end{cases} \qquad [3.79]$$

The area under the PSD at a given frequency is either zero or equal to the mean square value of the component if $f = n\, f_1$ (whereas, for a random signal, this area is always zero).

3.7. Case of a periodic signal superimposed onto random noise

Let us assume that:

$$\ell(t) = a(t) + p(t) \qquad [3.80]$$

$a(t)$ = random signal, of PSD $G_a(f)$ defined in [2.26]

$p(t)$ = periodic signal, of PSD $G_p(f)$ defined in the preceding section.

The PSD of $\ell(t)$ is equal to:

$$G_\ell(f) = G_a(f) + G_p(f) \qquad [3.81]$$

$$G_\ell(f) = G_a(f) + \sum_{n=0}^{\infty} \overline{\ell_n^2}\, \delta(f - f_n) \qquad [3.82]$$

where

$f_n = n\, f_1$

$n = \text{integer} \in (0, \infty)$

$f_1 =$ fundamental frequency of the periodic signal

$\overline{\ell_n^2} =$ mean square value of the nth component $\ell_n(t)$ of $\ell(t)$

The rms value of this composite signal is, as previously, equal to the square root of the area under $G_\ell(f)$.

Chapter 4

Practical Calculation of the Power Spectral Density

The analysis of a random vibration is carried out most of the time by assuming that it is stationary and ergodic. This assumption makes it possible to replace a study based on the statistical properties of a great number of signals with that of only one sample of finite duration T. Several approaches are possible for the calculation of the PSD of such a sample. The duration T being often small, a statistical error is necessarily introduced. This chapter shows how it can be estimated and reduced.

4.1. Sampling of signal

Sampling consists of transforming a vibratory signal that is continuous at the outset by a succession of sample points regularly distributed in time. If δt is the time interval separating two successive points, the sampling frequency is equal to $f_{samp.} = 1/\delta t$. In order for the digitized signal to be correctly represented, it is necessary that the sampling frequency is sufficiently high compared to the largest frequency of the signal to be analyzed.

A too low sampling frequency can thus lead to an aliasing phenomenon, characterized by the appearance of frequency components having no physical reality.

Example 4.1.

Figure 4.1 thus shows a component of frequency of 70 Hz artificially created by the sampling of 200 points/s of a sinusoidal signal of frequency of 350 Hz.

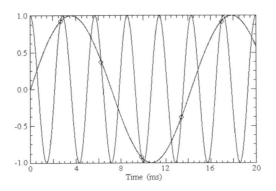

Figure 4.1. *Highlighting of the aliasing phenomenon due to under-sampling*

Example 4.2.

Consider a signal with a 30 s duration, defined in the bandwidth 2–500 Hz and sampled at 4 times its maximum frequency, i.e. at 2 KHz, with 60,416 points. Its PSD, calculated with a statistical error of about 13%, is shown in Figure 4.2. This signal was then decimated by only retaining one point in 3 (20,138 points). The sampling frequency is then equal to 666 Hz, leading to a maximum PSD frequency of 333 Hz.

The PSD of this shows an aliasing between 166 Hz and 333 Hz, a range in which the amplitude is doubled (so the rms value is retained). The 166 Hz frequency is symmetric by 500 Hz compared to 333 Hz.

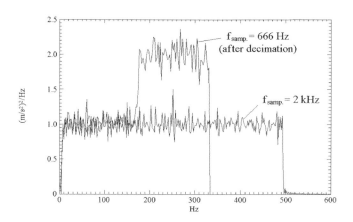

Figure 4.2. *PSD of the signal sampled at 2 kHz (60,416 points) and of the signal decimated at 20,136 points (sampling frequency: 666 Hz)*

Shannon's theorem (Volume 1, Chapter 1) indicates that *if a function contains no frequencies higher than f_{max} Hz, it is completely determined by its ordinates at a series of points spaced $1/2 f_{max}$ seconds apart [SHA 49]*.

This theorem helps in the determination of the minimum sampling frequency to retain all the information present in the signal. It is sufficient to correctly calculate the rms value of the signal and its PSD.

Example 4.3.

Consider a random vibration with a PSD close to $(1 \text{ m/s}^2)^2/\text{Hz}$ between 5 and 500 Hz.

It was sampled with a frequency respecting Shannon's theorem: 30,000 points over 30 s, i.e. 1,000 points/s, twice as big as the greatest signal frequency (500 Hz).

This signal is compared (Figure 4.3) with the same signal over-sampled at 13,000 Hz ("original signal").

156   Random Vibration

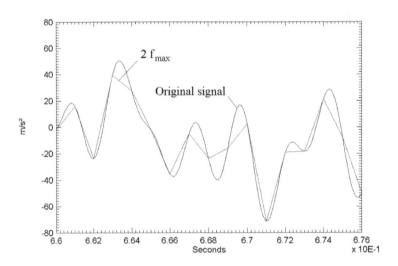

Figure 4.3. *Random sampled vibration with frequency equal to 1,000 Hz, superimposed to the original signal*

Figure 4.4 shows that PSDs of these three vibrations (calculated with the same statistical error) are very close. Digitization respecting Shannon's theorem does not lead to a slightly different PSD from the original PSD.

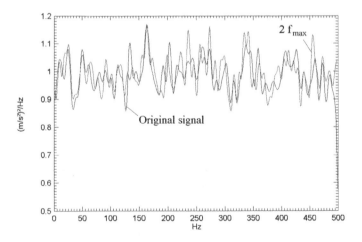

Figure 4.4. *PSD of sampled random vibration with frequency equal to 1,000 Hz, superimposed to the PSD of the original signal*

Figure 4.5 shows the rms value of a vibration calculated from the signal based on sampled time with several frequencies.

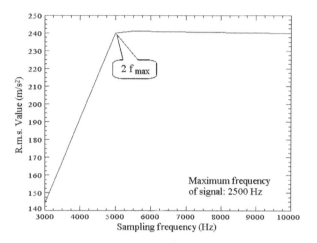

Figure 4.5. *Variation of rms value based on the sampling frequency of the signal*

We can verify that this rms value remains constant as long as the sampling frequency is highest than twice that of the maximum signal frequency.

NOTE.– *The response of a mechanical system calculated from a sampling signal according to Shannon's theorem can be different to the response that would be obtained from an analog signal. We have seen (Volume 1, Chapter 1) that it is possible to reconstruct the original signal.*

Given a signal which we wish to analyze up to the frequency f_{max}, it is thus appropriate to avoid aliasing:

– to filter it using a low-pass filter in order to eliminate frequencies higher than f_{max} (the high frequency part of the spectrum which can have a physical reality or noise);

– to sample it with a frequency at least equal to $2 f_{max}$ [CUR 87], [GIL 88], [PRE 90], [ROT 70].

NOTE.–

$f_{Nyquist} = f_{samp.} / 2$ is called Nyquist frequency.

We noted (Volume 1, Chapter 1) that low-pass filters used in practice are not perfect and they do not completely break the frequencies that are higher than the desired value. Because of this, we often impose a sampling frequency equal to approximately 2.6 times the highest frequency of the signal to analyze.

4.2. PSD calculation methods

Three methods can mainly be used to calculate a PSD:

– from the autocorrelation function;

– by signal filtering with Δf wide filters and calculation of the rms value of the filtered signals;

– with the help of Fourier transforms.

The most widely used method is the last one.

4.2.1. *Use of the autocorrelation function*

The calculation of the PSD can also be carried out by using relation [4.1], by evaluating the correlation in the time domain and by carrying out a Fourier transformation (Wiener-Khintchine method) (correlation analyzers) [MAX 86]:

$$G_{\ddot{x}}(\tau) = 4 \int_0^\infty R_{\ddot{x}}(f) \cos(2\pi f \tau) \, df \qquad [4.1]$$

4.2.2. *Calculation of the PSD from the rms value of a filtered signal*

Theoretical relation [2.26], which would assume an infinite duration T and a zero analysis bandwidth Δf, is replaced by the approximate relation [KEL 67] (Chapter 2):

$$G(f) = \frac{1}{T \, \Delta f} \int_0^T \ell_{\Delta f}^2(f, t) \, dt = \frac{\overline{\ell_{\Delta f}^2}}{\Delta f} \qquad [4.2]$$

where $\overline{\ell_{\Delta f}^2}$ is the mean square value of the sample of finite duration T, calculated at the output of a filter of central frequency f and non-zero width Δf [MOR 56].

NOTE.– *Given a random white noise vibration $\ell(t)$ and a perfect rectangular filter, the result of filtering is a signal having a constant spectrum over the width of the filter, which is zero elsewhere [CUR 64].*

The result can be obtained by multiplying the PSD G_0 of the input $\ell(t)$ by the square of the transmission characteristic of the filter (frequency-response characteristic) at each frequency (transfer function, defined as the ratio of the amplitude of the filter response to the amplitude of the sinewave excitation as a function of the frequency. If this ratio is independent of the excitation amplitude, the filter is said to be linear).

In practice, the filters are not perfectly rectangular. The mean square value of the response is equal to G_0 multiplied by the area squared under the transfer function of the filter. This surface is defined as the "rms bandwidth of the filter".

If the PSD of the signal to be analyzed varies with the frequency, the mean square response of a perfect filter divided by the width Δf of the filter gives a point on the PSD (mean value of the PSD over the width of the filter). With a real filter, this approximate value of the PSD is obtained by considering the ratio of the mean square value of the response to the rms bandwidth of the filter Δf, defined by [BEN 62], [GOL 53] and [PIE 64]:

$$\Delta f = \int_0^\infty \left| \frac{H(f)}{H_{max}} \right|^2 dt \qquad [4.3]$$

where $H(f)$ is the frequency response function of the (narrow) band-pass filter used and H_{max} its maximum value.

4.2.3. *Calculation of PSD starting from a Fourier transform*

The most used method consists of considering expressions [2.39] and [2.41]:

$$G_{\ell\ell}(f) = \lim_{T \to \infty} \frac{2}{T} E\left[|L(f, T)|^2 \right] \qquad [4.4]$$

NOTE.– *Knowing that the discrete Fourier transform can be written [KAY 81]*

$$L(m, T) = \frac{T}{N} \sum_{j=0}^{N-1} \ell_j \exp\left(-i \frac{2\pi jm}{N} \right) \qquad [4.5]$$

the expression of the PSD can be expressed for calculation in the form [BEN 71], [ROT 70]:

$$G(m\ \Delta f) = \frac{2}{N} \left| \sum_{j=0}^{N-1} \ell_j\ \exp\left(-i\ \frac{2\ \pi\ j\ m}{N}\right) \right|^2 \qquad [4.6]$$

where $0 < m \leq M$ and $\ell_j = j\ \delta t$.

4.3. PSD calculation steps

The calculation data are in general the following:
– the maximum frequency of the spectrum;
– the number of points of the PSD (or the frequency step Δf);
– the maximum statistical error tolerated.

4.3.1. *Maximum frequency*

Given an already sampled signal (frequency $f_{samp.}$) and taking into account the elements of section 1.4 and of Volume 1, Chapter 1, the PSD will be correct only for frequencies lower than $f_{max} = f_{samp.} / 2.6$.

4.3.2. *Extraction of sample of duration T*

Two approaches are possible for the calculation of the PSD:

– assuming that the signal is periodic and composed of the repetition of the sample of duration T;

– assuming that the signal has zero values at all the points outside the time corresponding to the sample.

These two approaches are equivalent [BEN 75]. In both cases, we are led to isolate by truncation a part of the signal which amounts to applying to it a rectangular temporal window $r(t)$ of amplitude 1 for $0 \leq t \leq T$ and zero elsewhere (*windowing*).

If $\ell(t)$ is the signal to be analyzed, the Fourier transform is thus calculated in practice with $f(t) = \ell(t)\ r(t)$.

Practical Calculation of the PSD   161

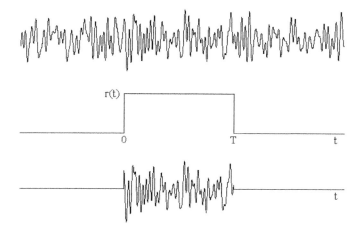

Figure 4.6. *Application of a temporal window*

This operation has consequences. In the frequency domain, the transform of a product is equal to the convolution of the Fourier transforms $L(\Omega)$ and $R(\Omega)$ of each term:

$$F(\Omega) = \int_0^\Omega L(\omega)\, R(\Omega - \omega)\, d\omega \qquad [4.7]$$

(ω is a variable of integration).

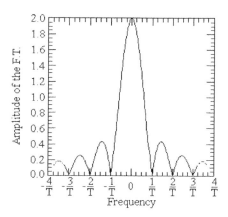

Figure 4.7. *Fourier transform of a square waveform*

The Fourier transform of a square temporal window appears as a principal central lobe surrounded by small lobes of decreasing amplitude (see Volume 2, Chapter 1). The transform cancels out regularly for Ω a multiple of $\frac{2\pi}{T}$ (i.e. a frequency f multiple of $\frac{1}{T}$). The effect of the convolution is to widen the peaks of the spectrum, the resolution, consequence of the width of the central lobe, not being able to better $\Delta f = \frac{1}{T}$.

Expression [4.7] shows that, for each point of the spectrum of frequency Ω (multiple of $\frac{2\pi}{T}$), the side lobes have a parasitic influence on the calculated value of the transform (*leakage*).

NOTE.– *In the case of a periodic function, leakage occurs when the truncation of a periodic function is done at some point other than a multiple of the period. The result is a discontinuity in the time domain which appears as side-lobes in the frequency domain.*

To reduce this influence and to improve the precision of calculation, their amplitude needs to be reduced.

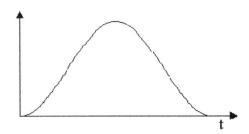

Figure 4.8. *The Hanning window*

This result can be obtained by considering a modified window which removes discontinuities of the beginning and end of the rectangular window in the time domain [PRE 92].

Many shapes of temporal windows are used [BLA 91], [DAS 89], [JEN 68], [NUT 81].

Practical Calculation of the PSD 163

One of best known and the most frequently used is the Hanning window [FER 99], which is represented by a versed sine function (Figure 4.8):

$$r(t) = \frac{1}{2}\left(1 - \cos\frac{2\pi t}{T}\right) \quad [4.8]$$

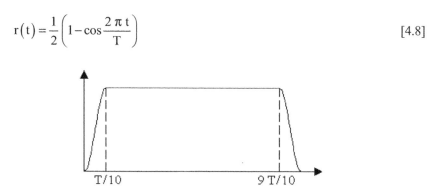

Figure 4.9. *The Bingham window [BIN 67]*

This shape is only sometimes used to constitute the rising and decaying parts of the window (the Bingham window, Figure 4.9).

The *weighting coefficient* of the window is the term given to the percentage of rise time (equal to the decay time) of the total length T of the window. This ratio cannot naturally exceed 0.5, corresponding to the case of the previously defined Hanning window.

Examples of windows

The advantages of the various windows have been discussed in the literature [BIN 67], [NUT 81]. These advantages are related to the nature of the signal to be analyzed. Actually, the most important point in the analysis is not the type of window, but rather the choice of the bandwidth [JEN 68]. The Hanning window is nevertheless recommended.

The replacement of the square window by a more smoothed shape modifies the signal actually treated through attenuation of its ends, which results in a reduction of the rms duration of the sample and consequently in a reduction of the resolution, depending on the width of the central lobe.

We should not forget to correct the result of the calculation of the PSD to compensate for the difference in area related to the shape of the new window. Given a temporal window defined by $r(t)$, having $R(f)$ for Fourier transform, the area intervening in the calculation of the PSD is equal to:

$$Q = \int_{-\infty}^{+\infty} |R(f)|^2 \, df \qquad [4.9]$$

From Parseval's theorem, this expression can be written in a form utilizing the N points of the digitized signal:

$$Q = \frac{1}{N} \sum_{i=0}^{N-1} r_i^2 \qquad [4.10]$$

The multiplicative compensation factor to apply in order to take account of the difference between this area and the unit area of a rectangular window is thus equal to 1/Q [DUR 72].

The choice of the window is not easy. It is still a compromise between the minimization of lateral lobes of the window, precision in amplitude and frequency and the increase in the main lobe size. It depends on the application considered and the frequency content of the signal involved.

Window type	Preferably used for	Frequency resolution	Amplitude accuracy	Spectral leakage
Barlett	Random	Good	Fair	Fair
Hanning	Random	Good	Fair	Good
Hamming	Random	Good	Fair	Fair
Kaiser-Bessel	Random	Fair	Good	Good
Tukey	Random	Good	Poor	Poor
Welch	Random	Good	Fair	Good
Blackman	Random or mixed	Poor	Good	Best
Rectangular	Transient or Synchronous sampling	Best	Poor	Poor
Flat top	Sinusoids	Poor	Best	Good

Table 4.1. *Time domain window shape characteristics*

For the analysis of continuous signals, the following windows can be used [GAD 87], [SHR 95], [SOH 84]:

Rectangular Weighting	Should only be used when analyzing special sinusoids, the frequencies of which coincide with the centre frequencies / lines in the analysis. This is often the situation when pseudo-random types of signal are analyzed, or when order tracking is applied.
Hanning Weighting	General purpose weighting and should be used in most cases. Hanning window with 66 $^2/_3$% or 75% overlap should be used when true real-time analysis is needed; makes it possible to achieve the best frequency resolution/amplitude resolution compromise.
Kaiser-Bessel Weighting	Very good selectivity and should be used for two-tone separation of harmonic signals with widely different levels.
Flat Top Weighting	Mainly designed for calibration and correct amplitude measurement; very good amplitude resolution, but very bad frequency resolution.

Table 4.2. *Windows for the analysis of continuous signals*

The following windows should be used for system analysis frequency response function measurements [HER 84a], [HER 84b]:

Rectangular Weighting	Both excitation and response signal when a pseudo-random excitation signal is used.
Hanning Weighting	Both excitation and response signals when a random excitation signal is used.
Hamming Weighting	Providing better frequency resolution at the expense of amplitude; used to separate close frequency components.
Kaiser-Bessel Weighting	Better than the Hamming technique for separating close frequencies because the filter has even less leakage into side bins; better amplitude resolution than the Hamming window, to the detriment of the frequency resolution.
Blackman-Harris Weighting	Good tool for frequency separation provides good amplitude accuracy.

Table 4.3. *Windows for system analysis frequency response function measurements*

Window type	Definition	Compensation factor
Bingham (Figure 4.9)	$r(t) = \dfrac{1}{2}\left\{1 + \cos\left[\dfrac{10\pi(t - 9T/10)}{T}\right]\right\}$ for $0 \leq t \leq \dfrac{T}{10}$ and $\dfrac{9T}{10} \leq t \leq T$	1/0.875
Blackman-Harris	3 terms $r(t) = 0.42323 - 0.49755 \cos\left(\dfrac{2\pi t}{T}\right) + 0.07922 \cos\left(\dfrac{4\pi t}{T}\right)$ for $0 \leq t \leq T$	1/0.30604
	4 terms $r(t) = 0.35875 - 0.48829 \cos\left(\dfrac{2\pi t}{T}\right)$ $+ 0.14128 \cos\cos\left(\dfrac{4\pi t}{T}\right) - 0.01168 \cos\left(\dfrac{6\pi t}{T}\right)$ for $0 \leq t \leq T$	1/0.25786
Hamming	$r(t) = 0.08 + 0.46\left(1 - \cos\dfrac{2\pi t}{T}\right)$ for $0 \leq t \leq T$	1/0.3974
Hanning	$r(t) = \dfrac{1}{2}\left[1 - \cos\left(\dfrac{2\pi t}{T}\right)\right]$ for $0 \leq t \leq T$	1/0.375
Parzen	$r(t) = 1 - 6\left(\dfrac{2t}{T} - 1\right)^2 + 6\left|\dfrac{2t}{T} - 1\right|^3$ for $\dfrac{T}{4} \leq t \leq \dfrac{3T}{4}$ $r(t) = 2\left[1 - \left|\dfrac{2t}{T} - 1\right|\right]^3$ for $0 \leq t \leq \dfrac{T}{4}$ and $\dfrac{3T}{4} \leq t \leq T$	1/0.269643

Table 4.4. *The principal windows*

Flat top	$r(t) = 1 - 1.933 \cos\left(\frac{2\pi}{T}t\right) + 1.286 \cos\left(2\frac{2\pi}{T}t\right)$ $- 0.388 \cos\left(3\frac{2\pi}{T}t\right) + 0.032 \cos\left(4\frac{2\pi}{T}t\right)$ for $0 \leq t \leq T$	1/3.7709265
Kaiser-Bessel	$r(t) = 1 - 1.24 \cos\left(\frac{2\pi}{T}t\right) + 0.244 \cos\left(2\frac{2\pi}{T}t\right)$ $- 0.00305 \cos\left(3\frac{2\pi}{T}t\right)$ for $0 \leq t \leq T$	1/1.798573

Table 4.4. *(Continued) The principal windows*

4.3.3. *Averaging*

We attempt in the calculations to obtain the best possible resolution with the data at our disposal, which results in trying to plot the PSD with the smallest possible frequency step.

The PSD is defined from the Fourier transform of the random vibration $\ddot{x}(t)$ sample with T duration studied with the help of the following relation:

$$G(f) = \lim_{T \to \infty} \frac{2}{T} |\ddot{X}(f)|^2 \qquad [4.11]$$

In practice, the sample duration is finite and:

$$G(f) = \frac{2}{T} |\ddot{X}(f)|^2 \qquad [4.12]$$

The Fourier transform gives the highest peak amplitude of the output signal from a filter (f, Δf) during T. It is only one value among all those that could be obtained from other similar signal samples.

It is therefore desirable to carry out *a priori* an average of several Fourier transforms to improve precision.

We also attempt to obtain the best resolution possible in the calculations with the data we have, leading us to try to draw the PSD with the lowest frequency step.

For a sample of duration T, this step cannot be lower than $\Delta f = 1/T$, but then we only calculate the Fourier transform on this single signal sample. With this resolution, the precision obtained is unacceptable. Several solutions are possible to improve it:

– carry out several measures of the phenomenon, calculate the PSD of each sample lasting T and carry out an average of the resulting spectrums;

– if we only have one sample with T duration, deliberately limit the resolution by accepting a higher Δf analysis step than $1/T$ and carry out *averaging* [BEN 71]. The sample of the random signal to study, duration, is broken down into K parts (blocks) lasting $\Delta T = T/K$ (we will see that this K number must be higher than 50) and we calculate a Fourier transform for each of these blocks (Figure 4.10) with a resolution equal to $1/\Delta T$.

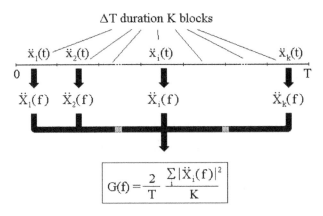

Figure 4.10. *PSD calculation from the Fourier transform*

The PSD is evaluated by considering the square mean of the modules of these transforms, multiplied by 2/T [BAR 55], [MAX 81]:

$$G(f) = \frac{1}{K} \sum_i \frac{2}{\Delta T} \left| \ddot{X}_i(f) \right|^2 = \frac{1}{K} \sum_i \hat{G}_i(f) \qquad [4.13]$$

$\hat{G}_i(f)$ being the evaluated PSD of block i.

Practical Calculation of the PSD 169

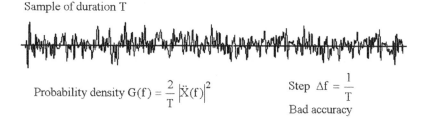

Figure 4.11. *PSD calculation by the Fourier transform of the complete sample*

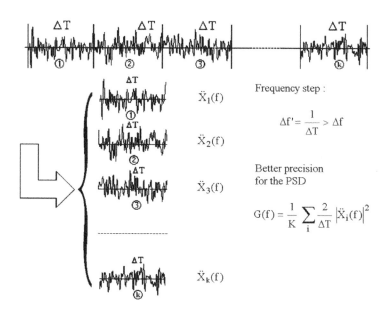

Figure 4.12. *Averaging process for PSD calculation*

Extracting blocks of duration ΔT consists of the implicit application of a temporal rectangular window to the signal which has some drawbacks. There are other forms of windows that are better suited (see section 4.3.2).

For a sample of duration T, this step cannot be lower than 1/T. With this resolution, the precision obtained is unacceptable. Several solutions are possible:

170   Random Vibration

– to carry out several measurements of the phenomenon, to calculate the PSD of each sample of duration T and to proceed to an average of the obtained spectra;

– if only one sample of duration T is available, to voluntarily limit the resolution by accepting an analysis step Δf larger than 1/T and to carry out an averaging [BEN 71]:

- either by calculating the average of several frequency components close to the considered spectrum component, separated by intervals 1/T, when the noise to be analyzed can be comparable to a white noise. If the average is carried out on K PSD, the average obtained is assigned to the central frequency of an interval of width equal to K/T (which characterizes the effective resolution of the PSD thus calculated),

- or by dividing up the initial sample of duration T into K subsamples (or blocks) of duration $\Delta T = T/K$ which will be used to calculate K spectra of resolution $1/\Delta T$ and their average [BAR 55], [MAX 81]:

$$\frac{1}{K}\sum_i G_i(f)$$

The results of these two approaches are identical for given duration T and given resolution [BEN 75]. It is the last procedure which is the most often used. The window, rectangular or not, is applied to each block.

NOTE.– *PSDs are usually calculated from measurements of the real environment with a linear averaging as that presented in the previous sections. Each instantaneous spectrum is counted with an identical weight in the average. However, for some specific applications, it may be necessary to use other types of averages, for example the exponential averaging [SHR 95]. The latest instantaneous spectrum entered has an equivalent weight to the n first instantaneous measured. It is useful in observing conditions that are changing slowly with respect to sampling time (steady-state process for example).*

4.3.4. *Addition of zeros*

The smallest interval Δf between two points of the PSD is related to the duration of the block considered by at least $\Delta f = \dfrac{1}{\Delta T}$. The calculation of the PSD is carried out at M points with distances of Δf between 0 and $f_{samp.}/2$ ($f_{samp.}$ = sampling frequency of the signal). As long as this condition is observed, it is said that the components of the spectrum are *statistically independent*.

Practical Calculation of the PSD 171

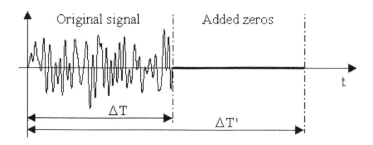

Figure 4.13. *Addition of zeros at end of the signal sample*

If the available signal sample is characterized by a small number of points, M can be small enough and lead to a PSD that is defined by only a few straight line segments.

As in the case of the Fourier transform of a shock, we can still add components to the spectrum to obtain a smoother curve by artificially increasing the number of points with zeros placed at the end of each block (leading to a new $\Delta T' > \Delta T$ duration) [FER 99]. This is called *padding* the signal with zeros.

We can however add components to the spectrum to obtain a more smoothed curve by artificially increasing the number of points using zeros placed at the end of the block (leading to a new duration $\Delta T' > \Delta T$).

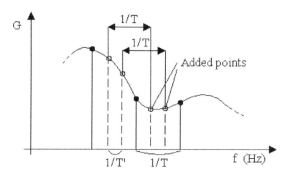

Figure 4.14. *The addition of zeros increases the number of points of the PSD*

172   Random Vibration

Although the components added are no longer statistically independent, the validity of each component remains whole.

The additional points of the PSD thus obtained lie between the original points corresponding to the duration ΔT and are on the continuous theoretical curve.

The resolution and the statistical error are unchanged [KAY 81]. All the components have an equal validity in the analysis [ENO 69]. We should attach no particular importance to the components spaced out at $\dfrac{1}{\Delta T}$, except that they constitute a range of independent components. An equivalent unit could be selected by considering the points at the frequencies $(1 + \delta f)/T$, $(2 + \delta f)/T$, etc., where δf is an increment ranging between 0 and 1 [BEN 75].

NOTES.–

1. *Similarly, we can add zeros to a PSD (at high frequency) to increase the sampling frequency of a signal to create from this PSD.*

2. *The addition of zeros must be carried out after the windowing to obtain the desired effect, thus returns to zero smoothed at the beginning and end of the sample of the signal considered. This would not be the case if the zeros were added before windowing [FER 99].*

Overlapping enables us to increase the number of PSD points by retaining the statistical error value. The resulting improvement being limited; however, adding zeros is mainly used after this operation to obtain smoother PSDs.

Example 4.4.

Let us take a random vibration measured on a truck, characterized by a 3.04 s signal sample with N = 4,096 points. The statistical error is set at 15%.

– maximum PSD frequency: $f_{max} = \dfrac{1}{2\,\delta t} = \dfrac{N}{2\,T} = \dfrac{4\,096}{2 \cdot 3.04} \approx 674 \text{ Hz}$;

– the number of PSD points necessary to respect the statistical error: $\dfrac{N\,\varepsilon^2}{2} = \dfrac{4096 \cdot 0.15^2}{2} \approx 46$, where the closest lower power of 2 M = 32;

– frequency step: $\Delta f = \dfrac{f_{max}}{M} = \dfrac{674}{32} \approx 21.06 \text{ Hz}$;

– number of blocks: $K = \dfrac{N}{2M} = \dfrac{4096}{2 \cdot 32} = 64$;

– number of points per block: $\Delta N = 2M = 2 \times 32 = 64$;

– duration of each block: $\Delta T = \dfrac{T}{K} = \dfrac{3.04}{64} \approx 0.0475$ s.

The PSD will thus only be defined by 32 points separated by 21 Hz (Figure 4.15).

Adding 192 zeros to each of the 64 blocks leads to a signal with 16,384 points, making it possible to calculate the PSD with 128 points (power of 2 immediately lower than $\dfrac{N\varepsilon^2}{2} = \dfrac{16\,384 \cdot 0.15^2}{2} \approx 184$). The resulting PSD is smoother (Figure 4.16).

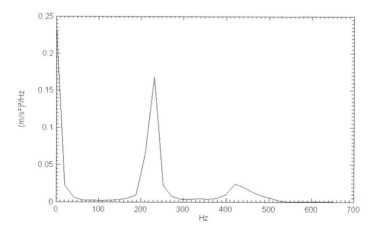

Figure 4.15. *PSD of the truck vibration calculated on 32 points (statistical error: 0.15)*

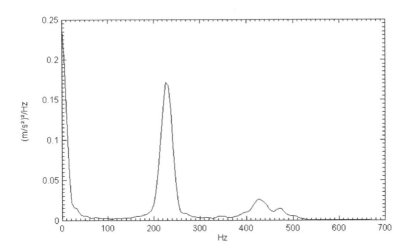

Figure 4.16. *PSD of the truck vibration calculated on 128 points with addition of zeros at each block (statistical error: 0.15)*

The PSDs calculated above are compared in Figure 4.17. Adding zeros results in intermediate points, but does not improve the statistical precision (Figures 4.18 and 4.19).

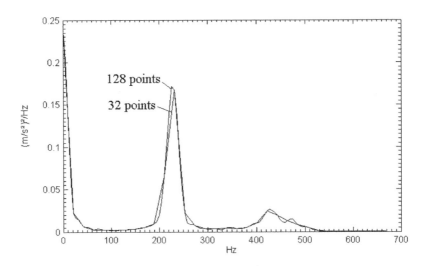

Figure 4.17. *Comparison of truck vibration PSDs calculated with 32 points (without additional zeros) and with 128 points (with zeros)*

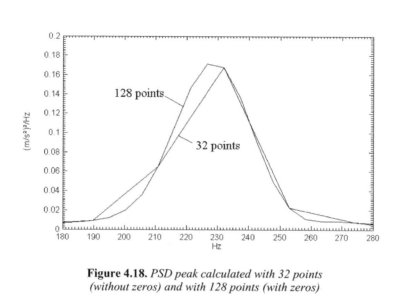

Figure 4.18. *PSD peak calculated with 32 points (without zeros) and with 128 points (with zeros)*

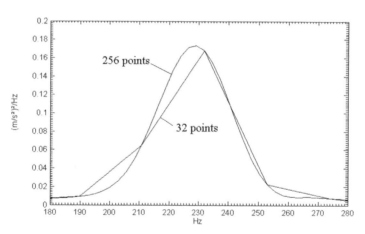

Figure 4.19. *PSD peak calculated with 32 points (without zeros) and with 256 points (with zeros)*

4.4. FFT

In 1965, J.W. Cooley and J. Tukey [COO 65] developed a method called the *Fast Fourier Transform* or *FFT*, making it possible to reduce considerably the calculation time of the Fourier transforms.

A FFT analyzer functions with a number of points which is [MAX 86]:

– a power of 2 for the Cooley-Tukey algorithms and those which derive from them;

– a product of integer powers of prime numbers (Vinograd's algorithm).

With the Cooley-Tukey algorithm, the calculation time of the transform of a signal defined by N points is proportional to $N \log_2 N$ instead of the theoretically necessary value N^2.

Calculations of PSD are done today primarily using the FFT, which also has applications for the calculation of coherence functions (square of the amplitude) [CAR 73] and of convolutions. This algorithm, which is based in practice on the discrete Fourier transform, leads to a frequency sampling of the Fourier transform and thus of the PSD [NEW 93].

Number of Points	Number of points of the Fourier transform	Speed ratio $\dfrac{N}{\log_2 N}$
256	128	32
512	256	56.9
1,024	512	102.4
2,048	1,024	186.2
4,096	2,048	341.3

Table 4.5. *Speed ratio for FFT calculation*

NOTES.–

1. *Whilst in theory equivalent, the FFT and the method using the correlation can in practice lead to different results, which can be explained by the non-cognisance of the theoretical assumptions due to the difficulties of producing the analyzers [MAX 86]. J. Max, M. Diot and R. Bigret showed that a correlation analyzer presents a certain number of advantages such as:*

– a greater flexibility in the choice of the frequency sampling step, facilitating the analysis of the periodic signals;

– a choice more adapted to the conditions of analysis of the signal.

2. *When these algorithms are used to calculate the Fourier transform of a shock, we should not forget to multiply the result by the duration T of the treated signal.*

4.5. Particular case of a periodic excitation

The PSD of a periodic excitation was defined by [3.69]:

$$G'(f) = \sum_{n=0}^{\infty} \overline{\ell_n^2} \, \delta(f - f_n) \qquad [4.14]$$

The PSD of such an excitation being characterized by very narrowbands centered on the frequencies f_n, the calculation of $G'(f)$ supposes that $\ell(t)$ is analyzed in sufficiently narrow filters Δf. The PSD is approximated by:

$$G'(f) = \sum_{n=0}^{\infty} \overline{\ell_n^2} \, \delta\left(f - f_n \pm \frac{\Delta f}{2}\right) \qquad [4.15]$$

$\overline{\ell_n^2}$ can be obtained either by direct calculation of:

$$\overline{\ell_n^2} = \frac{1}{T} \int_0^T L_n^2 \sin^2 2\pi f_n \, dt = \frac{1}{2} L_n^2 \qquad [4.16]$$

with $T = \dfrac{1}{f_n}$ or $T = \dfrac{k}{f_n}$, i.e. by calculation of the mean value:

$$\overline{\ell_n} = \frac{1}{T} \int_0^T L_n \sin 2\pi f_n \, dt = \frac{2}{\pi} L_n \qquad [4.17]$$

T having the same definition. It should be noted that:

$$\overline{\ell_n^2} = \frac{\pi^2}{8} (\overline{\ell})^2 \qquad [4.18]$$

T must be multiple of $\dfrac{1}{f_n}$. If this is not the case, the error is weaker the larger the number of selected periods. For a periodic excitation, the measurment or calculation accuracy is only related to the selected width Δf of the chosen filter (the signal being periodic and thus deterministic, there is no error of statistical origin related to the choice of T).

4.6. Statistical error

4.6.1. *Origin*

Let us consider a stationary random signal whose PSD we wish to calculate. Even if the measure duration were to be long (the signal remaining stationary), the PSD would only be determined from a signal sample lasting a few dozen seconds because of calculation time. The characteristic of such a signal being precisely to vary in a random way, the PSD obtained is different according to the moment at which it is calculated.

The PSD of an acceleration signal characterized by a sample of duration T is obtained by calculating the average of Fourier transform modules of K blocks for this sample, i.e. the average $\overline{G}$ of K PSD values $\hat{G}(f)$ of these blocks [4.13].

Because of the stochastic nature of the signal, the PSD value obtained at a given frequency varies according to the position over time of the signal sample chosen for the calculation.

Let us consider the PSD $\hat{G}(f)$ evaluated at frequency f starting from a sample of duration T chosen successively between the times t_0 and $t_0 + T$, then $t_0 + T$ and $t_0 + 2T$, and so on.

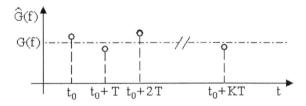

Figure 4.20. *PSD estimates at frequency f for various signal samples*

The values of $\hat{G}(f)$ thus calculated are all different from each other and different also from the exact value $G(f)$. We have:

$$G(f) = \lim_{K \to \infty} \frac{1}{K} \sum_{i=1}^{K} \hat{G}_i(f) = E\left[\hat{G}(f)\right] \qquad [4.19]$$

The true PSD is thus the mean value of the quantities $\hat{G}(f)$ estimated at various times, when their number tends towards the infinity. We could also define the standard deviation $\hat{s}$ of $\hat{G}(f)$. For K values,

$$\hat{s} = \sqrt{\frac{1}{K-1}\left\{\sum_{i=1}^{K}\hat{G}_i^2(f) - \frac{1}{K}\left[\sum_{i=1}^{K}\hat{G}_i(f)\right]^2\right\}} \qquad [4.20]$$

and, for $K \to \infty$

$$s = \lim_{K \to \infty} \hat{s} \qquad [4.21]$$

s, being the true standard deviation for a measurement $\hat{G}(f)$, is a description of uncertainty of this measure. In practice, we will make only one calculation of $\hat{G}(f)$ at the frequency f and we will try to estimate the error carried out according to the conditions of the analysis.

We can easily understand that the precision of this mean depends on the number of blocks.

This mean is only an estimation of the exact $G(f)$ value that would be obtained with a very large number of adequately sized blocks, or with a very long sample.

The true PSD is in theory the mean value of values $\hat{G}(f)$ evaluated at different moments when the number leans toward the infinite.

We should note that for a sample of duration T, when the number of K blocks increases, the $\Delta f = 1/\Delta t$ step between two consecutive PSD points also increases and the resolution decreases.

In practice, the number K is not infinite and we only have an estimation of the PSD at each frequency associated with a standard deviation. We can simply state, with a certain level of confidence, that the exact PSD is, for example, located between two limits made up of the estimated mean PSD ($\overline{\hat{G}}$) ± one standard deviation curves. We are then mistaken in the estimation of the PSD which is, *a priori*, a function of the length of the chosen sample.

4.6.2. Definition

The *statistical error* or *normalized rms error* is the quantity defined by the ratio:

$$\varepsilon = \frac{s_{\ell^2_{\Delta f}}}{\overline{\ell^2_{\Delta f}}} \qquad [4.22]$$

(variation coefficient) where $\overline{\ell^2_{\Delta f}}$ is the mean square value of the signal filtered in the filter of width Δf (quantity proportional to $\hat{G}(f)$) and $s_{\ell^2_{\Delta f}}$ is the standard deviation of the measurement of $\overline{\ell^2_{\Delta f}}$ related to the error introduced by taking a finite duration T.

NOTE.— *We are interested here in the statistical error related to calculation of the PSD. An error of comparable nature is also made during the calculation of other quantities such as coherence, transfer function, etc. (see section 4.17).*

4.7. Statistical error calculation

4.7.1. *Distribution of the measured PSD*

If the ratio $\varepsilon = \dfrac{s_{\ell^2_{\Delta f}}}{\overline{\ell^2_{\Delta f}}}$ is small, we can ensure with a high confidence level that a measurement of the PSD is close to the true average [NEW 75]. If, on the contrary, ε is large, the confidence level is small. We propose below to calculate the confidence level which can be associated with a measurement of the PSD when ε is known. The analysis is based on an assumption concerning the distribution of the measured values of the PSD.

The measured value of the mean square z^2 of the response of a filter Δf to a random vibration is itself a random variable. It is assumed in what follows that z^2 can be expressed as the sum of the squares of a certain number of Gaussian random variables statistically independent, zero average and of the same variance:

$$z^2 = \frac{1}{T}\left[\int_0^{T/n} \ddot{x}^2(t)\,dt + \int_{T/n}^{2T/n} \ddot{x}^2(t)\,dt + \cdots + \int_{T(1-1/n)}^{T} \ddot{x}^2(t)\,dt\right] \qquad [4.23]$$

We can indeed think that z^2 satisfies this assumption, but we cannot prove that these terms have an equal weight or that they are statistically independent. However, we note in experiments [KOR 66] that the measured values of z^2 roughly have the distribution which would be obtained if these assumptions were checked, namely a chi-square law, of the form:

$$\chi^2 = \chi_1^2 + \chi_2^2 + \chi_3^2 + \cdots + \chi_n^2 \qquad [4.24]$$

If it can be considered that the random signal follows a Gaussian law, it can be shown ([BEN 62], [BEN 71], [BLA 58], [GOL 53], [JEN 68], [NEW 75]) that measurements $\hat{G}(f)$ of the true PSD $G(f)$ are distributed as $G(f)\dfrac{\chi_n^2}{n}$ where χ_n^2 is the chi-square law with n degrees of freedom, mean n and variance 2 n (if the mean value of each independent variable is zero and their variance equal to 1 [BLA 58], [PIE 64]).

Figure 4.21 shows some curves of the probability density of this law for various values of n. We notice that, when n grows, the density approaches that of a normal law (a consequence of the central limit theorem).

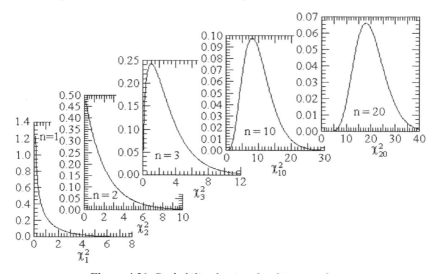

Figure 4.21. *Probability density: the chi-square law*

NOTE.– *Some authors [OSG 69] consider that measurements $\hat{G}(f)$ are distributed more like $G(f)\dfrac{\chi_{n-1}^2}{n-1}$, basing themselves on the following reasoning. From the* values $X_1, X_2, X_3, \cdots, X_n$ of a normally distributed population, of mean m (unknown value) and standard deviation s, we can calculate

$$\chi^2 = \frac{(X_1 - \bar{X})^2 + (X_2 - \bar{X})^2 + (X_3 - \bar{X})^2 + \cdots + (X_n - \bar{X})^2}{s^2} \quad [4.25]$$

where

$$\bar{X} = \frac{\sum X_i}{n} \quad [4.26]$$

(mean of the various values taken by variable X by each of the n elements). Let us consider the reduced variable

$$U_i = \frac{X_i - \bar{X}}{s} \quad [4.27]$$

The variables U_i are no longer independent, since there is a relationship between them: according to a property of the arithmetic mean, the algebraic sum of the deviations with respect to the mean is zero, therefore $\sum (X_i - \bar{X}) = 0$, and consequently, $\frac{\sum (X_i - \bar{X})}{s} = 0$ yielding:

$$\sum U_i = 0$$

In the sample of size n, only $n-1$ data are really independent, for if $n-1$ variations are known, the last variation results from this. If there is $n-1$ independent data, there are also $n-1$ degrees of freedom.

However, the majority of authors agree in considering that it is necessary to use a law with n degrees of freedom. This dissention has little incidence in practice, the number of degrees of freedom to be taken into account being necessarily higher than 90 so that the statistical error remains, according to the rulebook, lower than approximately 15%.

4.7.2. Variance of the measured PSD

The variance of $\hat{G}(f)$ is given by:

$$s^2_{\hat{G}(f)} = \text{var}\left[\frac{\hat{G}(f)\,\chi_n^2}{n}\right]$$

$$s^2_{\hat{G}(f)} = \left[\frac{G(f)}{n}\right]^2 \text{var}\left[\chi_n^2\right] \qquad [4.28]$$

However, the variance of a chi-square law is equal to twice the number of degrees of freedom:

$$\text{Var}\left(\chi^2\right) = 2\,n \qquad [4.29]$$

yielding

$$s^2_{\hat{G}(f)} = 2\,\frac{G^2(f)}{n^2}\,n = 2\,\frac{G^2(f)}{n} \qquad [4.30]$$

The mean of this law is equal to n.

4.7.3. Statistical error

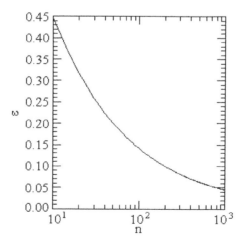

Figure 4.22. *Statistical error as a function of the number of degrees of freedom*

$$\overline{\hat{G}(f)} = \frac{G(f)}{n}\overline{\chi^2} = \frac{G(f)}{n}n$$

$$\hat{G}(f) = G(f) \qquad [4.31]$$

The statistical error is thus such that

$$\varepsilon^2 = \frac{s_{\hat{G}(f)}^2}{G(f)^2} = \frac{2}{n}$$

$$\varepsilon = \sqrt{\frac{2}{n}} \qquad [4.32]$$

ε is also called the *standard error*.

If this error is small, we can assume with a high level of confidence that the measure of the DSP is close to the real average [NEW 75]. If, on the contrary, ε is large, the result is not satisfactory.

The statistical error is therefore a quantity that makes it possible to estimate the precision with which a DSP has been calculated. It is important to configure its value on the traced curves.

4.7.4. *Relationship between number of degrees of freedom, duration and bandwidth of analysis*

This relation can be obtained either by using a series expansion of $E\left\{[\hat{G}(f) - G(f)]^2\right\}$ or starting from the autocorrelation function.

4.7.4.1. *From a series expansion*

It is shown that [BEN 61b], [BEN 62]:

$$E\left\{[\hat{G}(f) - G(f)]^2\right\} \approx \underbrace{\frac{G^2(f)}{T\,\Delta f}}_{\text{variability}} + \underbrace{\frac{\Delta f^2}{576}[G''(f)]^2}_{\text{bias}} \qquad [4.33]$$

Except when the slope of the PSD varies greatly with Δf, the bias is in general negligible. Then

$$\varepsilon^2 = \frac{E\left\{[\hat{G}(f) - G(f)]^2\right\}}{G^2(f)} \approx \frac{1}{T\,\Delta f}$$

This relation is a good approximation as long as ε is lower than approximately 0.2 (i.e. for $T\,\Delta f > 25$). The product $T\,\Delta f$, called the *bandwidth–time product*, is important in the estimation of spectra. For a given error, the resolution (Δf) cannot be reduced without increasing the duration of the sample of the processed signal.

$$\varepsilon \approx \frac{1}{\sqrt{T\,\Delta f}} \qquad [4.34]$$

The error is thus only a function of the duration T of the sample and of the width Δf of the analysis filter (always assumed to be ideal [BEA 72], [BEN 63], [NEW 75]).

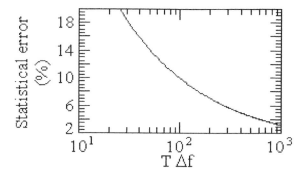

Figure 4.23. *Statistical error*

Figure 4.23 shows the variations of this quantity with the product $T\,\Delta f$. The number of events n represented by a record of a white noise signal, duration T, filtered by a filter of width Δf, is thus, starting from [4.32]:

$$n = 2\,\Delta f\,T \qquad [4.35]$$

Table 4.6 enables us to compare the approximate value given by [4.34] with the exact value for $n = 2\,T\,\Delta f$ between 2 and 65.

	Statistical error				
n	Exact	Approximate	n	Exact	Approximate
2	0.841	1.000	16	0.346	0.354
3	0.729	0.816	17	0.336	0.343
4	0.650	0.707	18	0.327	0.333
5	0.591	0.632	19	0.319	0.324
6	0.546	0.577	20	0.311	0.316
7	0.510	0.535	25	0.279	0.283
8	0.480	0.500	30	0.255	0.258
9	0.454	0.471	35	0.237	0.239
10	0.433	0.447	40	0.222	0.224
11	0.414	0.426	45	0.209	0.211
12	0.397	0.408	50	0.199	0.200
13	0.382	0.392	55	0.190	0.191
14	0.369	0.378	60	0.182	0.183
15	0.357	0.365	65	0.175	0.175

Table 4.6. *Comparison of the exact statistical error and the statistical error calculated from the approximate relation for a level of confidence equal to 68%*

Example 4.5.

In order for ε to be lower than 0.1, product $T \Delta f$ must be higher than 100, which can be done either with $T = 1$ s and $\Delta f = 100$ Hz or with $T = 100$ s and $\Delta f = 1$ Hz, for example.

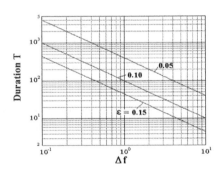

Figure 4.24. *Signal sample duration according to the frequency step, for a statistical error equal to 0.05, 0.10 and 0.15*

We will see later the incidence of the choice of these parameters in the PSD calculation (see section 4.8).

Definition

The quantity n = 2 Δf T is called the *number of degrees of freedom.*

NOTE.– *The expression degree of freedom is used differently in several sectors and should be clarified.*

In the studies on structure behavior during vibration, the number of degrees of freedom of a system is equal to the number of dimensions required to determine the state of this system at each moment. The simplest system, a hardware point, generally has three degrees of freedom: three coordinates are necessary to define its position in space at each moment. For a structure represented by a group of masses that can only be moved in one direction, springs and dampers, the number of degrees of freedom is equal to the number of masses. A solid in space has six degrees of freedom: its position is determined by three coordinates and three angles. The number of equations to characterize the movement of the system must be equal to the number of degrees of freedom;

We also use the expression "number of degrees of freedom" in control strategies for random vibration exciters to characterize the importance given to the last PSD measure of the signal generated in relation to the average of prior measures for correcting the control.

4.7.4.2. *From the autocorrelation function*

Let us consider $\ell(t)$ a vibratory signal response collected at the output of a filter of width Δf. The mean square value of $\ell(t)$ is given by [COO 65]:

$$\lambda_{\Delta f}^2 = \frac{1}{T} \int_0^T \ell^2(t) \, dt$$

Setting $\overline{\ell_{\Delta f}^2}$ the measured value of $\lambda_{\Delta f}^2$, we have, by definition:

$$\varepsilon = \frac{\left[\overline{\left(\lambda_{\Delta f}^2 - \overline{\ell_{\Delta f}^2} \right)^2} \right]^{1/2}}{\overline{\ell_{\Delta f}^2}} \qquad [4.36]$$

$$\varepsilon = \frac{\sqrt{\lambda_{\Delta f}^4 - 2\lambda_{\Delta f}^2 \overline{\ell_{\Delta f}^2} + \left(\overline{\ell_{\Delta f}^2}\right)^2}}{\overline{\ell_{\Delta f}^2}} = \frac{\sqrt{\overline{\lambda_{\Delta f}^4} - \left(\overline{\ell_{\Delta f}^2}\right)^2}}{\overline{\ell_{\Delta f}^2}}$$

However, we can write:

$$\overline{\lambda_{\Delta f}^4} - \left(\overline{\ell_{\Delta f}^2}\right)^2 = \overline{\frac{1}{T}\int_0^T \ell^2(u)\,du \cdot \frac{1}{T}\int_0^T \ell^2(v)\,dv} - \left(\overline{\ell_{\Delta f}^2}\right)^2$$

$$\overline{\lambda_{\Delta f}^4} - \left(\overline{\ell_{\Delta f}^2}\right)^2 = \frac{1}{T^2}\int_0^T du \int_0^T \left[\overline{\ell^2(u)\,\ell^2(v)} - \left(\overline{\ell_{\Delta f}^2}\right)^2\right]dv$$

i.e., while setting $t = u$ and $\tau = v - u = v - t$,

$$\overline{\lambda_{\Delta f}^4} - \left(\overline{\ell_{\Delta f}^2}\right)^2 = \frac{1}{T^2}\int_0^T dt \int_{-t}^{T-t}\left[\overline{\ell^2(t)\,\ell^2(t+\tau)} - \left(\overline{\ell_{\Delta f}^2}\right)^2\right]d\tau$$

yielding

$$\overline{\lambda_{\Delta f}^4} - \left(\overline{\ell_{\Delta f}^2}\right)^2 = \frac{2}{T^2}\int_0^T dt \int_{-t}^{T-t}\left(\overline{\ell_{\Delta f}^2}\right)^2 \rho^2(\tau)\,d\tau$$

where $\rho(\tau)$ is the autocorrelation coefficient. Given a narrowband random signal, we saw that the coefficient ρ is symmetric with regard to the axis $\tau = 0$ and that ρ decreases when $|\tau|$ becomes larger. If T is sufficiently large, as well as the majority of the values of t:

$$\varepsilon^2 = \frac{\overline{\lambda_{\Delta f}^4} - \left(\overline{\ell_{\Delta f}^2}\right)^2}{\left(\overline{\ell_{\Delta f}^2}\right)^2} = \frac{2}{T^2}\int_0^T dt \int_{-\infty}^{+\infty}\rho^2(\tau)\,d\tau$$

yielding the *standardized variance* ε^2 [BEN 62]:

$$\varepsilon^2 = \frac{2}{T}\int_{-\infty}^{+\infty}\rho^2(\tau)\,d\tau = \frac{4}{T}\int_0^\infty \rho^2(\tau)\,d\tau \qquad [4.37]$$

$$\varepsilon = 2\sqrt{\frac{1}{T}\int_0^\infty \rho^2(\tau)\,d\tau} \qquad [4.38]$$

Practical Calculation of the PSD 189

Particular cases

1. Rectangular band-pass filter

We saw [2.72] that in this case [MOR 58]:

$$\rho(\tau) = \frac{\cos 2\pi f_0 \tau \, \sin \pi \Delta f \tau}{\pi \tau \Delta f}$$

yielding

$$\varepsilon^2 \approx \frac{4}{T} \int_0^\infty \frac{\cos^2 2\pi f_0 \tau \, \sin^2 \pi \Delta f \tau}{\pi^2 \tau^2 \Delta f^2} \, d\tau$$

$$\varepsilon^2 \approx \frac{1}{T \Delta f}$$

and [BEN 62], [KOR 66], [MOR 63]:

$$\varepsilon \approx \frac{1}{\sqrt{T \Delta f}} \qquad [4.39]$$

Example 4.6.

For ε to be lower than 0.1, it is necessary that the product $T \Delta f$ be greater than 100, which can be achieved, for example, either with $T = 1$ s and $\Delta f = 100$ Hz, or with $T = 100$ s and $\Delta f = 1$ Hz. We will see, later on, the incidence of these choices on the calculation of the PSD.

---

2. Resonant circuit

For a resonant circuit:

$$\rho(\tau) = \cos 2\pi f_0 \tau \, e^{-\pi \tau \Delta f}$$

yielding

$$\varepsilon^2 \approx \frac{4}{T} \int_0^\infty \cos^2 2\pi f_0 \tau \, e^{-2\pi \tau \Delta f} \, d\tau$$

190   Random Vibration

$$\varepsilon^2 \approx \frac{2}{T} \int_0^\infty e^{-2\pi\tau\Delta f} \, d\tau$$

$$\varepsilon \approx \frac{1}{\sqrt{\pi T \Delta f}} \qquad [4.40]$$

4.7.5. Confidence interval

Uncertainty concerning $\hat{G}(f)$ can also be expressed in terms of the confidence interval. If the signal $\ell(t)$ has a roughly normal probability density function, the distribution of $\dfrac{\hat{G}(f)}{G(f)}$, for any f, is the same as $\dfrac{\chi^2}{n}$. Given an estimate $\hat{G}(f)$ obtained from a signal sample, for $n = 2 \Delta f\, T$ events, the confidence interval in which the true PSD $G(f)$ is located is, on the confidence level $(1-\alpha)$:

$$\frac{n\,\hat{G}(f)}{\chi^2_{n,\,1-\alpha/2}} \leq G(f) \leq \frac{n\,\hat{G}(f)}{\chi^2_{n,\,\alpha/2}} \qquad [4.41]$$

where $\chi^2_{n,\,\alpha/2}$ and $\chi^2_{n,\,1-\alpha/2}$ have n degrees of freedom.

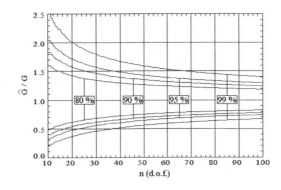

Figure 4.25. *Confidence intervals of $\hat{G}/G$ according to the number of degrees of freedom* [MOO 61]

Table 4.7 gives some values of $\chi^2_{n,\alpha}$ according to the number of degrees of freedom n for various values of α.

n\α	0.995	0.99	0.975	0.95	0.90	0.75	0.50	0.25	0.10	0.05	0.025	0.01	0.005
1	7.88	6.63	5.02	3.84	2.71	1.32	0.455	0.102	0.0158	0.0039	0.0010	0.0002	0.0000
2	10.6	9.21	7.38	5.99	4.61	2.77	1.39	0.575	0.211	0.103	0.0506	0.0201	0.0100
3	12.8	11.3	9.35	7.81	6.25	4.11	2.37	1.21	0.584	0.352	0.216	0.115	0.072
4	14.9	13.3	11.1	9.49	7.78	5.39	3.36	1.92	1.06	0.711	0.484	0.297	0.207
5	16.7	15.1	12.8	11.1	9.24	6.63	4.35	2.67	1.61	1.15	0.831	0.554	0.412
6	18.5	16.8	14.4	12.6	10.6	7.84	5.35	3.45	2.20	1.64	1.24	0.872	0.676
7	20.3	18.5	16.0	14.1	12.0	9.04	6.35	4.25	2.83	2.17	1.69	1.24	0.989
8	22.0	20.1	17.5	15.5	13.4	10.2	7.34	5.07	3.49	2.73	2.18	1.65	1.34
9	23.6	21.7	19.0	16.9	14.7	11.4	8.34	5.90	4.17	3.33	2.70	2.09	1.73
10	25.2	23.2	20.5	18.3	16.0	12.5	9.34	6.74	4.87	3.94	3.25	2.56	2.16
11	26.8	24.7	21.9	19.7	17.3	13.7	10.3	7.58	5.58	4.57	3.82	3.05	2.60
12	28.3	26.2	23.3	21.0	18.5	14.8	11.3	8.44	6.30	5.23	4.40	3.57	3.07
13	29.8	27.7	24.7	22.4	19.8	16.0	12.3	9.30	7.04	5.89	5.01	4.11	3.57
14	31.3	29.1	26.1	23.7	21.1	17.1	13.3	10.2	7.79	6.57	5.63	4.66	4.07
15	32.8	30.6	27.5	25.0	22.3	18.2	14.3	11.0	8.55	7.26	6.26	5.23	4.60
16	34.3	32.0	28.8	26.3	23.5	19.4	15.3	11.9	9.31	7.96	6.91	5.81	5.14
17	35.7	33.4	30.2	27.6	24.8	20.5	16.3	12.8	10.1	8.67	7.56	6.41	5.70
18	37.2	34.8	31.5	28.9	26.0	21.6	17.3	13.7	10.9	9.39	8.23	7.01	6.26
19	38.6	36.2	32.9	30.1	27.2	22.7	18.3	14.6	11.7	10.1	8.91	7.63	6.84
20	40.0	37.6	34.2	31.4	28.4	23.8	19.3	15.5	12.4	10.9	9.59	8.26	7.43
21	41.4	38.9	35.5	32.7	29.6	24.9	20.3	16.3	13.2	11.6	10.3	8.90	8.03
22	42.8	40.3	36.8	33.9	30.8	26.0	21.3	17.2	14.0	12.3	11.0	9.54	8.64
23	44.2	41.6	38.1	35.2	32.0	27.1	22.3	18.1	14.8	13.1	11.7	10.2	9.26
24	45.6	43.0	39.4	36.4	33.2	28.2	23.3	19.0	15.7	13.8	12.4	10.9	9.89
25	46.9	44.3	40.6	37.7	34.4	29.3	24.3	19.9	16.5	14.6	13.1	11.5	10.5
26	48.3	45.6	41.9	38.9	35.6	30.4	25.3	20.8	17.3	15.4	13.8	12.2	11.2
27	49.6	47.0	43.2	40.1	36.7	31.5	26.3	21.7	18.1	16.2	14.6	12.9	11.8
28	51.0	48.3	44.5	41.3	37.9	32.6	27.3	22.7	18.9	16.9	15.3	13.6	12.5
29	52.3	49.6	45.7	42.6	39.1	33.7	28.3	23.6	19.8	17.7	16.0	14.3	13.1
30	53.7	50.9	47.0	43.8	40.3	34.8	29.3	24.5	20.6	18.5	16.8	15.0	13.8
40	66.8	63.7	59.3	55.8	51.8	45.6	39.3	33.7	29.1	26.5	24.4	22.2	20.7
50	79.5	76.2	71.4	67.5	63.2	56.3	49.3	42.9	37.7	34.8	32.4	29.7	28.0
60	92.0	88.4	83.3	79.1	74.4	67.0	59.3	52.3	46.5	43.2	40.5	37.5	35.5
70	104.2	100.4	95.0	90.5	85.5	77.6	69.3	61.7	55.3	51.7	48.8	45.4	43.3
80	116.3	112.3	106.6	101.9	96.6	88.1	79.3	71.1	64.3	60.4	57.2	53.5	51.2
90	128.3	124.1	118.1	113.1	107.6	98.6	89.3	80.6	73.3	69.1	65.6	61.8	59.2
100	140.2	135.8	129.6	124.3	118.5	109.1	99.3	90.1	82.4	77.9	74.2	70.1	67.3

Table 4.7. Values of $\chi^2_{n,\alpha}$ as a function of the number N of degrees of freedom [SPI 74]

192 Random vibration

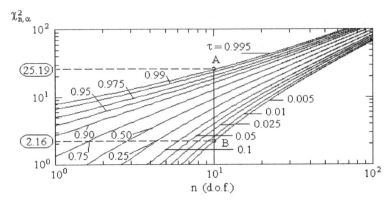

Figure 4.26. Values of $\chi^2_{n,\alpha}$ with respect to the number of degrees of freedom and of α

Figure 4.26 graphically represents the function χ^2_α with respect to n, parameterized by the probability α.

Example 4.7.

99% of the values lie between 0.995 and 0.005. We read from Figure 4.26, for n = 10, that the limits are χ^2 = 25.2 and 2.16.

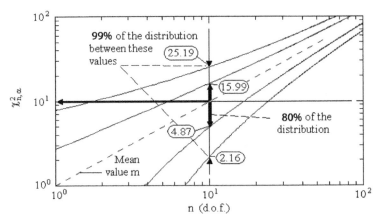

Figure 4.27. Example of use of the curves $\chi^2_{n,\alpha}(n)$

Example 4.8.

Figure 4.27 shows how in a particular case these curves can be used to numerically evaluate the limits of the confidence interval defined by relation [4.41].

Let us set n = 10. We note from this figure that 80% of the values are within the interval 4.87 and 15.99 with mean value m = 10. If the true value of the mean of the calculated PSD S_0 is m, it cannot be determined exactly, but it is known that

$$\frac{4.87}{10} < \frac{S_0}{m} < \frac{15.99}{10}$$

$$2.05\ S_0 > m > 0.625\ S_0$$

More specific tables or curves were published to directly provide the value of the limits [DAR 72], [MOO 61], [PIE 64]. For example, Table 4.8 gives the confidence interval defined in [4.41] for three values of $1 - \alpha$ [PIE 64].

NOTE.– When $n \geq 30$, $\sqrt{2\chi_n^2}$ follows a law close to a Gaussian law of mean $\sqrt{2n-1}$ and standard deviation 1 (Fisher's law). Let x be a normal reduced variable and α a value of the probability such that

$$prob\left[|x| < k(\alpha)\right] = 1 - \alpha \qquad [4.42]$$

where k is a constant function of the probability α.

For example:

α	90%	95%	99%
$k(\alpha)$	1.645	1.960	2.58

We have

$$prob\left[\sqrt{2n-1} - k(\alpha) \leq \sqrt{2\chi_n^2} \leq \sqrt{2n-1} + k(\alpha)\right] \qquad [4.43]$$

Degrees of freedom n	Confidence interval limits relating to a measured PSD $\hat{G}(f) = 1$							
	$(1-\alpha) = 0.68$		$(1-\alpha) = 0.90$		$(1-\alpha) = 0.95$		$(1-\alpha) = 0.99$	
	Lower limit	Higher limit	Lower limit	Higher limit	Lower limit	Higher limit	Lower limit	Higher limit
2	0.543	5.789	0.334	19.46	0.271	39.498	0.189	199.50
5	0.628	2.432	0.452	4.365	0.390	6.015	0.299	8.879
10	0.698	1.760	0.546	2.538	0.488	3.080	0.397	4.639
15	0.737	1.555	0.600	2.066	0.546	2.395	0.457	4.545
20	0.763	1.451	0.637	1.843	0.585	2.085	0.500	2.690
25	0.782	1.387	0.664	1.711	0.615	1.906	0.533	2.377
30	0.797	1.343	0.685	1.622	0.639	1.787	0.559	2.176
35	0.809	1.310	0.703	1.558	0.658	1.702	0.581	2.036
40	0.818	1.285	0.717	1.509	0.674	1.637	0.599	1.932
45	0.827	1.265	0.730	1.470	0.688	1.586	0.615	1.851
50	0.834	1.248	0.741	1.438	0.700	1.545	0.629	1.786
60	0.846	1.222	0.759	1.389	0.720	1.482	0.653	1.689
70	0.856	1.202	0.773	1.353	0.737	1.436	0.672	1.618
80	0.864	1.187	0.785	1.325	0.750	1.400	0.688	1.563
90	0.871	1.174	0.795	1.302	0.762	1.371	0.701	1.520
100	0.876	1.164	0.804	1.283	0.772	1.347	0.713	1.485
120	0.886	1.148	0.819	1.254	0.788	1.310	0.733	1.431
140	0.893	1.135	0.30	1.232	0.802	1.283	0.749	1.391
160	0.900	1.126	0.840	1.214	0.813	1.261	0.763	1.360
180	0.905	1.118	0.848	1.200	0.822	1.244	0.774	1.334
200	0.909	1.111	0.855	1.189	0.830	1.229	0.784	1.314

Multiply the lower and higher limits in the table by the measured value $\hat{G}(f)$ to obtain the limits of the confidence interval of the true value $G(f)$.

Table 4.8. *Confidence limits for the calculation of a PSD [PIE 64]*

yielding the approximate value of the limits of χ_n^2

$$\text{prob}\left[\frac{\left[\sqrt{2n-1}-k(\alpha)\right]^2}{2} \leq \chi_n^2 \leq \frac{\left[\sqrt{2n-1}+k(\alpha)\right]^2}{2}\right] = 1-\alpha \quad [4.44]$$

and that of the confidence interval limits of $\dfrac{G(f)}{\hat{G}(f)}$ (since the probability of $\dfrac{\hat{G}}{G}$ is the same as that of $\dfrac{\chi_n^2}{n}$):

$$\text{prob}\left[\frac{2n}{\left[\sqrt{2n-1}+k(\alpha)\right]^2} \leq \frac{G(f)}{\hat{G}(f)} \leq \frac{2n}{\left[\sqrt{2n-1}-k(\alpha)\right]^2}\right] = 1-\alpha \quad [4.45]$$

For large values of n $[n \geq 120]$, i.e. for ε small, it is shown that the chi-square law tends towards the normal law and that the distribution of the values of $\hat{G}(f)$ can itself be approximated by a normal law of mean n and standard deviation $\sqrt{2n}$ (law of large numbers). In this case,

$$\text{prob}\left[n-k(\alpha)\sqrt{2n} \leq \chi_n^2 \leq n+k(\alpha)\sqrt{2n}\right] = 1-\alpha \quad [4.46]$$

yielding

$$\text{prob}\left[\frac{n}{n+k(\alpha)\sqrt{2n}} \leq \frac{G(f)}{\hat{G}(f)} \leq \frac{n}{n-k(\alpha)\sqrt{2n}}\right] = 1-\alpha \quad [4.47]$$

196   Random Vibration

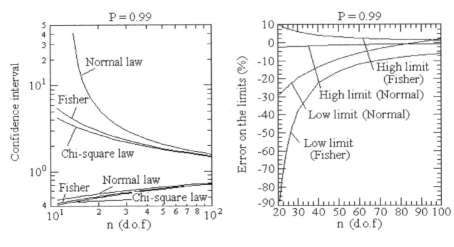

Figure 4.28. *Confidence interval for $P = 0.99$*

Figure 4.29. *Error related to the use of the normal or Fisher laws*

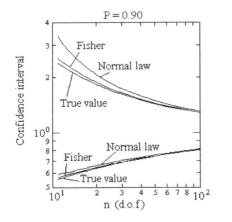 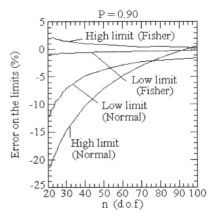

Figure 4.30. *Confidence interval for $P = 0.90$*

Figure 4.31. *Error related to the use of the normal or Fisher laws*

Figures 4.28 to 4.31 provide, for a confidence level of 99%, and then 90%:

– variations in the confidence interval limits depending to the number of degrees of freedom n, obtained using an exact calculation (chi-square law), by considering the Fisher and normal assumptions;

– the error made using each of these simplifying assumptions.

Practical Calculation of the PSD   197

These curves show that the Fisher assumption constitutes an approximation acceptable for n greater than approximately 30 (according to the confidence level), with relatively simple analytical expressions for the limits.

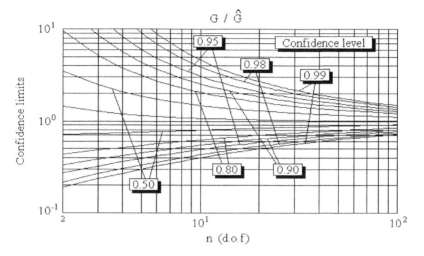

Figure 4.32. *Confidence limits* $(G/\hat{G})$

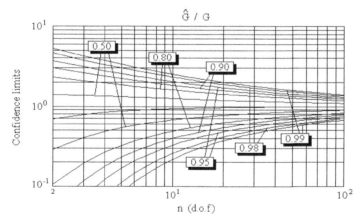

Figure 4.33. *Confidence limits* $(\hat{G}/G)$ *[MOO 61]*

The ratio $\dfrac{G}{\hat{G}}$ (or $\dfrac{\hat{G}}{G}$, depending on the case) is plotted in Figures 4.32 and 4.33 with respect to n, for various values of the confidence level.

Example 4.9.

Let us suppose that a PSD level $\hat{G} = 2$ has been measured with a filter of width $\Delta f = 2.5$ Hz and from a signal sample of duration $T = 10$ s. The number of degrees of freedom is $n = 2\,T\,\Delta f = 50$ (yielding $\varepsilon = \dfrac{1}{\sqrt{T\,\Delta f}} = 0.2$). Table 4.8 gives, for $1 - \alpha = 0.90$:

$$0.741\,\hat{G} < G < 1.44\,\hat{G}$$

i.e. $1.482 < G < 2.88$ if $\hat{G} = 2$.

This result can also be obtained from the curves in Figure 4.33. For $n = 50$:

$$\frac{\hat{G}}{G} > 0.69 \text{ on the confidence level } 5\%;$$

$$\frac{\hat{G}}{G} < 1.35 \text{ on the confidence level } 95\%.$$

With a confidence level of 90%, we thus have:

$$0.69 < \frac{\hat{G}}{G} < 1.35$$

i.e. $\dfrac{\hat{G}}{1.35} < G < \dfrac{\hat{G}}{0.69}$

$$0.74\,\hat{G} < G < 1.44\,\hat{G}$$

$$1.48 < G < 2.88$$

For $\varepsilon \leq 0.1$ [PIE 64], we can see that the relative error between the true PSD and the calculated PSD lies between $\pm s_{\hat{G}}$ with a confidence level of 68% (68.2689%), i.e. that during approximately 68% of the time, the exact PSD lies between $\hat{G}(f) \pm s_{\hat{G}}$:

$$\left| G(f) - \hat{G}(f) \right| < s_{\hat{G}} \qquad [4.48]$$

From this inequality, we can write [PIE 64]:

$$\frac{\hat{G}(f)}{1+\varepsilon} < G(f) < \frac{\hat{G}(f)}{1-\varepsilon} \qquad [4.49]$$

The confidence limits on the 68% level are plotted in Figure 4.34 for n ranging between 2 and 1,000, then ranging between 20 and 200.

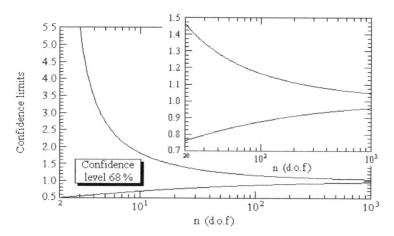

Figure 4.34. *Confidence limits at the 68% level*

NOTE.– *At confidence level $1 - \alpha = 68\%$, expressions [4.41] and [4.49] show that*

$$\begin{cases} \dfrac{1}{1+\varepsilon} = \dfrac{n}{\chi^2_{n,\,1-\alpha/2}} \\ \dfrac{1}{1-\varepsilon} = \dfrac{n}{\chi^2_{n,\,\alpha/2}} \end{cases} \qquad [4.50]$$

yielding

$$\begin{cases} \varepsilon = \dfrac{\chi^2_{n,\,1-\alpha/2}}{n} - 1 \\ \varepsilon = 1 - \dfrac{\chi^2_{n,\,\alpha/2}}{n} \end{cases} \qquad [4.51]$$

where, if $\varepsilon \leq 0.2$, $\varepsilon \approx \dfrac{1}{\sqrt{T\,\Delta f}}$, we deduce:

$$\chi^2_{n,\,1-\alpha/2} + \chi^2_{n,\,\alpha/2} = 2\,n \qquad [4.52]$$

This expression is applicable for any n for confidence level 68% and any α when n is large.

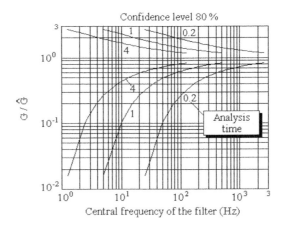

Figure 4.35. $G/\hat{G}$ *as a function of frequency of filter and length of analysis [CUR 64]*

Figure 4.35 shows the variations of:

$$\frac{G}{\hat{G}} = \frac{\text{true PSD (large T)}}{\text{measured PSD}}$$

with respect to the central frequency of the filter, for various lengths of analysis, at the confidence level of 80% and for a ratio $\dfrac{\text{central frequency}}{\Delta f} = 10$ [CUR 64].

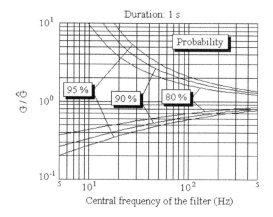

Figure 4.36. $G/\hat{G}$ *as a function of frequency of the filter and probability*

Figure 4.36 is parameterized, in the same axes, by the probability.

Figures 4.35 and 4.36 are deduced from Figure 4.34 as follows: for a given f, $\Delta f = \dfrac{f}{10}$ is calculated, then, for a given T, n = 2 Δf T, yielding $\dfrac{\hat{G}}{G}$ and $\dfrac{G}{\hat{G}}$.

Example 4.10.

We want to calculate a PSD with a statistical error less than 17.5% at a confidence level of 95%. At this level, we have ±1.96 times the standard error. The standard error should thus not exceed:

$$\varepsilon = \frac{17.5}{1.96} = 8.94\,\%$$

Knowing ε, the calculation conditions can be chosen from

$$\varepsilon = \frac{1}{\sqrt{T\,\Delta f}} = 8.94\ 10^{-2}.$$

Duration T (s)	60	30	20	10	5
Δf (Hz)	2.086	4.171	6.256	12.512	25.02

4.7.6. Expression for statistical error in decibels

While dividing, in [4.44], χ_n^2 by its mean value by n, we obtain

$$\text{prob}\left[\frac{[\sqrt{2n-1}-k(\alpha)]^2}{2n} \leq \frac{\chi_n^2}{n} \leq \frac{[\sqrt{2n-1}+k(\alpha)]^2}{2n}\right] = 1-\alpha \quad [4.53]$$

The error can be evaluated from $\frac{\hat{G}}{G}$, i.e. $\frac{\chi_n^2}{n}$, in the form

$$\varepsilon_{dB} = 10 \log_{10}\left[\frac{\chi_n^2}{n}\right] \quad [4.54]$$

It is raised, according to n, by

$$\varepsilon_{dB} = 10 \log_{10} \frac{[\sqrt{2n-1}+k(\alpha)]^2}{2n}$$

$$\varepsilon_{dB} = 10 \log_{10}\left[1 - \frac{1}{2n} + \frac{k(\alpha)\sqrt{2n-1}}{n} + \frac{k^2(\alpha)}{2n}\right] \quad [4.55]$$

Figure 4.37 shows the variations of ε_{dB} with the number of degrees of freedom n for a confidence level of 99%.

If $k(\alpha) = 1$, there is a 68.27% chance that the measured value is in the interval $\pm 1\, s_{\hat{G}(f)}$ and an 84.13% chance that it is lower than $1\, s_{\hat{G}(f)}$. Then:

$$\varepsilon_{dB} = 10 \log_{10}\left[1 + \frac{\sqrt{2n-1}}{n}\right]$$

If n is large compared to 1,

$$\varepsilon_{dB} \approx 10 \log_{10}\left(1 + \sqrt{\frac{2}{n}}\right)$$

$$\varepsilon_{dB} = 10 \log \left(1 + \frac{1}{\sqrt{T \, \Delta f}}\right) \qquad [4.56]$$

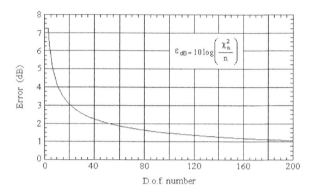

Figure 4.37. *Statistical error in dB (confidence level of 99%)*

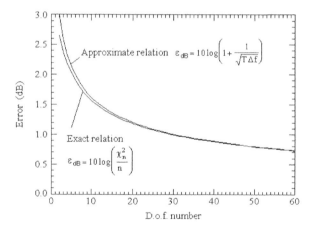

Figure 4.38. *Statistical error approximation (confidence level of 68%)*

The curves in Figure 4.38 allow comparison of exact relation [4.54] with approximate relation [4.56]: the approximation is very good for $n > 50$.

Example 4.11.

If it is required that $\varepsilon = \pm 0.5$ dB, i.e. that $\varepsilon = \pm 12.2\%$, it is necessary, at a confidence level of 84%, that $T \Delta f = 67.17$, or that the number of degrees of freedom is equal to $n = 2 T \Delta f \approx 135$.

If $\Delta f = 24$ Hz, $T = \dfrac{1}{(0.122)^2 \Delta f} 10^4 = 2.8$ s. At a confidence level of 90%, the variations of the PSD are, in the interval [BAN 78]:

n	Lower limit (dB)	Upper limit (dB)
50	-1.570	1.329
100	-1.077	0.958
250	-0.665	0.617

NOTE.– *From [4.56]*,

$$\varepsilon_{dB} \approx 10 \log_{10}\left(1 + \frac{1}{\sqrt{T \Delta f}}\right) \approx \frac{10}{\ln(10)} \ln\left(1 + \frac{1}{\sqrt{T \Delta f}}\right)$$

and, if $\dfrac{1}{\sqrt{T \Delta f}}$ is small, the decibel error can also be written as

$$\varepsilon_{dB} \approx \frac{10}{\ln(10)} \frac{1}{\sqrt{T \Delta f}} \approx \frac{4.34}{\sqrt{T \Delta f}}$$

4.7.7. *Statistical error calculation from digitized signal*

Let N be the number of sampling points of the signal $\ddot{x}(t)$ of duration T, M the number of points in frequency of the PSD, $f_{samp.}$ the sampling frequency of the signal, f_{max} the maximum frequency of the PSD, lower than or equal to $\dfrac{f_{samp.}}{2.6}$

Practical Calculation of the PSD   205

(modified Shannon's theorem, Volume 1, Chapter 1) and δt the time interval between two points.

We obtain:

$$T = N \delta t \qquad [4.57]$$

$$\Delta f = \frac{f_{samp.}}{2M} \qquad [4.58]$$

NOTE.– *M points separated by an interval Δf lead to a maximum frequency $f_{max} = M \Delta f = \frac{f_{samp.}}{2}$. To fulfill the condition of section 4.3.1, it is necessary to limit in practice the useful field of the PSD to $f_{max} \leq \frac{f_{samp.}}{2.6}$.*

If we need a PSD calculated based on M points, we need at least $\Delta N = 2M$ points per block. Since the signal is composed of N points, we will cut it up into $K = \frac{N}{2M}$ blocks of duration $\Delta T = \frac{T}{K}$.

Knowing that $f_{samp.} = \frac{1}{\delta t}$:

$$\Delta f = \frac{1}{2 M \delta t}$$

yielding

$$\varepsilon = \frac{1}{\sqrt{T \Delta f}} = \sqrt{\frac{2 M \delta t}{N \delta t}}$$

i.e. $\varepsilon = \sqrt{\frac{2M}{N}} = \frac{1}{\sqrt{K}} \qquad [4.59]$

Example 4.12.

$$N = 32{,}768 \text{ points} \quad M = 512 \text{ points} \quad T = 64 \text{ s}$$

yielding

$$2M = 1{,}024 \text{ points per sample}$$

$$K = \frac{N}{2M} = 32 \text{ samples (of 2 s)}$$

$$\Delta f = \frac{K}{T} = \frac{32}{64} = 0.5 \text{ Hz} \qquad f_{samp.} = \frac{N}{T} = \frac{32{,}768}{64} = 512 \text{ points/s}$$

$$\varepsilon = \sqrt{\frac{2 \times 512}{32{,}768}} = 0.1768$$

Even if $M \, \Delta f = 512 \; 0.5 \text{ Hz} = 256 \text{ Hz}$, we must have, in practice,

$$f_{max} \leq \frac{f_{samp.}}{2.6} = \frac{512}{2.6} \approx 197 \text{ Hz}.$$

Example 4.13.

From a signal sample of 5 seconds duration measured on a plane (Figure 4.39), several PSDs were calculated to show the influence of statistical error. The signal, sampled with a frequency equal to 2.6 times its maximum frequency, was initially defined by 32,302 points; zeros were added to each block to obtain a power of two (or 32,768 points).

The PSD in Figure 4.40 was obtained by considering a single block of equal duration as the sample, 5 s. The frequency resolution is great (0.2 Hz), but the statistical error is maximum. Since the duration of the sample is mandated, the only way to decrease the statistical error is to increase step Δf in frequency (and thus decrease resolution) and/or to use overlapping (trick to increase duration in the calculation).

Figures 4.41 to 4.46 show the PSDs traced for several Δf values included between 0.39 Hz and 12.62 Hz. The reduction in error leads to a smoother line.

Practical Calculation of the PSD 207

The curve is only truly "clean" for an error of approximately 13% (Figure 4.46), which confirms the rule (error lower than 15%).

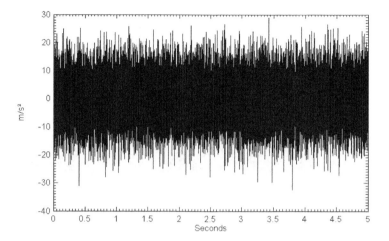

Figure 4.39. *Gaussian "airplane" vibration (rms value 7.58 m/s2, duration 5 s, 32,302 points)*

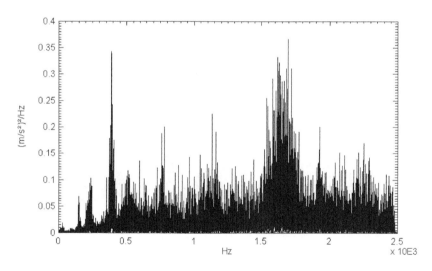

Figure 4.40. *PSD calculated with a single block (no averaging)*

208   Random Vibration

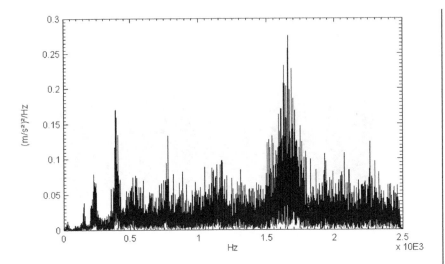

Figure 4.41. *PSD calculated with an error of approximately 71%
(2 blocks, Δf = 0.39 Hz, 8,192 points)*

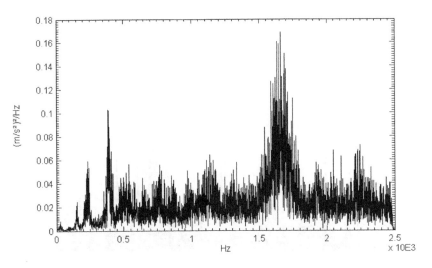

Figure 4.42. *PSD calculated with an error of approximately 50%
(4 blocks, Δf = 0.79 Hz, 4,096 points)*

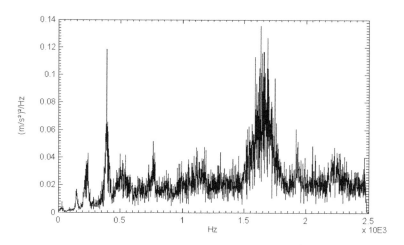

Figure 4.43. *PSD calculated with an error of approximately 36% (8 blocks, Δf = 1.58 Hz, 2,048 points)*

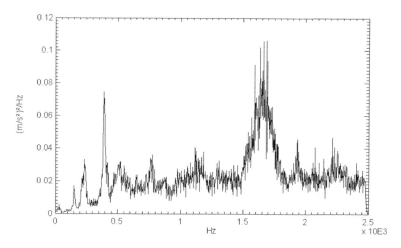

Figure 4.44. *PSD calculated with an error of approximately 25% (16 blocks, Δf = 3.15 Hz, 1,024 points)*

210   Random Vibration

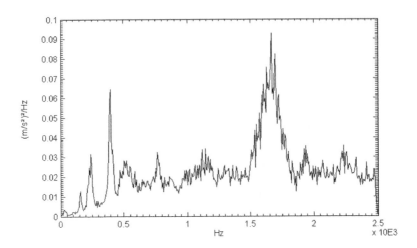

Figure 4.45. *PSD calculated with an error equal to 18%
(32 blocks, $\Delta f = 6.31$ Hz, 512 points)*

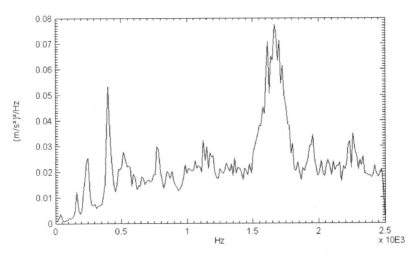

Figure 4.46. *PSD calculated with an error equal to 13%
(64 blocks, $\Delta f = 12.62$ Hz, 256 points)*

The number of PSD points can be increased to a constant statistical error:

– with overlapping that can reach 75%;

– by adding zeros to the signal before calculation (Figures 4.47 with 1,024 points and 4.48 with 512 points).

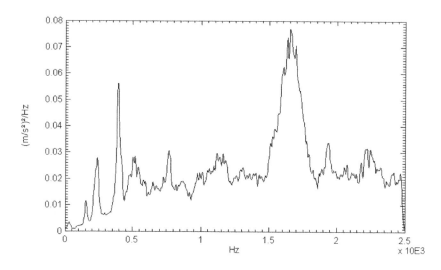

Figure 4.47. *PSD calculated with an error equal to 9.4% (Δf = 3.15 Hz, 1,024 points, 75% overlapping with zeros)*

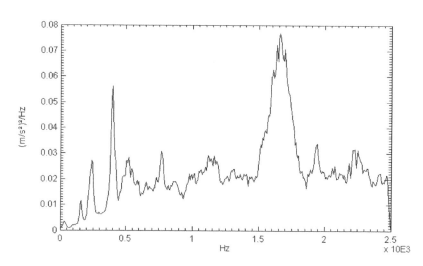

Figure 4.48. *PSD calculated with an error equal to 9.4% (Δf = 6.31 Hz, 512 points, 75% overlapping with zeros)*

4.8. Influence of duration and frequency step on the PSD

The signal studied below theoretically has a generally constant PSD up to 1,000 Hz (0.1 (m/s^2)^2/Hz). To highlight the influence of each parameter studied, the PSD traced in the following figures was artificially broken into 3 or 4 frequency intervals. Each band shows a part of the PSD calculated with a different value from the parameter involved.

4.8.1. *Influence of duration*

PSDs are calculated with a frequency of 512 points, up to 1,000 Hz (a frequency step approximately equal to 1.95 Hz, therefore constant resolution). Each part of the curve shows the PSD obtained with a signal sample of 100 s, 20 s and 5 s durations respectively. With 100 s, the statistical error is low, equal to 7.2%: the PSD remains close to 0.1. When duration decreases, the statistical error increases and the curve becomes increasingly dispersed around 0.1.

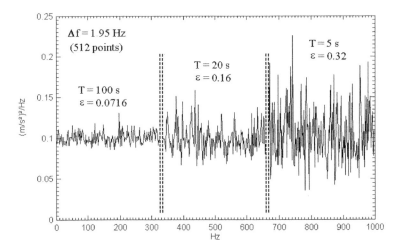

Figure 4.49. *Influence of sample duration on the PSD*

For a given ε statistical error, we can say, with a level of confidence equal to 68%, that the true PSD is included between the calculated PSD divided by 1 + ε and the calculated PSD divided by 1 − ε.

Figure 4.50 shows the sector delimited by these two curves in the previous calculation hypotheses. The indetermination decreases with the statistical error.

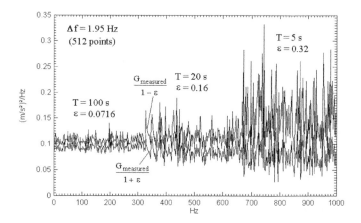

Figure 4.50. *Influence of duration on the precision of the PSD calculation. Confidence intervals for 3 duration values (at constant Δf)*

4.8.2. *Influence of the frequency step*

We now review the case where we have a signal sample with a given duration (20 s) and we examine the influence the Δf frequency step (i.e. of the number of points).

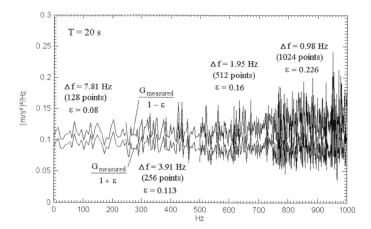

Figure 4.51. *Influence of frequency step on precision d PSD calculation Confidence intervals for 4 Δf values (at constant sample duration)*

214   Random Vibration

We can observe that when the resolution increases (Δf smaller), the statistical error increases, the resulting PSD is less smoothed with higher peaks. The greater the resolution, the greater the uncertainty.

To the left of the curve, the statistical error is small, precision is good, but the resolution is only of 7.81 Hz. The PSD is smooth.

4.8.3. *Influence of duration and of constant statistical error frequency step*

When duration T of the sample and the frequency step Δf vary to retain the same statistical error value (constant T Δf product), we obtain a spectral curve that is defined with more or less points (since the frequency step varies) and where the amplitude of variations around the true value (0,1) remains generally constant (linked to the statistical error).

Example 4.14.

Figures 4.52 to 4.55 show a PSD calculated with a statistical error equal to 0.16, for different values of duration T of the sample (1.25 s to 20 s) and frequency step (31.24 Hz to 1.95 Hz).

The curve is more detailed the smaller the frequency step.

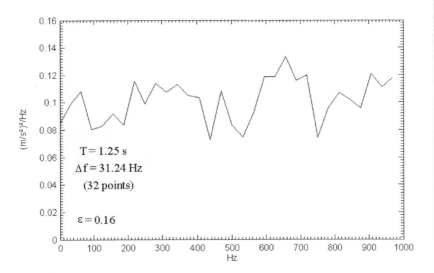

Figure 4.52. *PSD calculated for T = 1.25 s and Δf = 31.24 Hz (ε = 0.16)*

Practical Calculation of the PSD 215

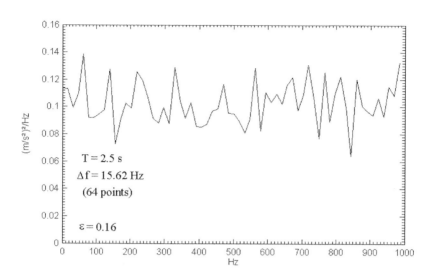

Figure 4.53. *PSD calculated for T = 2.5 s and Δf = 15.62 Hz (ε = 0.16)*

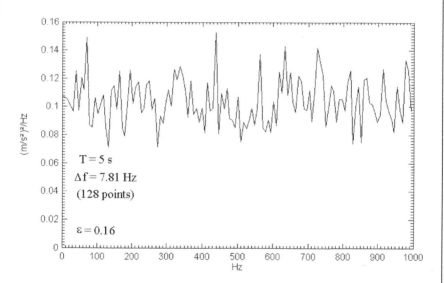

Figure 4.54. *PSD calculated for T = 5 s and Δf = 7.31 Hz (ε = 0.16)*

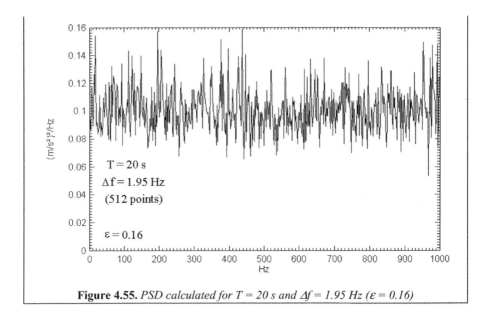

Figure 4.55. *PSD calculated for T = 20 s and Δf = 1.95 Hz (ε = 0.16)*

4.9. Overlapping

4.9.1. *Utility*

We can carry out an overlapping of blocks for three reasons:

– to limit the loss of information related to the use of a window on sequential blocks, which results in ignorance of a significant part of the signal because of the low values of the window at its ends [GAD 87];

– to reduce the length of analysis (interesting for real time analyses) [CON 95], [SHR 95];

– to reduce the statistical error when the duration T of the signal sample cannot be increased. We saw that this error is related to the number of blocks taken in the sample of duration T. If all the blocks are sequential, the maximum number K of blocks of fixed duration ΔT (arising from the frequency resolution desired) is equal to the integer part of $T/\Delta T$ [WEL 67]. An overlapping makes it possible to increase this number of blocks whilst preserving their size ΔT.

Overlapping rate

The *overlapping rate* R is the ratio of the duration of the block overlapped by the following block over the total duration of the block.

This rate is in general limited to the interval between 0 and 0.75.

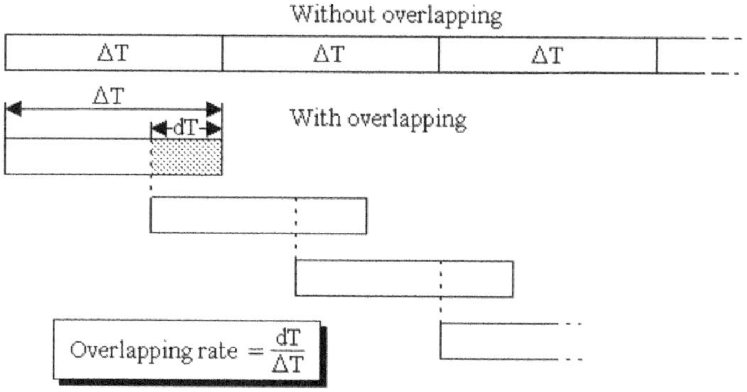

Figure 4.56. *Overlapping of blocks*

Overlapping in addition makes it possible to minimize the influence of the side lobes of the windows [CAR 80], [NUT 71], [NUT 76].

4.9.2. *Influence on the number of degrees of freedom*

Let N be the number of points of the signal sample, N' (> N) the number of points necessary to respect the desired statistical error with K blocks of size ΔN ($N' = K \Delta N$). The difference $N'-N$ must be distributed over $K-1$ possible overlappings [NUT 71]:

$$N'-N = (K-1) R \Delta N$$

yielding

$$R = \frac{N'-N}{\Delta N (K-1)} = \frac{N'-N}{N'-\Delta N} \qquad [4.60]$$

For R to be equal to 0.5 for example, it is necessary that $N' = 2N - \Delta N$.

Overlapping modifies the number of degrees of freedom of the analysis since the blocks can no longer be regarded as independent and non-correlated. The estimated value of the PSD no longer obeys a single chi-square law. The variance of the PSD measured from an overlapping is less than that calculated from contiguous blocks [WEL 67]. However, R. Potter and J. Lortscher [POT 78] showed that, when K is sufficiently large, the calculation could still be carried out on the assumption of non-overlapping, on the condition that the result could still be corrected by a reduction factor depending on the type of window and the selected overlapping rate. The correlation as a function of overlapping can be estimated using the coefficient:

$$c(R) = \frac{\int_0^{\Delta T} r(t) \, r[t + (1-R)\Delta T] \, dt}{\int_0^{\Delta T} r^2(t) \, dt} \qquad [4.61]$$

Window	Correlation coefficient c			Coefficient μ	
	R = 25%	R = 50%	R = 75%	R = 50%	R = 75%
Rectangle	0.25000	0.50000	0.75000	0.66667	0.36364
Bingham	0.17143	0.45714	0.74286	0.70524	0.38754
Hamming	0.02685	0.23377	0.70692	0.90147	0.47389
Hanning	0.00751	0.16667	0.65915	0.94737	0.51958
Parzen	0.00041	0.04967	0.49296	0.999509	0.67071
Flat top	0.00051	-0.01539	0.04553	0.99953	0.99540
Kaiser-Bessel	0.00121	0.07255	0.53823	0.98958	0.62896

Table 4.9. *Reduction factor*

4.9.3. *Influence on statistical error*

When the blocks are statistically independent, the number of degrees of freedom is equal to $n = 2K = 2T \Delta f$ whatever the window. With overlappings of K blocks, the effective number of blocks to consider in order to calculate the statistical error is given [HAR 78], [WEL 67]:

– for R = 50% by:

$$K_{50} = \frac{1}{\frac{1 + 2c_{50\%}^2}{K} - \frac{2c_{50\%}^2}{K^2}} \approx \frac{K}{1 + 2c_{50\%}^2} = \mu_{50} K \qquad [4.62]$$

– for R = 75% by:

$$K_{75\%} = \cfrac{1}{\cfrac{1 + 2\,c_{75\%}^2 + 2\,c_{50\%}^2 + 2\,c_{25\%}^2}{K} - 2\,\cfrac{c_{75\%}^2 + c_{50\%}^2 + 3\,c_{25\%}^2}{K^2}}$$

$$K_{75\%} \approx \frac{K}{1 + 2\,c_{75\%}^2 + 2\,c_{50\%}^2 + 2\,c_{25\%}^2} = \mu_{75}\,K \qquad [4.63]$$

(the approximation being acceptable for $K > 10$). Under these conditions, the statistical error is no longer equal to $1/\sqrt{K}$, but to:

$$\varepsilon = \frac{1}{\sqrt{\mu\,K}} \qquad [4.64]$$

The coefficient μ being less than 1, the statistical error is, for a given K, all the larger as overlapping is greater. However, with an overlapping, the total duration of the treated signal is smaller, which makes it possible to carry out the analyses quickly more in real time (control of the test facilities). The time saving can be calculated from [4.60]:

$$R = \frac{N'-N}{N'-\Delta N} = \frac{T'-T}{T'-\Delta T}$$

(ΔT = duration of a block). To avoid a confusion of notations, we will let T_O be the duration of the signal to be treated with an overlapping and T be the duration without overlapping. We then have:

$$R = \frac{N'-N}{N'-\Delta N} = \frac{T-T_O}{T-\Delta T} \qquad [4.65]$$

yielding

$$T_O = T(1-R) + R\,\Delta T \qquad [4.66]$$

Since $R < 1$ and $\Delta T \ll T$, we have in general $T_O \approx T(1-R)$. The time saving is thus approximately equal to $\dfrac{T_O}{T} \approx (1-R)$.

Example 4.15.

Consider a PSD calculated from a vibration of duration $T = 25$ s with a $\Delta f = 4$ Hz frequency step (or $K = T \Delta f = 100$) leading to a statistical error $\varepsilon_0 = 0.1$ (without overlapping).

With overlapping rate equal to $R = 0.75$ and a Hann window, the coefficient μ is approximately equal to 0.52, yielding $\varepsilon = 1/\sqrt{0.52 \times 25 \times 4} \approx 0.139$. However, this result is obtained with a signal of duration $T_R \approx (1-0.75) \, 25 \approx 6.25$ s.

If we now consider a sample of given duration T, overlapping makes it possible to define a greater number of blocks. This K' number can be deducted from [4.60]:

$$N' = \frac{N - R \, \Delta N}{1 - R}$$

yielding, if $N' = K' \Delta N$

$$K' = \frac{K - R}{1 - R} \qquad [4.67]$$

The increase in the number of blocks makes it possible to reduce the statistical error which becomes equal to:

$$\varepsilon = \frac{1}{\sqrt{\mu \dfrac{K-R}{1-R}}} \approx \sqrt{\frac{1-R}{\mu K}} = \varepsilon_0 \sqrt{\frac{1-R}{\mu}} \qquad [4.68]$$

Example 4.16.

With the data of the above example, the statistical error would be equal to

$$\varepsilon \approx \varepsilon_0 \sqrt{\frac{1 - 0.75}{0.52}} \approx 0.693 \, \varepsilon_0 = 0.0693 \, .$$

4.9.4. *Choice of overlapping rate*

The calculation of the PSD uses the square of the signal values to be analyzed. In this calculation, the square of the function describing the window for each block thus intervenes in an indirect way, by taking account of the selected overlapping rate R. For a linear average, this leads to an effective weighting function $r_{rms}(t)$ such that [GAD 87]:

$$r_{rms}^2(R) = \frac{1}{K}\sum_{i=1}^{K} r^2\left[t - i(1-R)T\right] \qquad [4.69]$$

where T is the duration of the window used (duration of the block), i is the number of the window in the sum and K is the number of windows at time t.

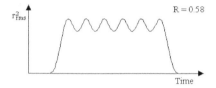
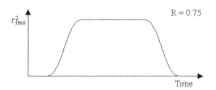

Figure 4.57. *Ripple on the Hanning window (R = 0.58)*   **Figure 4.58.** *Hanning window for R = 0.75*

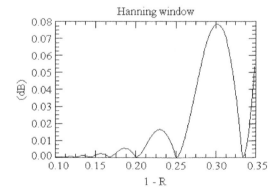

Figure 4.59. *Ripple amplitude versus 1 − R*

With the Hanning window, one of the most frequently used windows, it can be observed (Figure 4.57) that there is a ripple on $r_{rms}^2(t)$, except when $1-R$ is of the form $1/p$ where p is an integer equal to or higher than 3 (Figure 4.58). The ripple has a negligible amplitude when $1-R$ is small (lower than $1/3$) [CON 95], [GAD 87]. This property can be observed in Figure 4.59, which represents the variations of the ratio of the maximum and minimum amplitudes of the ripple (in dB) with respect to $1-R$.

This makes it possible to justify the use, in practice, of an overlapping equal to 0.75, which guarantees a constant weighting on a broad part of the window (the other possible values, 2/3, 3/4, 4/5, etc., are less used, because they do not lead, like 3/4, to an integer number of points when the block size is a power of two).

4.10. Information to provide with a PSD

A lot of information should be included with the PSD. Imperatively:

– the *PSD rms value*, providing a global idea of the severity of the vibration;

– the *statistical error* during the PSD calculation, linked to the duration of the signal sample and to the frequency step of the PSD. The parameter is very important, since it is a characteristic of the validity of the PSD (the rule cited lays down that its value be lower than 15%).

Ideally:

– the frequency step (or the number of points of PSD calculation);

– the *signal rms value*. If the average is zero, the difference between the rms value of the PSD and of the signal shows that we have ignored part of the frequency content of the signal during the calculation of its PSD, generally of high frequency (section 4.11);

– *standard deviation* of the signal sample used to calculate the PSD. The rms value of the signal and standard deviation enable the calculation of the mean;

– *skewness* and *kurtosis of the signal*, making it possible to determine the Gaussian character of signal instantaneous values and in particular to detect the presence of faults in the signal (or shock) sample.

4.11. Difference between rms values calculated from a signal according to time and from its PSD

The rms value of a vibratory signal can be calculated:

Practical Calculation of the PSD   223

– directly from the signal itself, by considering the square root of the sum of the squares of point amplitudes defining the signal (square root of the quadratic average);

– by integration of the PSD, theoretically between zero and infinite, between two frequencies f_1 and f_2 in practice (square root of the surface under the spectral curve between two frequencies).

In the first case, the rms value $\ddot{x}_{r.m.s.}$ obtained takes into consideration all the points in the signal and thus its frequency content, as well as its mean m if it is not zero.

We know that the rms value is linked to the standard deviation s by:

$$\ddot{x}^2_{r.m.s.} = s^2 + m^2 \qquad [4.70]$$

In the second case, we calculate the rms value of the signal in the frequency band f_1, f_2 without taking into consideration the mean (which would correspond to zero frequency).

Both results are identical if the band f_1, f_2 covers the frequency content of the signal (or if f_1 is equal to zero and f_2 is infinite).

The data of the rms value of the signal and its standard deviation with the PSD is therefore useful, since it allows for the calculation of the average value of the signal. The comparison of rms values calculated from the signal and PSD makes it possible to assess the importance of the part of the spectrum that may be ignored during PSD calculation.

4.12. Calculation of a PSD from a Fourier transform

The PSD can be calculated from a Fourier transform:

– by raising to the square the amplitude of the Fourier transform and by multiplying this square by the Δf frequency step of the Fourier transform:

$$PSD = |TF|^2 \, \Delta f \qquad [4.71]$$

– or, which leads to the same result, by dividing the square of the Fourier transform amplitude by duration T of the original signal (this is the equivalent because Δf is equal to $1/T$):

$$\text{PSD} = \frac{|\text{TF}|^2}{T} \qquad [4.72]$$

The resulting PSD is obtained with a very small frequency step and thus a very large resolution, but with very bad precision on amplitude (see section 4.7.7).

The expression of error $\varepsilon = \dfrac{1}{\sqrt{T\,\Delta f}}$ is no longer correct here, since it is only valid for $\varepsilon < 0.2$ (see section 4.7.4). The number of degrees of freedom, equal to product 2 T Δf, is equal to 2, which leads to a statistical error of 83% with exact relations. In other words, with a level of confidence of 68%, we can simply state that the true PSD is between 0.54 and 5.79 times that of the resulting PSD. This result therefore has no practical value.

Example 4.17.

The acceleration signal studied is that in Figure 4.60. It is an approximate white noise of duration 5 s, defined by 5,005 points (rms value: 15.7 m/s^2).

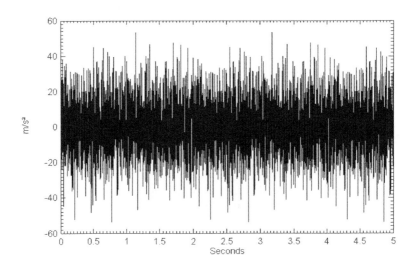

Figure 4.60. *Acceleration signal according to the time studied (rms value: 15.7 m/s^2)*

The amplitude of its Fourier transform is given in Figure 4.61 and the resulting PSD by using relation [4.72] in Figure 4.62.

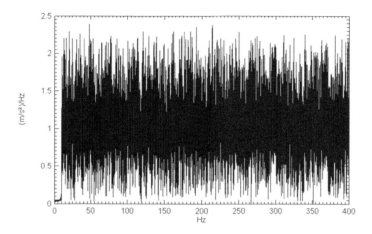

Figure 4.61. *Amplitude of the Fourier transform of signal in Figure 4.60*

The Fourier transform is calculated on 4,096 points, with a Δf frequency step equal to 0.122 Hz.

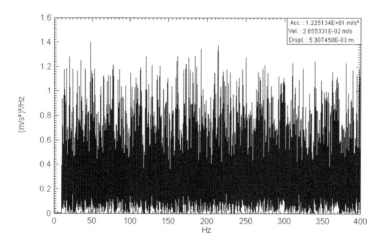

Figure 4.62. *PSD deducted from the Fourier transform*

The resulting PSD has the same frequency step. Its rms value is equal to 12.25 m/s². This PSD can be compared to PSDs calculated with a statistical error equal to 0.114, i.e. with 4,096 points ($\Delta f = 0.09$ Hz) or 512 points ($\Delta f = 0.98$ Hz).

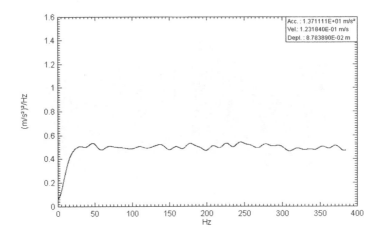

Figure 4.63. *PSD calculated with a statistical error equal to 0.114 (4,096 points)*

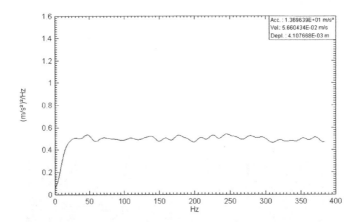

Figure 4.64. *PSD calculated with a statistical error equal to 0.114 (512 points)*

PSDs calculated with a statistical error of 0.114 can be superimposed, with an rms value approximately equal to 13.7 m/s², and are very different from the PSD obtained from the Fourier transform for the complete signal sample, obtained with a very high and unacceptable statistical error.

4.13. Amplitude based on frequency: relationship with the PSD

The first vibration test standards established between 1945 and 1965 [FOL 72b], [SCH 65] specified vibrations defined by amplitude based on the frequency, established from measures in the real, albeit random, environment.

The main reason was the lack of power in the test facilities back then, which did not allow for random vibrations. The only possibility was sinusoid vibrations, swept or otherwise.

Analyzing vibratory signals, filtering in particular was done in analog ways; the results were shown on an oscilloscope.

Today, it is not always easy to find the exact conditions of reduction of data used then. An example is given by R.W. Hager, R.L. Partington and R.J. Leistikow [HAG 62], in which the vibratory signal to analyze, available on magnetic tape, was consecutively filtered by several narrowband filters made up of a low-pass and a high-pass filter such that the output to input ratio was reduced by 6 dB from the central frequency and 24 dB from an octave on each side.

Although it is sometimes clearly indicated [FOL 69, MAG 78], the width of the filter, which can vary with its central frequency, is most often not mentioned [GRA 62], even though the result greatly depends on it. Because of this, it is not certain today that we can compare the severity of several random vibrations from this type of curve used in the literature.

The maximum amplitude of the filtered signal was used on a diagram (acceleration 0 – peak) according to the central frequency of the filter represented in abscissas. The ordinate axis sometimes indicated the displacement corresponding to this acceleration as if the filtered signal was sinusoidal.

The curves drawn by linking the points of the line spectrum thus obtained [HAG 62] were used to carry out severity comparisons [GEN 68], [SCH 65].

Specifications were determined by enveloping, with straight line segments, the scatter plot resulting from the analysis of a large number of signals collected from different points in a single vehicle or different types of vehicles (constant displacement, generally peak-to-peak, constant velocity and constant acceleration in logarithmic axes) [TOL 63].

Some authors presented the results in statistical form, with several curves representing the value of the acceleration around P% of accelerations measured by the frequency, which comes down to representing the peak value as well as the distribution of accelerations below the peak at each frequency [OST 67], [OST 79]. Sometimes, the curves also provided amplitude according to probability of occurrence.

Relationship with the PSD

The acceleration peak can be calculated from the PSD by choosing, for a given G value of the PSD, a 1 Hz bandwidth, leading to an rms value equal to $\ddot{x}_{rms} = \sqrt{G \cdot 1} = \sqrt{G}$ and by taking into consideration the peak value between 3 and 5 times the rms value for a random signal [OST 79].

For example, of the PSD has an amplitude of G = 9 (m/s²)²/Hz at 200 Hz, the amplitude is equal to $\sqrt{G}$ = 3 m/s². To obtain the rms value of the whole spectrum, we have to integrate the amplitude to the square of this value. We should note that this amplitude–frequency spectrum is generally drawn in logarithmic axes and that we have to consider it for the integration.

Another possibility is to use the product of G by frequency step Δf of the PSD at each frequency. However, the result is then a function of the step width.

4.14. Calculation of the PSD for given statistical error

4.14.1. *Case study: digitization of a signal is to be carried out*

Given a vibration $\ell(t)$, we set out to calculate its PSD between 0 and f_{max} with M points (M must be a power of 2), for a statistical error not exceeding a selected value ϵ. The procedure is summarized in Table 4.10 [BEA 72], [LEL 73], [NUT 80].

The signal of total duration T (to be defined) will be cut out in K blocks of unit duration ΔT, under the following conditions:	
$f_{samp.} \geq 2.6 f_{max}$	Condition to avoid the aliasing phenomenon (modified Shannon's theorem).
$f_{Nyquist} = \dfrac{f_{samp.}}{2}$	Nyquist frequency [PRE 90].
$\Delta f = \dfrac{f_{Nyquist}}{M}$	Interval between two points of the PSD (this interval limits the possible precision of the analysis starting from the PSD).

$\delta t = \dfrac{1}{f_{samp}}$	Temporal step (time interval between two points of the signal), if the preceding condition is observed.
$\Delta N = 2\,M$	Number of points per block.
$N = \dfrac{2\,M}{\varepsilon^2}$	Minimum number of signal points to analyze in order to respect the statistical error.
$T = N\,\delta t$	Minimum total duration of the sample to be treated.
$K = \dfrac{N}{2\,M}$	Number of blocks.
$\Delta T = \dfrac{T}{K}\;(= 2\,M\,\delta t = \dfrac{1}{\Delta f})$	Duration of one block.
Calculation of $\dfrac{2}{\Delta T}\lvert L(f)\rvert^2$ for each point of the PSD, where $f = m\,\Delta f\;(0 < m \le M)$	Calculation from the FFT of each block.
$\dfrac{1}{K}\sum_{i=1}^{K}\dfrac{2}{\Delta T}\lvert L_i(f)\rvert^2$	Averaging of the spectra obtained for each of the K blocks (stationary and ergodic process).

Table 4.10. *Calculation process of a PSD starting from a non-digitized signal*

With these conditions, the maximum frequency of the calculated PSD is equal to $f'_{max} = f_{Nyquist}$, but it is preferable to consider the PSD only in the interval $(0, f_{max})$.

NOTE.– *It is assumed here that the signal has frequency components greater than f_{max} and that it was thus filtered by a low-pass filter to avoid aliasing. If it is known that the signal has no frequency beyond f_{max}, this filtering is not necessary and $f'_{max} = f_{max}$.*

4.14.2. *Case study: only one sample of an already digitized signal is available*

If the signal sample of duration T has already been digitized with N points, we can use the value of the statistical error to calculate the number of points M of the PSD (i.e. the frequency interval Δf), which is thus no longer to be freely selected (but it is nevertheless possible to increase the number of points of the PSD by overlapping and/or addition of zeros).

Data: the digitized signal, f_{max} and ε.	
$f'_{max} = \dfrac{f_{samp.}}{2}$	Theoretical maximum frequency of the PSD (see preceding note).
$f_{max} = \dfrac{f_{samp.}}{2.6}$	Practical maximum frequency.
$\delta t = \dfrac{1}{f_{samp.}}$	Temporal step (time interval between two points of the signal).
$N = \dfrac{T}{\delta t}$	Number of signal points of duration T.
$M = \dfrac{N \varepsilon^2}{2}$	Number of points of the PSD necessary to respect the statistical error (we will take the number immediately beneath that equal to the power of 2).
$f_{Nyquist} = \dfrac{f_{samp.}}{2}$	Nyquist frequency.
$\Delta f = \dfrac{f_{Nyquist}}{M}$	Interval between two points of the PSD.
$\Delta N = 2M$	Number of points per block.
$K = \dfrac{N}{2M}$	Number of blocks.

Table 4.11. *Calculation process of a PSD starting from an already digitized signal*

If the number of points M of the PSD to be plotted is itself imposed, it would be necessary to have a signal defined by N' points instead of N given points (N < N'). We can avoid this difficulty in two complementary ways:

– either by using an overlapping of the blocks (of 2 M points). We will set the overlapping rate R equal to 0.5 and 0.75 while taking the smallest of these two values (for a Hanning window) which satisfies the inequality:

$$\sqrt{\frac{1-R}{\mu} \frac{2M}{N}} \leq \varepsilon$$

When it is possible, overlapping chosen in this manner makes it possible to use K' blocks with $K' = \frac{N'}{2M}$, where [4.60] $N' = \frac{N - 2MR}{1 - R}$;

– or, if overlapping does not sufficiently reduce the statistical error, by fixing this rate at 0.75 to benefit as much as possible from its effect and then to evaluate the size of the blocks which would make it possible, with this rate, to respect the statistical error, using:

$$\sqrt{\frac{1-R}{\mu} \frac{\Delta N}{N}} \leq \varepsilon$$

The value ΔN thus obtained is lower than the number 2 M necessary to obtain the desired resolution on the PSD. Under these conditions, the number of items used for the calculation of the PSD is equal to:

$$N' = \frac{N - 0.75 \Delta N}{1 - 0.75}$$

and the numbers of blocks to $K' = N'/\Delta N$. We can then add zeros to each block to increase the number of calculation points of the PSD and to make it equal to 2 M. For each block, this number is equal to $\frac{2MK' - N'}{K'}$. This is however only an artifice, the information contained in the initial signal not evidently increasing with the addition of zeros.

4.15. Choice of filter bandwidth

4.15.1. *Rules*

It is important to recall that the precision of calculation of the PSD depends, for given T, on the width Δf of the filter used [RUD 75]. The larger the width Δf of the filter, the smaller the statistical error ε and the better the precision of calculation of $G(f)$. However, this width cannot be increased limitlessly [MOO 61]. The larger Δf, the fewer details on the curve are obtained, which is smoothed. The resolution being weaker, the narrow peaks of the spectrum are no longer shown [BEN 63]. A compromise must thus be found.

Figure 4.65 shows as an example three spectral curves obtained starting from the same vibratory signal with three widths of filter (3.9 Hz, 15.625 Hz and 31.25 Hz). These curves were plotted without the amplitude being divided by Δf, as is normally the case for a PSD.

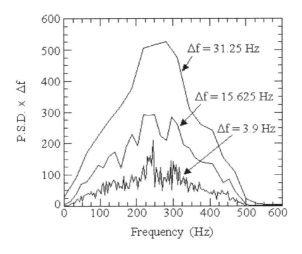

Figure 4.65. *Influence of width of filter*

We observe in these conditions that the area under the curve calculated for $\Delta f = 15.625$ Hz is approximately half of that obtained for $\Delta f = 31.25$ Hz. In the case of a true PSD, division by Δf gives the same area, whatever the value of Δf.

We note in addition on these curves that the spectrum obtained for $\Delta f = 15.6$ Hz is very much smoothed; in particular, the peak observed for $\Delta f = 3.9$ Hz has disappeared. In order to choose the value of Δf, it would be necessary to satisfy two requirements:

1) The filter should not be broader than a quarter of the width of the narrowest resonance peak expected [BEN 61b], [BEN 63], [FOR 64], [MOO 61], [WAL 81].

2) The statistical error should remain small, with a value not exceeding approximately 15%.

If the first condition is observed, the precision of the PSD calculation is proportional to the width of the filter. If, on the contrary, resonances are narrower than the filter, the precision of the estimated PSD is proportional to the width of the resonance of the specimen and not to the width of the filter. To solve this problem, C.T. Morrow [MOR 58], and then R.C. Moody [MOO 61] suggested making two analyses, by using the narrowest filter first of all to emphasize resonances, then by making a second analysis with a broader filter in order to improve the precision of the PSD estimate.

Other more complicated techniques have been proposed (H. Press and J.W. Tukey, for example, [BLA 58], [NEW 75]).

NOTE.– *Additionally, the curves in Figure 4.65 show the interest of dividing the mean square value of the response from each filter by its width Δf. We can verify that, in the conditions of this example, the area below the curve calculated for $\Delta f = 15.625$ Hz is approximately half of that obtained for $\Delta f = 31.25$ Hz. Dividing by Δf during PSD calculation leads to the same area regardless of Δf.*

4.15.2. *Bias error*

Let us consider a random signal $\ell(t)$ with a constant PSD (white noise) $G_\ell(f) = G_{\ell_0}$ applied to a linear system with transfer function with one degree of freedom [PIE 93], [WAL 81]:

$$H(f) = \frac{1}{\sqrt{\left[1 - \left(\frac{f}{f_0}\right)^2\right]^2 + \left(\frac{f}{Q f_0}\right)^2}} \quad [4.73]$$

(f_0 being the natural frequency and Q the quality factor of the system). The response $u(t)$ of this system has the following PSD [THO 08]:

$$G_u(f) = |H(f)|^2 \, G_\ell(f)$$

$$G_u(f) = \frac{G_\ell(f)}{\left[1 - \left(\frac{f}{f_0}\right)^2\right]^2 + \left(\frac{f}{Q f_0}\right)^2} \quad [4.74]$$

234   Random Vibration

Let us analyze this PSD, which presents a peak at $f = f_0$, using a rectangular filter of width ΔF centered on f_c, with transfer function [FOR 64]:

$$\begin{cases} H_A = 1 & \text{for } f_c - \dfrac{\Delta F}{2} \leq f \leq f_c + \dfrac{\Delta F}{2} \\ H_A = 0 & \text{elsewhere} \end{cases} \quad [4.75]$$

We propose calculating the *bias error* made over the width between the half-power points of the peak of the PSD response and on the amplitude of this peak when using an analysis filter ΔF of non-zero width. For a given f_c, the PSD calculated with this filter has a value of:

$$G_F(f_c) = \frac{1}{\Delta F} \int_0^\infty |H_A|^2 \, G_u(f) \, df \quad [4.76]$$

i.e.

$$G_F(f_c) = \frac{1}{\Delta F} \int_{f_c - \Delta F/2}^{f_c + \Delta F/2} \frac{G_{\ell 0}}{\left[1 - \left(\dfrac{f}{f_0}\right)^2\right]^2 + \left(\dfrac{f}{Q f_0}\right)^2} \, df \quad [4.77]$$

It is known [LAL 94] (Appendix A6) that the integral

$$A = \int \frac{dh}{\left(1 - h^2\right)^2 + h^2/Q^2} \quad \text{is equal to:}$$

$$A = \frac{1}{8\sqrt{1-\xi^2}} \ln \frac{h^2 + 2h\sqrt{1-\xi^2} + 1}{h^2 - 2h\sqrt{1-\xi^2} + 1} + \frac{1}{4\xi} \left(\arctan \frac{h + \sqrt{1-\xi^2}}{\xi} + \arctan \frac{h - \sqrt{1-\xi^2}}{\xi} \right)$$

Consequently,

$$\Delta F \frac{G_F(f_c)}{G_{\ell 0}} = \frac{f_0}{8\sqrt{1-\xi^2}} \left[\ln \frac{h^2 + 2h\sqrt{1-\xi^2} + 1}{h^2 - 2h\sqrt{1-\xi^2} + 1} \right]_{h_1}^{h_2}$$

$$+\frac{1}{4\xi}\left(\arctan\frac{h+\sqrt{1-\xi^2}}{\xi}+\arctan\frac{h-\sqrt{1-\xi^2}}{\xi}\right)\bigg|_{h_1}^{h_2} \quad [4.78]$$

$$\frac{G_F(f_c)}{G_{\ell 0}} \approx \frac{f_0 \, Q}{2 \, \Delta F}\left[\arctan 2Q \frac{f_c - f_0 + \frac{\Delta F}{2}}{f_0} - \arctan 2Q \frac{f_c - f_0 - \frac{\Delta F}{2}}{f_0}\right] \quad [4.79]$$

For $f_c = f_0$:

$$\frac{G_F(f_0)}{G_{\ell 0}} \approx \frac{f_0 \, Q}{2 \, \Delta F}\left[\arctan\left(Q \frac{\Delta F}{f_0}\right) - \arctan\left(Q \frac{(-\Delta F)}{f_0}\right)\right] \quad [4.80]$$

i.e.

$$\frac{G_F(f_0)}{G_{\ell 0}} \approx \frac{f_0 \, Q}{\Delta F}\arctan\frac{Q \, \Delta F}{f_0} \quad [4.81]$$

However, by definition, the bandwidth between the half-power points is equal to $\Delta f = \frac{f_0}{Q}$, yielding:

$$\frac{G_F(f_0)}{G_{\ell 0}} \approx \frac{f_0^2}{\Delta f \, \Delta F}\arctan\frac{\Delta F}{f} \quad [4.82]$$

At the half-power points, the calculated spectrum has a value:

$$G_F(f_c) = \frac{1}{2} G_F(f_0) \quad [4.83]$$

where $f_c = f_0 + \frac{\Delta f_F}{2}$. We deduce that:

$$\frac{G_F}{G_{\ell 0}} = Q \frac{f_0}{2 \, \Delta F}\left[\arctan\frac{\Delta f_F + \Delta F}{\Delta f} - \arctan\frac{\Delta f_F - \Delta F}{\Delta f}\right] \quad [4.84]$$

From [4.82], [4.83] and [4.84], we obtain:

$$\arctan\frac{\Delta F}{\Delta f} = \left[\arctan\frac{\Delta f_F + \Delta F}{\Delta f} - \arctan\frac{\Delta f_F - \Delta F}{\Delta f}\right] \quad [4.85]$$

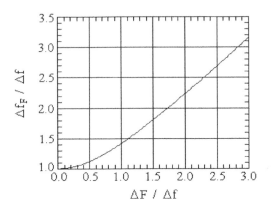

Figure 4.66. *Width of the peak at half-power versus width of the analysis filter (from [FOR 64])*

The curve in Figure 4.66 gives the variations of $\dfrac{\Delta f_F}{\Delta f}$ against $\dfrac{\Delta F}{\Delta f}$ after numerical resolution. It is noted that the measured value Δf_f of the width of the peak at half-power is obtained with an error lower than 10% so long as the width of the analysis filter is less than half the true value Δf.

Setting $x = \dfrac{\Delta F}{\Delta f}$ and $y = \dfrac{\Delta f_F}{\Delta f}$.

$$\arctan x = \arctan(y + x) - \arctan(y - x)$$

Knowing that $\arctan a - \arctan b = \arctan \dfrac{a - b}{1 + a b}$, we have:

$$\arctan x = \arctan \dfrac{2x}{1 + y^2 - x^2}.$$

This yields $x = \dfrac{2x}{1 + y^2 - x^2}$ and $y^2 = x^2 + 1$, i.e.:

$$\dfrac{\Delta f_F}{\Delta f} = \left[\left(\dfrac{\Delta F}{\Delta f}\right)^2 + 1\right]^{1/2} \qquad [4.86]$$

Practical Calculation of the PSD 237

In addition, the peak of the PSD occurs for $f = f_0$:

$$^PG = Q^2 = \left(\frac{f_0}{\Delta f}\right)^2 \qquad [4.87]$$

yielding the relationship between the measured value of the peak and the true value:

$$\frac{G_F(f_0)}{^PG} = \frac{\Delta f}{\Delta F} \arctan \frac{\Delta F}{\Delta f} \qquad [4.88]$$

Figure 4.67 shows the variations of this ratio versus $\Delta F/\Delta f$. If $\Delta F = \frac{1}{4}\Delta f$ according to the rule previously suggested,

$$\frac{\Delta f_F}{\Delta f} = \sqrt{1+\left(\frac{1}{4}\right)^2} = 1.0308$$

and

$$\frac{G_F}{^PG} = 4 \arctan \frac{1}{4} \approx 0.98$$

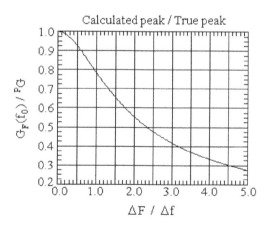

Figure 4.67. *Amplitude of peak versus filter width*

238 Random Vibration

Under these conditions, the error is about 3% of Δf and 2% of the peak.

Example 4.18.

Let us consider a one-degree-of-freedom system of natural frequency $f_0 = 100$ Hz and quality factor $Q = 10$, excited by a white noise. The error of measure of the PSD response peak is given by the curve in Figure 4.67.

If $\Delta F = 5$ Hz:

$$\frac{Q}{f_0} \Delta F = \frac{10}{100} 5$$

$$\frac{Q}{f_0} \Delta F = \frac{1}{2}$$

yielding:

$$\frac{G_F}{{}^P G} = 0.92$$

For $f_0 = 50$ Hz and $Q = 10$, we would similarly obtain $\dfrac{Q}{f_0} \Delta F = \dfrac{10}{50} 5 = 1$ and $\dfrac{G_F}{{}^P G} = 0.78$.

4.15.3. *Maximum statistical error*

When the phenomenon to be analyzed is of short duration, it can be difficult to obtain a good resolution (small ΔF) whilst preserving an acceptable statistical error.

Example 4.19.

Figure 4.68 shows, as an example, the PSDs of the same signal duration of 22.22 seconds, calculated respectively with ΔF equal to 4.69 Hz; 2.34 Hz and 1.17 Hz (i.e. with a statistical error equal to 0.098, 0.139 and 0.196).

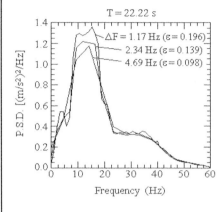 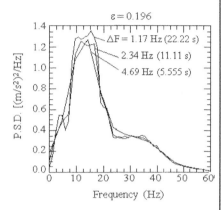

Figure 4.68. *Influence of analysis filter width for a sample of a given duration*

Figure 4.69. *Influence of analysis filter width for constant statistical error*

We observe that, the more detailed the curve (ΔF small), the larger the statistical error. Although the duration of the sample is longer than 20 s, a resolution of the order of 1 Hz can be obtained only with an error close to 20%.

A constant statistical error with different durations T and widths ΔF can lead to appreciably different results. Figure 4.69 shows three calculated PSDs of the same signal all three for $\varepsilon = 19.6\%$, with respectively:

$T = 22.22$ s   and $\Delta F = 1.17$ Hz

$T = 11.11$ s   and $\Delta F = 2.34$ Hz

$T = 5.555$ s   and $\Delta F = 4.69$ Hz

The choice of ΔF must thus be a compromise between the resolution and the precision. In practice, we try to comply with the two following rules: ΔF less than a quarter of the width of the narrowest peak of the PSD, which limits the width measurement error of the peak and its amplitude to less than 3%, and a statistical error of less than 15% (which corresponds to a number of degrees of freedom n equal to approximately 90). Certain applications (calculation of random transfer functions for example) can justify a lower value of the statistical error.

Taking into account the importance of these parameters, the filter width used for the analysis and the statistical error should always be specified on the PSD curves.

4.15.4. *Optimum bandwidth*

A.G. Piersol [PIE 93] defines the optimum bandwidth ΔF_{op} as the value of ΔF minimizing the total mean square error, the sum of the squares of the bias error and of the statistical error:

$$\varepsilon^2 = \varepsilon_{bias}^2 + \varepsilon_{stat}^2 \qquad [4.89]$$

The bias error calculated from [4.88] is equal to:

$$\varepsilon_{bias} = \frac{\Delta f}{\Delta F} \arctan\left(\frac{\Delta F}{\Delta f}\right) - 1 \qquad [4.90]$$

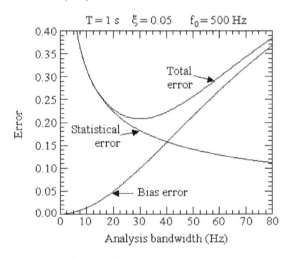

Figure 4.70. *Total mean square error*

where Δf is the width between the half-power points of the peak. Whence:

$$\varepsilon^2 \approx \left[\frac{\Delta f}{\Delta F} \arctan\left(\frac{\Delta F}{\Delta f}\right) - 1\right]^2 + \frac{1}{\Delta F\, T} \qquad [4.91]$$

Figure 4.70 shows the variations of bias error, statistical error and ε with ΔF.

Error ε has a minimum at $\Delta F = \Delta F_{op}$. The optimum bandwidth ΔF_{op} is thus obtained by canceling the derivative of ε^2 with respect to ΔF. This research is carried out numerically.

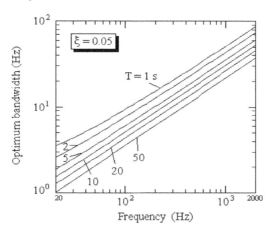

Figure 4.71. *Optimum bandwidth versus peak frequency and duration of the sample*

The curves in Figure 4.71 show ΔF_{op} versus f_0, for $\xi = 0.05$ and for some values of duration T.

If $\Delta F/\Delta f < 0.4$, the bias error can be approximated by:

$$\varepsilon_{bias} \approx -\frac{1}{3}\left(\frac{\Delta F}{\Delta f}\right)^2 \qquad [4.92]$$

Then,

$$\varepsilon^2 \approx \frac{\Delta F^4}{9\, \Delta f^4} + \frac{1}{T\, \Delta F} \qquad [4.93]$$

Whence, by canceling the derivative,

$$\Delta F_{op} \approx \left(\frac{9\, \Delta f^4}{4\, T}\right)^{1/5} \approx 2\,\frac{(\xi\, f_0)^{4/5}}{T^{1/5}} \qquad [4.94]$$

Figure 4.72 shows the error made versus the natural frequency f_0, for various values of T.

242   Random Vibration

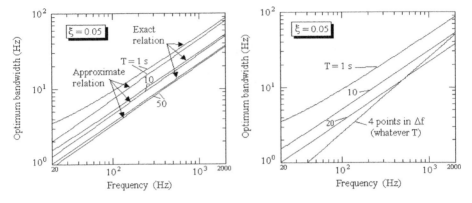

Figure 4.72. *Comparison of approximate and exact relations for calculation of optimum bandwidth*

Figure 4.73. *Comparison of the optimum bandwidth with the standard rules*

It can be interesting to compare the values resulting from these calculations with the standard rules which require four points in the half-power interval (Figure 4.73). It should be noted that this rule of four points generally leads to a smaller bandwidth. The calculation method of optimum width must be used with prudence, for it can lead to a much too large statistical error (Figure 4.74, plotted for $\xi = 0.05$).

To confine this error to the low resonance frequencies, A.G. Piersol [PIE 93] suggested limiting the optimum band to 2.5 Hz, which leads to the curves in Figure 4.75.

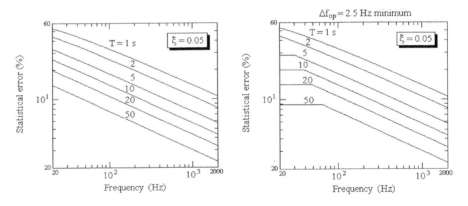

Figure 4.74. *Statistical error obtained using the optimum bandwidth*

Figure 4.75. *Statistical error obtained using the optimum bandwidth limited to 2.5 Hz*

By plotting the variations of $\Delta f / \Delta F_{op}$, we can also evaluate, with respect to f_0, the number of points in Δf which determines this choice of ΔF_{op}, in order to compare this number with the four points of the empirical rule. Figures 4.76 and 4.77 show the results obtained, for several values of T, with and without limitation of the ΔF_{op} band.

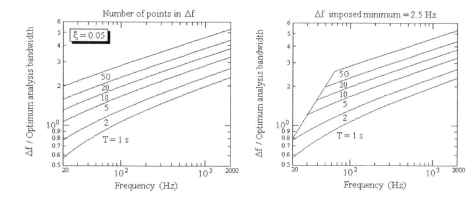

Figure 4.76. *Number of points in Δf resulting from choice of optimum bandwidth*

Figure 4.77. *Number of points in Δf resulting from the choice of optimum bandwidth limited to 2.5 Hz*

4.16. Probability that the measured PSD lies between ± one standard deviation

We saw that the approximate relation $\varepsilon = \dfrac{1}{\sqrt{T\,\Delta f}}$ is acceptable as long as $\varepsilon < 0.20$. In this same range, the error on the measured PSD $\hat{G}$ (or on $\hat{G}/G$) has a roughly Gaussian distribution [MOO 61], [PRE 56a]. Let us set $\hat{s} = s_{\hat{G}}$ to simplify the notations. The probability that the measured PSD is false by a quantity greater than $\theta\,\hat{s}$ (error in the positive sense) is [MOR 58]:

$$P = \frac{1}{\hat{s}\sqrt{2\pi}} \int_{\theta\hat{s}}^{\infty} e^{-\frac{a^2}{2\hat{s}^2}}\, da \qquad [4.95]$$

If we set $v = \dfrac{a}{\hat{s}}$, P takes the form:

$$P = \dfrac{1}{\sqrt{2\pi}} \int_{\theta}^{\infty} e^{-v^2/2} \, dv$$

Knowing that:

$$\operatorname{erf}(x) = \dfrac{2}{\sqrt{\pi}} \int_{0}^{x} e^{-t^2} \, dt \qquad [4.96]$$

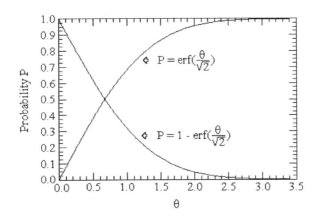

Figure 4.78. *Probability that the measured PSD lies between ±1 standard deviation*

P can be also written, to facilitate its numerical calculation (starting from the approximate expressions given in Appendix A4.1):

$$P = \dfrac{1}{2}\left[1 - \operatorname{erf}\left(\dfrac{\theta}{\sqrt{2}}\right)\right] \qquad [4.97]$$

The probability of a negative error is identical. The probability of an error outside the range $\pm \theta \hat{s}$ is thus equal to:

$$P = 1 - \text{erf}\left(\frac{\theta}{\sqrt{2}}\right) \qquad [4.98]$$

Example 4.20.

$\theta = 1 \qquad P = 68.26\%$

$\theta = 2 \qquad P = 95\%$

4.17. Statistical error: other quantities

The statistical error related to the estimate of the mean and mean square value is, according to the case, given by [BEN 80].

	Mean estimation	Error	Estimate of the mean square value	Error
Ensemble averages	$\dfrac{1}{N}\sum_{i=1}^{N} x_i$	$\dfrac{s_x}{\mu_x \sqrt{N}}$	$\dfrac{1}{N}\sum_{i=1}^{N} x_i^2$	$\sqrt{\dfrac{2}{N}}$
Temporal averages	$\dfrac{1}{T}\int_0^T x(t)\,dt$	$\dfrac{s_x}{\mu_x \sqrt{2\,T\,\Delta f}}$	$\dfrac{1}{T}\int_0^T x^2(t)\,dt$	$\dfrac{1}{\sqrt{T\,\Delta f}}$

Table 4.12. *Statistical error of the mean and the mean square value*

Calculations of the quantities defined in this chapter are carried out in practice on samples of short duration T, subdivided into K blocks of duration ΔT [BEN 71] [BEN 80], by using filters of non-zero width Δf. These approximations lead to the errors in Table 4.13. The expressions of these errors are established in the hypothesis of a constant spectrum in interval Δf.

Contrary to the autospectrum case, the statistical error linked to the determination of the cross-spectrum is a function of the frequency.

	Quantity	Error ε_r
Direct PSD	$\hat{G}_{xx}$ or $\hat{G}_{yy}$	$\dfrac{1}{\sqrt{T\,\Delta f}}$
Cross-PSD	$\left\|\hat{G}_{xy}(f)\right\|$	$\dfrac{1}{\left\|\gamma_{xy}(f)\right\|\sqrt{T\,\Delta f}}$
Coherence	$\hat{\gamma}_{xy}(f)$	$\dfrac{\sqrt{2}\left[1-\gamma_{xy}^2(f)\right]}{\left\|\gamma_{xy}(f)\right\|\sqrt{T\,\Delta f}}$
PSD of v(t) such that y(t) = v(t) + n(t) y(t) = measured output signal n(t) = output noise	$\hat{G}_{vv}(f) = \hat{\gamma}_{xy}^2(f)\,\hat{G}_{yy}(f)$	$\dfrac{\sqrt{2-\gamma_{xy}^2(f)}}{\left\|\gamma_{xy}(f)\right\|\sqrt{T\,\Delta f}}$
Transfer function	$\left\|\hat{H}_{xy}(f)\right\|$	$\dfrac{\sqrt{1-\gamma_{xy}^2(f)}}{\left\|\gamma_{xy}(f)\right\|\sqrt{2\,T\,\Delta f}}$
	$\left\|\hat{H}_{xy}(f)\right\|^2$	$\dfrac{\sqrt{2}\sqrt{1-\gamma_{xy}^2(f)}}{\left\|\gamma_{xy}(f)\right\|\sqrt{T\,\Delta f}}$

Table 4.13. *Other statistical errors*

Example 4.21.

The statistical error linked to the cross-power spectral density calculation in Figure 2.7 slightly increases with the frequency.

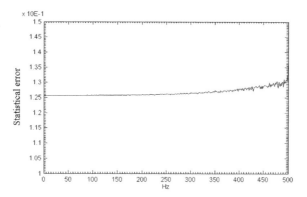

Figure 4.79. *Statistical error made during cross-spectrum calculation in Figure 2.7*

The same goes for the error made during the calculation of a crossed transfer function.

Example 4.22.

We go back to Example 2.4. Figure 4.80 shows the error made during the calculation of the transfer function with the cross-spectrum in the frequency band where the PSD has a significant value.

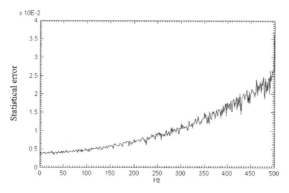

Figure 4.80. *Statistical error in the transfer function calculated from the cross-spectrum*

The expressions of Table 4.13 can be used with an estimated value $\hat{\gamma}_{xy}(f)$ of the coherence function coefficient instead of $\gamma_{xy}(f)$ (unknown); we then obtain approximate values of ε_r, when ε_r is small (i.e. $\varepsilon_r < 0.20$), which can be limited at the 95% confidence level using:

$$\hat{Z}(1 - 2\varepsilon_r) \leq Z \leq \hat{Z}(1 + 2\varepsilon_r)$$

where Z is the true value of the parameter and $\hat{Z}$ its estimated value.

Figure 4.81 shows the variations of the error made during the calculation of the transfer function $|\hat{H}_{xy}(f)|$, given by:

$$\varepsilon_r \approx \frac{\left(1 - \gamma_{xy}^2\right)^{1/2}}{|\gamma_{xy}| \sqrt{2\, T\, \Delta f}} \quad [4.99]$$

for various values of $n_d = T\, \Delta f$.

$$\hat{H}(f) = \frac{G_{xy}(f)}{G_x(f)} = |\hat{H}(f)|\, e^{j\hat{\phi}(f)}$$

$\hat{H}(f)$ = measured transfer function

$H(f)$ = exact function [BEN 63], [GOO 57].

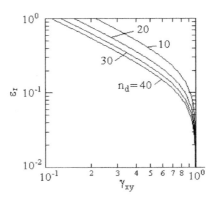

Figure 4.81. *Statistical error related to the calculation of the transfer function*

It is shown that if

$$P = \text{prob}\left[\left|\frac{\hat{H}(f)-H(f)}{H(f)}\right| < \sin \varepsilon_r \text{ and } \left|\hat{\phi}(f)-\phi(f)\right| < \varepsilon_r\right]$$

$$P = 1 - \left[\frac{1-\gamma_{xy}^2(f)}{1-\gamma_{xy}^2(f)\cos^2\varepsilon}\right]^n \qquad [4.100]$$

where n is the number of degrees of freedom, equal to $2\,T\,\Delta f$.

$$n = \frac{\ln(1-P)}{\ln\left(\dfrac{1-\gamma_{xy}^2}{1-\gamma_{xy}^2\cos^2\varepsilon_r}\right)} \qquad [4.101]$$

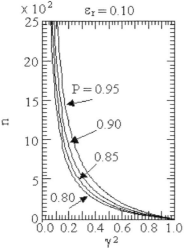
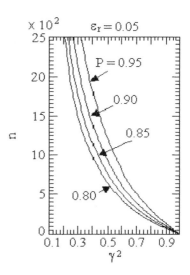

Figure 4.82. *Number of degrees of freedom necessary for the statistical error on the transfer function to be lower than 0.10 with probability P*

Figure 4.83. *Number of degrees of freedom necessary for the statistical error on the transfer function to be lower than 0.05 with probability P*

The statistical error resulting from the calculation of the autocorrelation R_x is given by [VIN 72]:

250 Random Vibration

$$\varepsilon_x = \frac{1}{\sqrt{2\,T\,\Delta f}} \left[1 + \frac{R_x^2(0)}{R_x^2(\tau)} \right]^{1/2} \qquad [4.102]$$

A reasonable value of T for the calculation of R_x is $T = \dfrac{1}{\Delta f\,\varepsilon_x^2}$. For the cross-correlation R_{xy}:

$$\varepsilon_{xy} = \frac{1}{\sqrt{2\,T\,\Delta f}} \left[1 + \frac{R_{xy}^2(0)}{R_{xy}^2(\tau)} \right]^{1/2} \qquad [4.103]$$

4.18. Peak hold spectrum

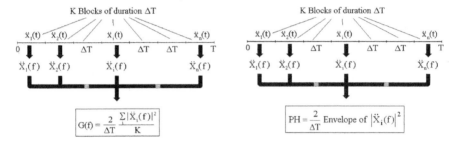

Figure 4.84. *Principle of PSD calculation*   **Figure 4.85.** *Principle of peak hold spectrum calculation.*

We go back to the definition of the PSD. This spectrum is obtained from a stationary signal by calculating the square mean of amplitudes from Fourier transforms of K blocks from a sample of duration T for this signal (Figure 4.84). Using this mean is justified by the necessity of obtaining an acceptable statistical error, leading to approximately 50 sub-samples.

When the vibration is not stationary, the calculation of this mean, and thus the PSD, makes no sense (average of a population with no similar statistical properties).

In order to attempt to characterize this type of environment, we sometimes replace the traditional PSD by an envelope of PSDs calculated in each sub-sample K (*peak hold spectrum*) (Figure 4.85) [GIR 04].

This spectrum provides information on the severity of the vibration. It should, however, be used very carefully, knowing that:

– a PSD calculated in a single sample presents a large statistical error (see Example 4.13). The amplitude is calculated with great inaccuracy. The probability of these amplitudes is very small, all the more so when the signal is not stationary or when there are transitory components. This spectrum is statistically very poor at best and at worst of no value;

– the level of confidence defined for the calculation of the traditional PSD (linked to the statistical error) has no value for the peak hold spectrum. No level of confidence can be attributed to it;

– the difference between the traditional PSD and the peak hold spectrum can be large, particularly in the case of a non-stationary signal or with transitory components.

Example 4.23.

Figure 4.86 enables us to compare the PSD of a stationary signal and the peak hold spectrum.

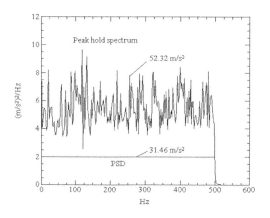

Figure 4.86. *Comparison between the PSD and the peak hold spectrum*

Despite the stationarity of the signal, we notice a large gap between both spectrums and the "broken" appearance of the peak hold spectrum linked to the statistical error.

252   Random Vibration

The combined use of the PSD and the peak hold spectrum can provide an indication on the non-stationary or on the transitory condition by comparison of the peak and mean spectral values [SHR 95], even though the interpretation of results is not easy. Peak hold is sometimes used to observe the highest frequency domain values, for example during startups and coastdowns. The presence of beats in the data can also be detected from a comparison between the PSD values and the peak hold values [CRA 92].

This spectrum sometimes (wrongly) replaces the PSD, but generally, it is used to increase it. It is not a PSD and should not be used directly as a test specification, because it can lead to very important conservatism, particularly in the specific cases discussed.

If we still want to rely on the peak hold spectrum to define a test specification, we must find a PSD producing the same peak hold spectrum as the one in the environment measured (2 to 3 times smaller value). However, another problem would be to specify a duration for this PSD.

4.19. Generation of random signal of given PSD

Several methods are used to obtain a random signal according to a given PSD. We will list two, the first one consisting in a sum of sinusoids with a random phase, and the second using the inverse Fourier transform.

4.19.1. *Random phase sinusoid sum method*

4.19.1.1. *Principle*

The method of generation of a random signal varying with time of given duration T from a PSD of maximum frequency f_{max} includes the following stages:

– calculation of the temporal step $\delta t = \dfrac{1}{f_{samp.}} = \dfrac{1}{2.6\, f_{max}}$;

– choice of the number M of points of definition of the PSD (power of two);

– calculation of the number of signal points: $N = \dfrac{T}{\delta t}$;

– possibly, modification of N (and thus of the duration) and/or of M in order to respect a maximum statistical error ε_0 (for a future PSD calculation of the generated signal), starting from the relation $\dfrac{M}{N} \leq \dfrac{\varepsilon_0^2}{2}$, maintaining M equal to a power of two;

– calculation of the frequency interval between 2 points of the PSD
$\Delta f = \dfrac{f_{samp.}}{2\,M}$;

– for each M points of the PSD, calculation at every time $t = k\,\delta t$ (k = constant integer between 1 and N) of a "sinusoid":

- of the form: $^m x(t) = {}^m x_{max}\sin(2\pi f t + \varphi_m)$,

- of duration T,

- of frequency $f_m = m\,df$ (m integral such that $1 \leq m \leq M$),

- of amplitude $\sqrt{2\,G(f_m)\,\Delta f}$ (where $G(f_m)$ is the value of the given PSD at the frequency f_m, the amplitude of a sinusoid being equal to twice its rms value,

- of random phase φ_m, whose expression is a function of the specified distribution law for the instantaneous values of the signal,

– sum of the M sinusoids at each time.

4.19.1.2. *Expression for phase*

Normal law

It is shown that we can obtain a normal distribution of the signal's instantaneous values when the phase is equal to [KNU 98]

$$\varphi_m = 2\pi\sqrt{-2\ln r_1}\,\cos(2\pi r_2) \qquad [4.104]$$

or

$$\varphi_m = 2\pi\sqrt{-2\ln r_1}\,\sin(2\pi r_2) \qquad [4.105]$$

In these expressions, r_1 and r_2 are two random numbers obeying a rectangular distribution in the interval [0, 1].

Definition

A random variable r has a uniform or rectangular distribution in the interval [a, b] if its probability density obeys

$$p(r) = \begin{cases} \dfrac{1}{b-a} & \text{for } a \leq r \leq b \\ 0 & \text{for } r < a \text{ or } r > b \end{cases} \qquad [4.106]$$

If a random variable is uniformly distributed about [0, 1], the variable $y = a + (b - a)\, r$ is uniformly distributed about [a, b], having a mean of $\dfrac{a+b}{2}$ and standard deviation $s = \dfrac{b-a}{2\sqrt{3}}$.

Other laws

Here we want to create a signal whose instantaneous values obey a given distribution law $F(X)$. This function not decreasing, the probability that $x \leq X$ is equal to [DAH 74]:

$$P(x \leq X) = P[F(x) \leq F(X)] \qquad [4.107]$$

Let us set $F(x) = r$ where r is a random variable uniformly distributed about [0,1]. It then becomes:

$$P[F(x) \leq F(X)] = P[r \leq F(X)] \qquad [4.108]$$

From definition of the uniform distribution, $P(r \leq R) = R$ where R is an arbitrary number between 0 and 1, yielding $P(x \leq X) = P[r \leq F(X)] = F(X)$. To create a signal of distribution $F(X)$, it is thus necessary that:

$$F(x) = r \qquad [4.109]$$

The problem can also be solved by setting:

$$F(x) = 1 - r \qquad [4.110]$$

Examples

1. Signal of exponential distribution: the distribution is defined by (Appendix A1.3)

$$F(X) = 1 - e^{-\lambda X} \qquad [4.111]$$

From [4.110],

$$1 - e^{-\lambda x} = 1 - r \qquad [4.112]$$

yielding

$$x = -\frac{\ln r}{\lambda} \qquad [4.113]$$

and

$$\varphi_m = -2\pi \frac{\ln r}{\lambda} \qquad [4.114]$$

2. Signal with Weibull distribution: from (Appendix A1.7)

$$F(X) = \begin{cases} 1 - \exp\left[-\left(\dfrac{X-\varepsilon}{v-\varepsilon}\right)^{\alpha}\right] & X > \varepsilon \\ 0 & X < \varepsilon \end{cases} \qquad [4.115]$$

it is shown in a similar way that we must have:

$$x = \varepsilon + (v - \varepsilon)(-\ln r)^{1/\alpha} \qquad [4.116]$$

$$\varphi_m = 2\pi\left[\varepsilon + (v - \varepsilon)(-\ln r)^{1/\alpha}\right] \qquad [4.117]$$

4.19.2. *Inverse Fourier transform method*

The PSD of a random signal x(t) is calculated by breaking a sample of this signal of duration T into K blocks of duration ΔT and by averaging the PSD Ks of these blocks. For a given block, the PSD is given by relation [2.20]:

$$G(f) = \frac{2}{\Delta T} |X(f)|^2$$

where X(f) is the block's Fourier transform.

Conversely, if we give a PSD G(f), it is possible to create a random signal respecting this PSD according to the following process:

– calculation, from the maximum PSD frequency, of sampling frequency $f_{samp} = 2 f_{max}$ and temporal step $\delta t = \dfrac{1}{2 f_{max}}$;

– calculation from the Δf frequency step of this same PSD of signal duration ΔT (block) which will be obtained from the inverse Fourier transform;

– multiplication of the PSD G(f) by $e^{j\varphi}$, where φ is a random phase;

– calculation of the Fourier transform of a block with the help of relation [2.20] by:

$$X(f) = \sqrt{\frac{\Delta T}{2} G(f)}\, e^{j\varphi}$$

– calculation of the X(f) inverse Fourier transform, to obtain a complex temporal variable of duration ΔT. Signal x(t) is the real part of this complex quantity;

– to verify the validity of the signal obtained by comparison of its PSD with the original PSD with a given ε statistical error, we must generate a signal with a total duration of $T = \dfrac{1}{\varepsilon^2 \, \Delta f}$, or in other words, repeat K times the last three steps (where K is the closest integer to $\dfrac{T}{\Delta T}$).

4.20. Using a window during the creation of a random signal from a PSD

The signal developed from the method from section 4.19.2 is constituted by serializing the real parts of the complex quantity thus calculated, for different values of the phase φ, until the desired duration is reached.

If the basic signals are simply connected end-to-end without precautions, each connection is done with a very high slope straight line segment creating high frequencies in the response and modifies in particular the extreme response spectrum (ERS) to greater frequencies than the PSD definition range.

To avoid this problem, each basic signal can be filtered by a window, a Bingham window for example (Figure 4.9), softening the extremities for much smoother connection.

Example 4.24.

Figure 4.87 shows one of the discontinuities present in the signal when it is developed by simple end-to-end connection of the real Fourier transform parts.

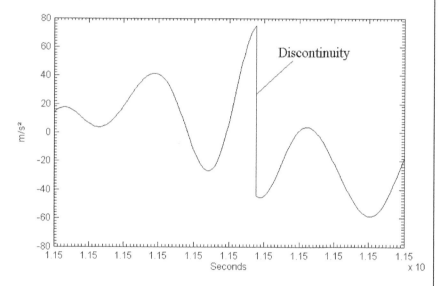

Figure 4.87. *Discontinuity of connection of two basic signals*

This discontinuity triggers a significant transitory response in systems with a frequency that is higher than the maximum frequency of the PSD used to create the signal (Figure 4.88).

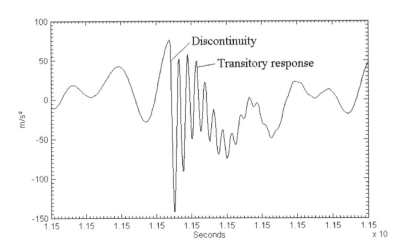

Figure 4.88. *Effect of the discontinuity on the response of a linear system at 1 ddl (2,000 Hz, Q = 10)*

This leads to an abnormally high extreme response spectrum (ERS) at high frequency (Figure 4.89) (see definition of ERS in Volume 5).

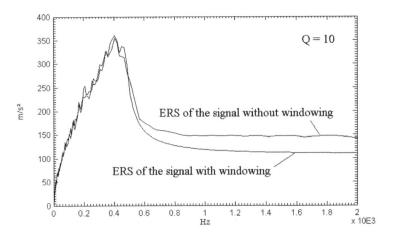

Figure 4.89. *ERS of the signal with and without windowing*

Chapter 5

Statistical Properties of Random Vibration in the Time Domain

The analysis of the statistical properties of the instantaneous values of a random signal $\ell(t)$ is based primarily on the work of S.O. Rice [RIC 44] and of S.H. Crandall [CRA 63]. We are more particularly interested in the study of the probability density of the instantaneous values of the signal and in that of the peaks (positive and negative maximum amplitude).

This study results in considering simultaneously at a given time $\ell(t)$ and its derivatives $\dot{\ell}(t)$ and $\ddot{\ell}(t)$ which respectively represent the value of the signal, its slope and its curvature at the time t. These parameters are in particular associated with a multidimensional normal probability density function of the form [BEN 58]:

$$p(\ell_1, \ell_2, \cdots, \ell_n) = (2\pi)^{-n/2} |M|^{-n/2} \exp\left(-\frac{1}{2|M|} \sum_{i,j=1}^{n} M_{ij} \ell_i \ell_j\right) \quad [5.1]$$

for the research of the distribution law of the peak values.

5.1. Distribution of instantaneous values

The distribution of the instantaneous values of the parameter describing the random phenomenon can very often be represented by a Gaussian law [MOR 75]. There can of course be particular cases where this assumption is not justified, for

260 Random Vibration

example, for vibrations measured on the axle of a vehicle whose suspension has just compressed an elastic thrust after deflection of the dampers (non-linear behavior in compression only).

5.2. Properties of derivative process

Let us consider a stationary random vibration $\ell(t)$ and its derivative $\dot{\ell}(t)$, defined by:

$$\dot{\ell}(t) = \lim_{\Delta t \to 0} \left[\frac{\ell(t + \Delta t) - \ell(t)}{\Delta t} \right] \quad [5.2]$$

with the condition that

$$\lim_{\Delta t \to 0} E\left\{ \left[\frac{\ell(t + \Delta t) - \ell(t)}{\Delta t} - \dot{\ell}(t) \right]^2 \right\} = 0 \quad [5.3]$$

Average value of the derivative process

This is

$$E[\dot{\ell}] = \lim_{\Delta t \to 0} E\left[\frac{\ell(t + \Delta t) - \ell(t)}{\Delta t} \right] \quad [5.4]$$

If the process is stationary,

$$E[\ell(t + \Delta t)] = E[\ell(t)]$$

yielding

$$E[\dot{\ell}] = 0 \quad [5.5]$$

NOTE.– *The autocorrelation $R_\ell(\tau)$ presents an absolute maximum for $\tau = 0$. We thus obtain $R''_\ell(0) < 0$.*

$$E[\ell \, \dot{\ell}] = E[\ell(t) \, \dot{\ell}(t)] = \lim_{\Delta t \to 0} E\left[\ell(t) \, \frac{\ell(t + \Delta t) - \ell(t)}{\Delta t} \right] \quad [5.6]$$

$$E[\ell \dot{\ell}] = R'_\ell(0) \quad [5.7]$$

The derivative of the autocorrelation function of a derivable process is:

– continuous and derivable at any point;

– even.

It is thus canceled for $\tau = 0$, yielding

$$E[\ell \dot{\ell}] = 0 \quad [5.8]$$

There is no correlation between a stationary process $\ell(t)$ and the derivative process $\dot{\ell}(t)$ (whatever the distribution law).

Mean square of the derivative

$$E\left\{\left[\frac{\ell(t+\Delta t) - \ell(t)}{\Delta t}\right]^2\right\} = 2\frac{R_\ell(0) - R_\ell(-\Delta t)}{(\Delta t)^2}$$

$$E\left\{[\dot{\ell}(t)]^2\right\} = -R''_\ell(0) \quad [5.9]$$

A stationary process $\ell(t)$ is thus derivable in the mean square sense if and only if its correlation function $R_\ell(\tau)$ contains a continuous second derivative.

Correlation function of the process and its derivative

1. By definition [1.64], $R_{\dot{\ell}\ell}(\tau) = E[\dot{\ell}(t)\,\ell(t+\tau)]$

$$R_{\dot{\ell}\ell}(\tau) = \lim_{\substack{\Delta t_1 \to 0 \\ \Delta t_2 \to 0}} E\left[\frac{\ell(t+\Delta t_1) - \ell(t)}{\Delta t_1} \cdot \frac{\ell(t+\tau+\Delta t_2) - \ell(t+\tau)}{\Delta t_2}\right]$$

$$R_{\dot{\ell}\ell}(\tau) = \lim_{\substack{\Delta t_1 \to 0 \\ \Delta t_2 \to 0}} \frac{R_\ell(\tau+\Delta t_2 - \Delta t_1) - R_\ell(\tau - \Delta t_1) - R_\ell(\tau + \Delta t_2) - R_\ell(\tau)}{\Delta t_1\,\Delta t_2}$$

$$R_{\dot{\ell}\ell}(\tau) = \lim_{\Delta t_1 \to 0} \frac{R''_\ell(t-\Delta t_1) - R'_\ell(\tau)}{\Delta t_1}$$

$$R_{\ddot{\ell}\ell}(\tau) = -\frac{d^2 R_\ell(\tau)}{d\tau^2} = -R''_\ell(\tau) \tag{5.10}$$

2. $R_{\ell\dot{\ell}}(\tau) = E[\ell(t)\,\dot{\ell}(t+\tau)]$

$$R_{\ell\dot{\ell}}(\tau) = \lim_{\Delta t \to 0} E\left[\ell(t)\,\frac{\ell(t+\tau+\Delta t)-\ell(t+\tau)}{\Delta t}\right]$$

$$R_{\ell\dot{\ell}}(\tau) = \lim_{\Delta t \to 0}\left[\frac{R_\ell(\tau+\Delta t)-R_\ell(\tau)}{\Delta t}\right]$$

$$R_{\ell\dot{\ell}}(\tau) = \frac{dR_\ell(\tau)}{d\tau} = R'_\ell(\tau) \tag{5.11}$$

In the same way:

$$R_{\dot{\ell}\ell}(\tau) = -\frac{dR_\ell(\tau)}{d\tau} = -R'_\ell(\tau) \tag{5.12}$$

In a more general way, if $\ell^{(m)}$ and $u^{(n)}$ are the mth derivative processes of $\ell(t)$ and n^{th} of $u(t)$, if the successive derivatives exist, we obtain,

$$R_{\ell^{(m)} u^{(n)}} = (-1)^m \frac{d^{m+n} R_{\ell u}(\tau)}{d\tau^{m+n}} \tag{5.13}$$

Variance of the derivative process

$$E[\dot{\ell}^2] = R_{\dot{\ell}\dot{\ell}}(0) = -R''(0) \tag{5.14}$$

Since $E[\dot{\ell}] = 0$, the variance $s_{\dot{\ell}}^2$ is equal to

$$s_{\dot{\ell}}^2 = E[\dot{\ell}^2] - [E(\dot{\ell})]^2 \tag{5.15}$$

PSD of the derivative process

By definition [2.48]:

$$R_\ell(\tau) = \int_{-\infty}^{+\infty} S_\ell(\Omega)\, e^{i\Omega t}\, d\Omega$$

Knowing that:

$$R_{\dot{\ell}}(\tau) = -R''_{\ell}(\tau)$$

$$R_{\dot{\ell}}(\tau) = -\int_{-\infty}^{+\infty} (i\,\Omega)^2\, S_{\ell}(\Omega)\, e^{i\Omega\tau}\, d\Omega \qquad [5.16]$$

This yields

$$S_{\dot{\ell}}(\Omega) = \Omega^2\, S_{\ell}(\Omega) \qquad [5.17]$$

$$E\left[\dot{\ell}^2\right] = \int_{-\infty}^{+\infty} S_{\dot{\ell}}(\Omega)\, d\Omega = \int_{-\infty}^{+\infty} \Omega^2\, S_{\ell}(\Omega)\, d\Omega \qquad [5.18]$$

and in the same way [MOR 56], [NEW 75], [SVE 80]:

$$E\left[\ddot{\ell}^2\right] = \int_{-\infty}^{+\infty} S_{\ddot{\ell}}(\Omega)\, d\Omega = \int_{-\infty}^{+\infty} \Omega^4\, S_{\ell}(\Omega)\, d\Omega \qquad [5.19]$$

$$S_{\ddot{\ell}}(\Omega) = \Omega^4\, S_{\ell}(\Omega) \qquad [5.20]$$

$$R_{\ddot{\ell}\ddot{\ell}}(\tau) = -R_{\dot{\ell}\dot{\ell}}(\tau) = \frac{d^4 R_{\ell}(\tau)}{d\tau^4} \qquad [5.21]$$

NOTE.– *The autocorrelation functions of the derivative processes of $\ell(t)$ depend only on τ. The derivatives of a stationary process are stationary functions. However, the integral of a stationary function is not necessarily stationary.*

The result obtained shows the existence of a transfer function $H(\Omega)$ between $\ell(t)$ and its derivatives:

$$S_{\dot{\ell}}(\Omega) = |H(\Omega)|^2\, S_{\ell}(\Omega) \qquad [5.22]$$

$$S_{\ddot{\ell}}(\Omega) = |H(\Omega)|^4\, S_{\ell}(\Omega) \qquad [5.23]$$

where

$$H(\Omega) = i\,\Omega$$

5.3. Number of threshold crossings per unit time

Let us consider a stationary and ergodic random vibration $\ell(t)$, and $p(\ell)$, the probability density function of the instantaneous values of $\ell(t)$. Let us seek to determine the number of times per unit time n_a^+ the signal crosses a threshold chosen *a priori* with a positive slope.

Let us set by n_a the number of occasions per unit time that the signal crosses the interval a, a + da with a positive or negative arbitrary slope, da being a very small interval corresponding to the time increment dt. We have, on average,

$$n_a^+ = \frac{n_a}{2} \qquad [5.24]$$

Let us set by n_0^+ the number of occasions per unit time that the signal crosses the threshold a = 0 with a positive slope (n_0^+ gives an indication of the average frequency of the signal). Let us finally set by $\dot{\ell}(t)$ the derivative of the process $\ell(t)$ and by b the value of $\dot{\ell}(t)$ when $\ell = a$. Let us suppose that the time interval dt is sufficiently small that the variation of the signals between t and t + dt is linear. To cross the threshold a, the process must have a velocity $\dot{\ell}(t)$ greater than $\frac{a - \ell(t)}{dt}$.

The probability of crossing is related to the joint probability density $p(\ell, \dot{\ell})$ between ℓ and $\dot{\ell}$. Given a threshold a, the probability that:

$$a < \ell(t) \leq a + da$$

and

$$b < \dot{\ell}(t) \leq b + db \qquad [5.25]$$

is thus, in a time unit,

$$p(a, b)\, da\, db = P[a < \ell(t) \leq a + da,\ b < \dot{\ell}(t) \leq b + db] \qquad [5.26]$$

Setting t_a the time spent in the interval da:

$$t_a = \frac{da}{|b|} \qquad [5.27]$$

(t_a being a primarily positive quantity). The number of passages per unit time in the interval a, a + da for $\dot{\ell}(t) = b$ is thus:

$$\frac{p(a, b) \, da \, db}{t_a} = |b| \, p(a, b) \, da \qquad [5.28]$$

and the average total number of crossings of the threshold a, per unit time, for all the possible values of $\dot{\ell}(t)$ is written:

$$n_a = \int_{-\infty}^{+\infty} |b| \, p(a, b) \, db = 2 \, n_a^+ \qquad [5.29]$$

where

$$n_a^+ = \left| \int_0^\infty p(\ell, \dot{\ell}) \, d\ell \, d\dot{\ell} \right|_{\ell=a} \qquad [5.30]$$

This expression is sometimes called the *Rice formula*. The only assumption considered is that of the stationarity. We deduce from [5.30], for a = 0,

$$n_0 = 2 \, n_0^+ = \int_{-\infty}^{+\infty} |b| \, p(a, b) \, db \qquad [5.31]$$

and

$$\frac{n_a}{n_0} = \frac{n_a^+}{n_0^+} = \frac{\int_{-\infty}^{+\infty} |b| \, p(a, b) \, db}{\int_{-\infty}^{+\infty} |b| \, p(0, b) \, db} \qquad [5.32]$$

These expressions can be simplified since the signals $\ell(t)$ and $\dot{\ell}(t)$ are statistically independent:

$$p(\ell, \dot{\ell}) = p(\ell) \, \pi(\dot{\ell}) \qquad [5.33]$$

Then,

$$n_a = p(a) \int_{-\infty}^{+\infty} |b|\, \pi(b)\, db \qquad [5.34]$$

and

$$\frac{n_a}{n_0} = \frac{n_a^+}{n_0^+} = \frac{p(a)}{p(0)} \qquad [5.35]$$

Lastly, if $\pi(\dot{\ell})$ is an even function of b,

$$\pi(b) = \pi(-b)$$

yielding

$$n_a = 2\, p(a) \int_0^{\infty} b\, \pi(b)\, db \qquad [5.36]$$

Particular case

If the function $\dot{\ell}(t)$ has instantaneous values distributed according to a Gaussian law, zero mean and variance $\dot{\ell}_{rms}^2$, such that

$$\pi(\dot{\ell}) = \frac{1}{\dot{\ell}_{rms}\sqrt{2\pi}}\, e^{-\frac{\dot{\ell}^2}{2\dot{\ell}_{rms}^2}} \qquad [5.37]$$

starting from [5.34], it results that:

$$n_a = p(a) \int_{-\infty}^{+\infty} \frac{|b|}{\dot{\ell}_{rms}\sqrt{2\pi}}\, e^{-\frac{b^2}{2\dot{\ell}_{rms}^2}}\, db \qquad [5.38]$$

or since $\pi(\dot{\ell})$ is even,

$$n_a = \frac{2\,\dot{\ell}_{rms}}{\sqrt{2\pi}}\, p(a) \qquad [5.39]$$

Statistical Properties of Random Vibration in the Time Domain 267

If the instantaneous acceleration is itself distributed according to a Gaussian law $(0, \ell_{rms})$:

$$p(\ell) = \frac{1}{\ell_{rms}\sqrt{2\pi}} e^{-\frac{\ell^2}{2\ell_{rms}^2}} \quad [5.40]$$

$$p(\ell,\dot{\ell}) = \frac{1}{2\pi\,\ell_{rms}\,\dot{\ell}_{rms}} e^{-\frac{1}{2}\left(\frac{\ell^2}{\ell_{rms}^2}+\frac{\dot{\ell}^2}{\dot{\ell}_{rms}^2}\right)} \quad [5.41]$$

and [LEY 65], [LIN 67], [NEW 75], [PRE 56a], [THR 64], [VAN 75]:

$$n_a = \frac{1}{\pi}\frac{\dot{\ell}_{rms}}{\ell_{rms}} e^{-\frac{a^2}{2\ell_{rms}^2}} \quad [5.42]$$

$$n_0 = \frac{1}{\pi}\frac{\dot{\ell}_{rms}}{\ell_{rms}} \quad [5.43]$$

$$n_a = n_0\, e^{-\frac{a^2}{2\ell_{rms}^2}} \quad [5.44]$$

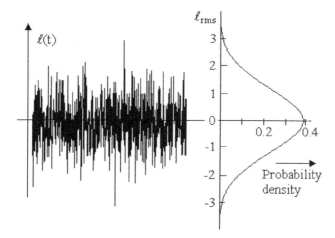

Figure 5.1. *Probability density of instantaneous values of a random signal*

Since

$$\ell^2_{rms} = \int_0^\infty G(\Omega) \, d\Omega = R(0) \qquad [5.45]$$

$$\dot{\ell}^2_{rms} = \int_0^\infty \Omega^2 \, G(\Omega) \, d\Omega = -R''(0) \left[-R^{(2)}(0)\right] \qquad [5.46]$$

$$\ddot{\ell}^2_{rms} = \int_0^\infty \Omega^4 \, G(\Omega) \, d\Omega = R^{(4)}(0) \qquad [5.47]$$

this results in [DEE 71], [VAN 75]:

$$n_a = 2 \, n_a^+ = \frac{1}{\pi} \left[\frac{\int_0^\infty \Omega^2 \, G(\Omega) \, d\Omega}{\int_0^\infty G(\Omega) \, d\Omega} \right]^{\frac{1}{2}} \exp\left(-\frac{a^2}{2 \, \ell^2_{rms}}\right) \qquad [5.48]$$

n_a is the mean number of crossings of the threshold a per unit time.

n_a^+ is the mean number of crossings of the threshold a with positive slope and per unit time.

NOTE.– *In a time interval Δt, there are on average N_a occurrences of value a:*

$$N_a = n_a \, \Delta t = n_0 \, \Delta t \, \exp\left(-\frac{a^2}{2 \, \ell^2_{rms}}\right) \qquad [5.49]$$

If the vibration is narrowband, $n_0 \approx 2 \, f_0$ and

$$N_a \approx 2 \, f_0 \, \Delta t \, \exp\left(-\frac{a^2}{2 \, \ell^2_{rms}}\right) \qquad [5.50]$$

The average time interval between occurrences, obtained for $N_a = 1$, is thus equal to [AND 11]:

$$\Delta t \approx \frac{1}{2 f_0} \exp\left(\frac{a^2}{2 \ell_{rms}^2}\right) \qquad [5.51]$$

5.4. Average frequency

Let us set [PAP 65], [PRE 56b]:

$$n_0 = \frac{1}{\pi} \left[\frac{\int_0^\infty \Omega^2 G(\Omega) d\Omega}{\int_0^\infty G(\Omega) d\Omega}\right]^{\frac{1}{2}} = \frac{1}{\pi} \sqrt{-\frac{R^{(2)}(0)}{R(0)}} \qquad [5.52]$$

Depending on f, n_0 becomes [BEN 58], [BOL 84], [CRA 63], [FUL 61], [HUS 56], [LIN 67], [POW 58], [RIC 64], [SJÖ 61], [SWA 63]:

$$n_0 = 2 n_0^+ = 2 \left[\frac{\int_0^\infty f^2 G(f) df}{\int_0^\infty G(f) df}\right]^{\frac{1}{2}} \qquad [5.53]$$

The quantity n_0^+ (*average* or *expected* frequency) can be regarded as the frequency at which energy is most concentrated in the spectrum (apparent frequency of the spectrum).

Band-limited white noise

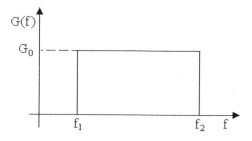

Figure 5.2. *PSD of a band-limited white noise*

270 Random Vibration

If the PSD is defined by

$$\begin{cases} G(f) = G_0 & \text{for } f_1 \leq f \leq f_2 \\ G(f) = 0 & \text{elsewhere} \end{cases},$$

we have

$$n_0^+ = \left[\frac{f_2^3 - f_1^3}{3(f_2 - f_1)}\right]^{\frac{1}{2}} \qquad [5.54]$$

$$n_0^+ = \sqrt{\frac{f_1^2 + f_1 f_2 + f_2^2}{3}} \qquad [5.55]$$

Ideal low-pass filter

If $f_1 = 0$,

$$n_0^+ = \frac{f_2}{\sqrt{3}} = 0.577 \, f_2 \qquad [5.56]$$

Case of a narrowband noise

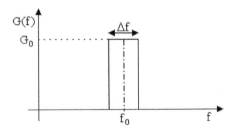

Figure 5.3. *PSD of a narrowband noise*

Let us consider a random vibration of constant PSD $G(\Omega) = G_0$ in the interval $\Delta\Omega$, zero elsewhere. We have [COU 70], [NEW 75]:

$$\dot{\ell}^2_{rms} = \int_0^\infty (2\pi f)^2 \, G(\Omega) \, d\Omega$$

$$\ddot{\ell}^2_{rms} = \int_0^\infty (2\pi f)^4 \, G(\Omega) \, d\Omega$$

$$\dot{\ell}^2_{rms} = G_0 \int_0^\infty \Omega^2 \, d\Omega = G_0 \, \omega_0^2 \, \Delta\Omega \qquad [5.57]$$

$$\ddot{\ell}^2_{rms} = G_0 \int_0^\infty \Omega^4 \, d\Omega = G_0 \, \omega_0^4 \, \Delta\Omega \qquad [5.58]$$

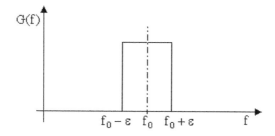

Figure 5.4. *PSD of a narrowband noise*

Let us set $\Delta f = 2\varepsilon$. We have $f_1 = f_0 - \varepsilon$ and $f_2 = f_0 - \varepsilon$, yielding

$$n_0^+ = \left[\frac{(f_0 - \varepsilon)^2 + (f_0 - \varepsilon)(f_0 + \varepsilon) + (f_0 + \varepsilon)^2}{3} \right]^{\frac{1}{2}}$$

and

$$n_0^+ = f_0 \sqrt{1 + \frac{\varepsilon^2}{3 f_0^2}} = f_0 \sqrt{1 + \frac{1}{12}\left(\frac{\Delta f}{f_0}\right)^2} \qquad [5.59]$$

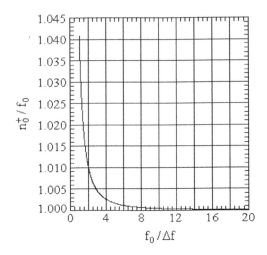

Figure 5.5. *Ratio of average frequency/central frequency of a narrowband noise*

n_0^+ tends towards f_0 when Δf tends towards zero. For any value of f_0, n_0^+ is equal to or higher than f_0.

In the case of the response of a linear slightly damped one-degree-of-freedom system, n_0^+ will thus generally be close to the natural frequency f_0 of the system.

5.5. Threshold level crossing curves

Threshold level crossing curves give, depending on the threshold a, the number of crossings of this threshold with positive slope. These curves can be plotted:

– either from the time history signal by effective counting of the crossings with positive slope over a duration T. For a given signal, the result is deterministic;

– or from the power spectral density of the vibration, by supposing that the distribution of the instantaneous values of the signal follows a Gaussian law to zero mean. We obtain here the expected value of the number of threshold crossings a over the duration T [LEA 69], [RIC 64]:

$$N_a^+ = n_a^+ T = n_0^+ T e^{-\frac{a^2}{2 \ell_{rms}^2}} \qquad [5.60]$$

Statistical Properties of Random Vibration in the Time Domain 273

with n_0^+ = expected frequency defined in [5.53]:

$$n_0^{+2} = \frac{\int_0^\infty f^2 \, G(f) \, df}{\int_0^\infty G(f) \, df}$$

The knowledge of $G(f)$ makes it possible to calculate n_0^+ and ℓ_{rms}, then to plot N_a^+ as a function of the threshold value a. In practice, we generally represent a with respect to N_a^+, the first value of N_a^+ being higher or equal to 1. For $N_a^+ = 1$,

$$a_0 = \ell_{rms} \sqrt{2 \ln N_0^+} = \ell_{rms} \sqrt{2 \ln n_0^+ \, T} \qquad [5.61]$$

a_0 is, on average, the strongest value of the signal observed over a duration T.

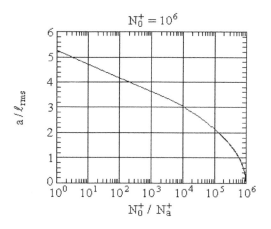

Figure 5.6. *Example of threshold level crossing curve for a Gaussian signal*

The curve in Figure 5.6 shows the variations of $\dfrac{a}{\ell_{rms}}$ with respect to $\dfrac{N_a^+}{N_0^+}$, plotted starting from the expression:

$$\frac{a}{\ell_{rms}} = \sqrt{2 \frac{N_a^+}{N_0^+}}$$

274   Random Vibration

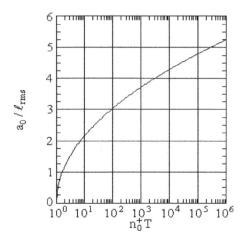

Figure 5.7. *Largest peak, on average, over a given duration*

The variations of $\dfrac{a_0}{\ell_{rms}}$ as a function of the product $n_0^+ T$ are represented in Figure 5.7:

$$\frac{a_0}{\ell_{rms}} = \sqrt{2 \ln n_0^+ T}$$

It is observed that it is possible to obtain, in very realistic situations, combinations of n_0^+ and T such that the ratio $\dfrac{a_0}{\ell_{rms}}$ is equal to or higher than 5. For this, it is necessary that:

$$n_0^+ T \geq e^{25/2}$$

$$n_0^+ T \geq 2.7 \; 10^5$$

For   T = 600 s   it is necessary that $n_0^+ \geq 447$ Hz

T = 3,600 s $n_0^+ \geq 74.5$ Hz

T = 4 hours $n_0^+ \geq 18.6$ Hz

The stress in a part subjected to a Gaussian random vibration may exceed 5 to 6 times the rms stress. The rule sometimes used, which requires to size the parts for 3 times the rms value, is insufficient [MET 03], [SCH 06], [WIJ 09].

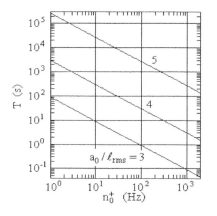

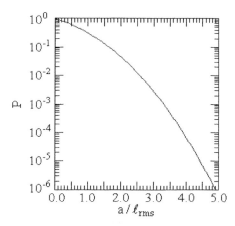

Figure 5.8. *Time necessary to obtain, on average, a given maximum level, versus the average frequency*

Figure 5.9. *Probability of crossing a given threshold, versus the threshold value*

Figure 5.8 indicates the duration T necessary to obtain a given ratio a_0 / ℓ_{rms}, as a function of n_0^+.

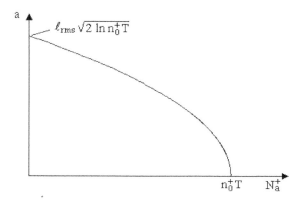

Figure 5.10. *Noteworthy points on the threshold level crossings curve*

276   Random Vibration

For a = 0,

$$N_a^+ = N_0^+ = n_0^+ T$$

The probability that the signal crosses the level a with a positive slope and that $a \leq \ell \leq a + \Delta a$ is equal to $N_a^+ \Delta a / N_0^+ \ell_{rms}$.

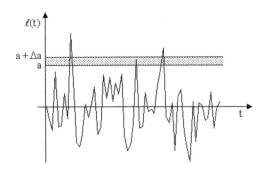

Figure 5.11. *Values of the signal in the interval a, a + da, after crossing threshold a with a positive slope*

The probability that $\ell(t)$ is higher than a is equal to:

$$P = \frac{1}{N_0^+ \ell_{rms}} \int_a^\infty N_a^+ \, da \qquad [5.62]$$

$$P = \frac{1}{\ell_{rms}} \int_a^\infty e^{-\frac{a^2}{2\ell_{rms}^2}} \, da$$

Knowing that the error function can be written:

$$\text{erf}\left(\frac{u}{\sqrt{2}}\right) = \sqrt{\frac{2}{\pi}} \int_0^u e^{-\frac{u^2}{2}} \, du \qquad [5.63]$$

and that:

$$\int_{-\infty}^{+\infty} e^{-\frac{u^2}{2}} \, du = \sqrt{2\pi}$$

resulting in, if $u = a/\ell_{rms}$,

$$P = \left[\sqrt{\frac{\pi}{2}} - \int_0^u e^{-\frac{u^2}{2}} du\right]$$ [5.64]

This yields, after standardization:

$$P\left(u > \frac{a}{\ell_{rms}}\right) = \left[1 - \text{erf}\left(\frac{a}{\ell_{rms}\sqrt{2}}\right)\right]$$ [5.65]

Figure 5.9 shows the variations of $P(u > a/\ell_{rms})$ for a/ℓ_{rms} ranging between 0 and 5.

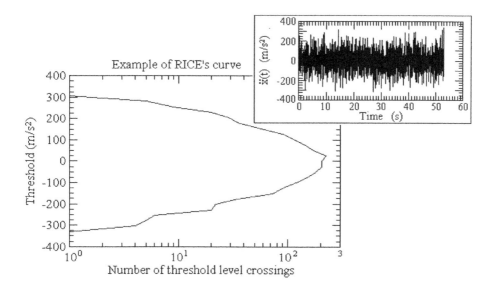

Figure 5.12. *Example of threshold level crossing curve*

278 Random Vibration

Example 5.1.

Let us consider a random acceleration defined over a duration T = 1 hr by its PSD G(f):

$G(f) = G_1 = 0.1 \text{ m}^2\text{s}^{-4}/\text{Hz}$ from 10 Hz to 50 Hz

$G(f) = G_2 = 0.2 \text{ m}^2\text{s}^{-4}/\text{Hz}$ from 50 Hz to 100 Hz

$G(f) = 0$ elsewhere

$\ddot{x}_{rms}^2 = (50+10)\ 0.1+(100-50)\ 0.2 = 14\ (m/s^2)^2$

$\ddot{x}_{rms} = 3.74\ m/s^2$

From [5.53]:

$$n_0^{+2} = \frac{0.1\ (50^3 - 10^3) + 0.2\ (100^3 - 50^3)}{3\ \ddot{x}_{rms}^2}\ \text{Hz}^2$$

$n_0^+ = 66.8\ \text{Hz}$

and [5.60]:

$$N_a^+ = 66.8\ 3{,}600\ e^{-\frac{a^2}{2\ 14}} = 2.4\ 10^5\ e^{-\frac{a^2}{28}}$$

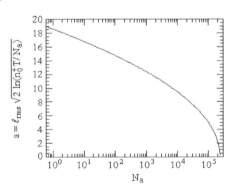

Figure 5.13. *Example of curve threshold level crossings*

The threshold which is only exceeded once on average over the duration T has an amplitude

$a_0 = 3.74\ \sqrt{2\ \ln\ 66.8\ 3{,}600} = 18.62\ m/s^2$

These threshold level crossings curves make it possible to verify the symmetry of the load and its Gaussian character [BEN 04]. They were used to compare the severity of several random vibrations [KAZ 70], to evaluate their damage potential or to reduce the test duration. This method can be justified if the treated signal is the stress applied to a part of a structure, with just one reserve, which is the non-immediate relationship between the number of peaks and the number of threshold level crossings; it is not, on the other hand, usable starting from the input signal of acceleration. The threshold crossings curve of the excitation $\ddot{x}(t)$ is not representative of the damage undergone by a part which responds at its natural frequency with its Q factor. In random mode, we cannot directly associate a peak of the excitation with a peak of the response.

NOTES.–

1. *All the relations of the preceding sections can be applied either to the vibration input on the specimen or to the response of the specimen.*

2. *ONERA proposed, in 1961 [COU 66], a method of calculation of the PSD G(f) of a stationary and Gaussian random signal starting from the average number of zero level crossings, its derivative and the rms value of the signal. The process can be extended to non-Gaussian processes.*

5.6. Moments

Many important statistical properties of the signal considered (excitation or response) can be obtained directly from the power spectral density $G(\Omega)$ and in particular *the moments* [VAN 79].

Definition

Given a random signal $\ell(t)$, the *moment of order* n (close to the origin) is the quantity:

$$M_n = E\left\{\left[\frac{d^{n/2}\ell(t)}{dt^{n/2}}\right]^2\right\} = \lim_{T\to\infty} \frac{1}{2T} \int_{-T}^{+T} \left[\frac{d^{n/2}\ell(t)}{dt^{n/2}}\right]^2 dt \qquad [5.66]$$

(if the derivative exists). The moment of order zero is none other than the square of the rms value ℓ_{rms}:

280   Random Vibration

$$M_0 = E[\ell^2(t)] = \lim_{T \to \infty} \frac{1}{2T} \int_{-T}^{+T} \ell^2(t) \, dt = \overline{\ell^2(t)} = \ell_{rms}^2$$

$$M_0 = R(0) = \int_0^\infty G(\Omega) \, d\Omega = \int_0^\infty G(f) \, df$$

The moment of order two is equal to:

$$M_2 = E\left[\left(\frac{d\ell}{dt}\right)^2\right] = \lim_{T \to \infty} \frac{1}{2T} \int_{-T}^{+T} \left(\frac{d\ell}{dt}\right)^2 dt = \overline{\dot{\ell}^2(t)} \quad [5.67]$$

However, by definition,

$$R(\tau) = E[\ell(t) \, \ell(t+\tau)]$$

$$R(\tau) = \int_0^\infty G(\Omega) \, \cos \Omega\tau \, d\Omega \quad [5.68]$$

If we set:

$$S(\tau) = E[\dot\ell(t) \, \dot\ell(t+\tau)] \quad [5.69]$$

$$S(\tau) = \int_0^\infty \Omega^2 \, G(\Omega) \, \cos \Omega\tau \, d\Omega = -\frac{d^2 R(\tau)}{d\tau^2} \quad [5.70]$$

(if $\ddot R$ exists). We have, for $\tau = 0$,

$$S(\tau) = \int_0^\infty \Omega^2 \, G(\Omega) \, d\Omega \quad [5.71]$$

In the same way, if:

$$T(\tau) = [\ddot\ell(t) \, \ddot\ell(t+\tau)] \quad [5.72]$$

$$T(\tau) = \frac{d^4 R(\tau)}{d\tau^4} = \int_0^\infty \Omega^4 \, G(\Omega) \, \cos \Omega\tau \, d\tau \quad [5.73]$$

it results that, for $\tau = 0$,

$$T(0) = \int_0^\infty \Omega^4 \ G(\Omega) \ d\Omega \qquad [5.74]$$

yielding [KOW 69]:

$$M_2 = -\ddot{R}(0) = \int_0^\infty \Omega^2 \ G(\Omega) \ d\Omega = (2\pi)^2 \int_0^\infty f^2 \ G(f) \ df = \dot{\ell}_{rms}^2 \qquad [5.75]$$

$$M_4 = R^{(4)}(0) = \int_0^\infty \Omega^4 \ G(\Omega) \ d\Omega = (2\pi)^4 \int_0^\infty f^4 \ G(f) \ df = \ddot{\ell}_{rms}^2$$

$$[5.76]$$

More generally, the nth moment can be defined as [CHA 72], [CHA 85], [DEE 71], [PAR 64], [SHE 83], [SWA 63], [VAN 72], [VAN 75], [VAN 79]:

$$M_n = \int_0^\infty \Omega^n \ G(\Omega) \ d\Omega$$

or

$$M_n = (2\pi)^n \int_0^\infty f^n \ G(f) \ df \qquad [5.77]$$

(n integer) while

$$R^{(2n)}(0) = (-1)^n \int_0^\infty \Omega^{2n} \ G(\Omega) \ d\Omega = (-1)^n \ M_{2n} \qquad [5.78]$$

M_n are the moments of the PSD $G(\Omega)$ with respect to the vertical axis $f = 0$.

Application

We deduce from the preceding relations [CRA 68], [CHA 72], [LEY 65], [PAP 65], [SHE 83]:

$$n_0^+ = \frac{1}{2\pi} \left(\frac{M_2}{M_0} \right)^{\frac{1}{2}} \qquad [5.79]$$

$$n_a^+ = n_0^+ \, e^{-\frac{a^2}{2 M_0}} \qquad [5.80]$$

NOTE.– *Some authors [CHA 85], [FUL 61], [KOW 69], [VAN 79], [WIR 73], [WIR 83] define M_n by:*

$$M_n = \int_0^\infty f^n \, G(f) \, df \qquad [5.81]$$

which leads to [BEN 58], [CHA 85]:

$$n_0^+ = \left(\frac{M_2}{M_0}\right)^{\frac{1}{2}} \qquad [5.82]$$

(sometimes noted Ω_ℓ) [VAN 79].

5.7. Average frequency of PSD defined by straight line segments

5.7.1. *Linear-linear scales*

$$n_0^+ = \frac{1}{2\pi} \sqrt{\frac{M_2}{M_0}}$$

with

$$M_0 = \int_{f_1}^{f_2} G(f) \, df$$

where

$$G(f) = a\,f + b$$

$$a = \frac{G_2 - G_1}{f_2 - f_1} \text{ and } b = \frac{f_2\, G_1 - f_1\, G_2}{f_2 - f_1}$$

$$M_0 = \left[\frac{a}{2}\left(f_2^2 - f_1^2\right) + b\left(f_2 - f_1\right)\right] \tag{5.83}$$

$$M_2 = (2\pi)^2 \int_{f_1}^{f_2} f^2 \, G(f) \, df$$

$$M_2 = (2\pi)^2 \left[\frac{a}{4}\left(f_2^4 - f_1^4\right) + \frac{b}{3}\left(f_2^3 - f_1^3\right)\right] \tag{5.84}$$

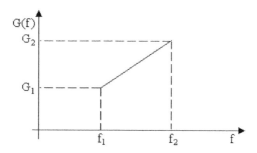

Figure 5.14. *PSD defined by a straight line segment on linear axes*

This yields, after having replaced a and b with their value according to f_1, f_2, G_1 and G_2 [BEN 62],

$$n_0^{+2} = \frac{(G_2 - G_1)\dfrac{f_2^4 - f_1^4}{4} + \dfrac{(f_2 G_1 - f_1 G_2)}{3}\left(f_2^3 - f_1^3\right)}{\dfrac{G_2 - G_1}{2}\left(f_2^2 - f_1^2\right) + (f_2 G_1 - f_1 G_2)(f_2 - f_1)} \tag{5.85}$$

Particular cases

$G_1 = G_2 = G_0 = $ constant

$$n_0^{+2} = \frac{f_2^3 - f_1^3}{3(f_2 - f_1)} = \frac{f_1^2 + f_1 f_2 + f_2^2}{3} \tag{5.86}$$

284 Random Vibration

If $f_1 = 0$ and if $G = G_0$ until f_2, it results that:

$$n_0^{+2} = \frac{f_2^2}{3} \qquad [5.87]$$

i.e.:

$$n_0^+ = 0.577 \, f_2 \qquad [5.88]$$

If the PSD is a narrowband noise centered around f_0, we can set $f_1 = f_0 - \frac{\varepsilon}{2}$ and $f_1 = f_0 + \frac{\varepsilon}{2}$ [BEN 62], yielding:

$$n_0^{+2} = f_0 + \frac{\varepsilon^2}{12} \qquad [5.89]$$

If $\varepsilon \to 0$,

$$n_0^+ \to f_0$$

5.7.2. Linear-logarithmic scales

In this case, the PSD is represented by:

$$\ln G = a \, f + b \qquad [5.90]$$

$$M_0 = \int_{f_1}^{f_2} e^{a \, f + b} \, df = \frac{1}{a} \left(e^{a \, f + b} \right)_{f_1}^{f_2} \qquad [5.91]$$

$$M_2 = (2\pi)^2 \int_{f_1}^{f_2} f^2 \, e^{a \, f + b} \, df$$

After integration by parts, it results that, if $a \neq 0$,

$$M_2 = (2\pi)^2 \frac{e^{a \, f + b}}{a^3} \left(a^2 \, f^2 - 2 \, a \, f + 2 \right) \qquad [5.92]$$

yielding

$$n_0^{+2} = \frac{e^{a f_2 + b}\left(a^2 f_2^2 - 2 a f_2 + 2\right) - e^{a f_1 + b}\left(a^2 f_1^2 - 2 a f_1 + 2\right)}{a^2 \left(e^{a f_2 + b} - e^{a f_1 + b}\right)} \qquad [5.93]$$

the constants a and b being calculated starting from the coordinates of the points f_1, G_1, f_2 and G_2.

Particular case

$$G_2 = G_1,$$

(a = 0)

$$M_2 = 2\pi \frac{G}{3}\left(f_2^3 - f_1^3\right) \qquad [5.94]$$

$$M_0 = G\left(f_2 - f_1\right) \qquad [5.95]$$

and

$$n_0^{+2} = \frac{f_2^3 - f_1^3}{3\left(f_2 - f_1\right)} = \frac{f_1^2 + f_1 f_2 + f_2^2}{3} \qquad [5.96]$$

5.7.3. *Logarithmic-linear scales*

$$G(f) = a \ln f + b \qquad [5.97]$$

$$M_0 = \int_{f_1}^{f_2} \left(a \ln f + b\right) df$$

$$M_0 = a\left(f \ln f\right)\big|_{f_1}^{f_2} + \left(f_2 - f_1\right)(b - a) \qquad [5.98]$$

$$M_2 = (2\pi)^2 \int_{f_1}^{f_2} f^2 \left(a \ln f + b\right) df$$

$$M_2 = (2\pi)^2 \left\{ \frac{f_2^3}{3} \left[a \left(\ln f_2 - \frac{1}{3} \right) + b \right] - \frac{f_1^3}{3} \left[a \left(\ln f_1 - \frac{1}{3} \right) + b \right] \right\} \qquad [5.99]$$

$$n_0^{+2} = \frac{\frac{f_2^3}{3} \left[a \left(\ln f_2 - \frac{1}{3} \right) + b \right] - \frac{f_1^3}{3} \left[a \left(\ln f_1 - \frac{1}{3} \right) + b \right]}{a \left(f_2 \ln f_2 - f_1 \ln f_1 \right) + \left(f_2 - f_1 \right) \left(b - a \right)} \qquad [5.100]$$

Particular case

$$G_1 = G_2 = G_0 = \text{constant}$$

In this case, $a = 0$ and $b = G_0$, yielding:

$$M_2 = (2\pi)^2 \frac{G_0}{3} \left(f_2^3 - f_1^3 \right)$$

$$M_0 = G_0 \left(f_2 - f_1 \right)$$

and

$$n_0^{+2} = \frac{f_2^3 - f_1^3}{3 \left(f_2 - f_1 \right)} = \frac{f_1^2 + f_1 f_2 + f_2^2}{3} \qquad [5.101]$$

5.7.4. *Logarithmic-logarithmic scales*

$$G(f) = G_1 \left(\frac{f}{f_1} \right)^b \qquad [5.102]$$

the constant b being such that $b = \dfrac{\ln G_2 / G_1}{\ln f_2 / f_1}$.

$$M_0 = \int_{f_1}^{f_2} G_1 \left(\frac{f}{f_1} \right)^b df = \frac{G_1}{f_1^b} \frac{1}{b+1} \left(f^{b+1} \right)_{f_1}^{f_2}$$

(if b ≠ −1):

$$M_2 = (2\pi)^2 \int_{f_1}^{f_2} f^2 \, G(f) \, df = (2\pi)^2 \frac{G_1}{f_1^b} \left(\frac{f^{b+3}}{b+3} \right)_{f_1}^{f_2}$$

(if b ≠ −3). It yields:

$$n_0^{+2} = \frac{b+1}{b+3} \frac{f_2^{b+3} - f_1^{b+3}}{f_2^{b+1} - f_1^{b+1}} \qquad [5.103]$$

If b = −1:

$$M_0 = G_1 \, f_1 \, \ln \frac{f_2}{f_1}$$

and

$$M_2 = (2\pi)^2 \, G_1 \, f_1 \, \frac{f_2^2 - f_1^2}{2}$$

$$n_0^{+2} = \frac{f_2^2 - f_1^2}{2 \ln f_2/f_1} \qquad [5.104]$$

If b = −3:

$$M_0 = \frac{G_1 \, f_1^3}{2} \left(\frac{1}{f_1^2} - \frac{1}{f_2^2} \right)$$

$$M_2 = (2\pi)^2 \, G_1 \, f_1^3 \, \ln \frac{f_2}{f_1}$$

$$n_0^{+2} = 2 \, f_1^2 \, f_2^2 \, \frac{2 \ln f_2/f_1}{f_2^2 - f_1^2} \qquad [5.105]$$

NOTE.— *If the PSD is made up of n straight line segments, the average frequency n_0^+ is obtained from:*

$$n_0^{+2} = \frac{1}{(2\pi)^2} \frac{\sum_{i=1}^{n} M_{2i}}{\sum_{i=1}^{n} M_{0i}} \qquad [5.106]$$

5.8. Fourth moment of PSD defined by straight line segments

The interest of this parameter lies in its participation, with M_0 and M_2 already studied, in the calculation of n_p^+ and r.

5.8.1. *Linear-linear scales*

By definition,

$$M_4 = (2\pi)^4 \int_0^\infty f^4 \, G(f) \, df$$

$$G(f) = a\,f + b$$

yielding:

$$M_4 = (2\pi)^4 \left[\frac{a}{6}\left(f_2^6 - f_1^6\right) + \frac{b}{5}\left(f_2^5 - f_1^5\right) \right] \qquad [5.107]$$

where $a = \dfrac{G_2 - G_1}{f_2 - f_1}$ and $b = \dfrac{f_2 G_1 - f_1 G_2}{f_2 - f_1}$.

Particular cases

1. $G_1 = G_2 = G_0$ =constant, i.e. $a = 0$ and $b = G_0$:

$$M_4 = (2\pi)^4 \left[\frac{G_0}{5}\left(f_2^5 - f_1^5\right) \right] \qquad [5.108]$$

2. $f_1 = 0$

$$M_4 = (2\pi)^4 \left[\frac{a}{6} f_2^6 + \frac{b}{5} f_2^5 \right] \qquad [5.109]$$

3. $f_1 = 0$ and $G_0 =$ constant

$$M_4 = (2\pi)^4 \frac{G_0}{5} f_2^5 \qquad [5.110]$$

5.8.2. Linear-logarithmic scales

$$G(f) = e^{af+b}$$

$$a = \frac{\ln G_2/G_1}{f_2 - f_1}$$

$$b = \frac{f_2 \ln G_1 - f_1 \ln G_2}{f_2 - f_1}$$

$$M_4 = (2\pi)^4 \int_{f_1}^{f_2} f^4 \, G(f) \, df = (2\pi)^4 \int_{f_1}^{f_2} f^4 \, e^{a f + b} \, df \qquad [5.111]$$

After several integrations by parts, we obtain, if $a \neq 0$,

$$M_4 = \frac{(2\pi)^4}{a} \left\{ e^{a f_2 + b} \left(f_2^4 - \frac{4 f_2^3}{a} + \frac{12 f_2^2}{a^2} - \frac{24 f_2}{a^3} + \frac{24}{a^4} \right) \right.$$

$$\left. - e^{a f_1 + b} \left(f_1^4 - \frac{4 f_1^3}{a} + \frac{12 f_1^2}{a^2} - \frac{24 f_1}{a^3} + \frac{24}{a^4} \right) \right\} \qquad [5.112]$$

Particular cases

1. $G_1 = G_2 = G_0 =$ constant. Then, $a = 0$ and $b = \ln G_0$

$$M_4 = (2\pi)^4 G_0 \frac{f_2^5 - f_1^5}{5} \qquad [5.113]$$

290 Random Vibration

2. $f_1 = 0$ and $a \neq 0$

$$M_4 = \frac{(2\pi)^4}{a}\left[e^{a f_2 + b}\left(f_2^4 - \frac{4}{a}f_2^3 + \frac{12}{a^2}f_2^3 - \frac{24}{a^3}f_2 + \frac{24}{a^4}\right) - \frac{24}{a^4}e^b\right] \quad [5.114]$$

and, if $G_1 = G_2 = G_0$

$$M_4 = (2\pi)^4 G_0 \frac{f_2^5}{5} \quad [5.115]$$

5.8.3. *Logarithmic-linear scales*

$G = a \ln f + b$

$$M_4 = (2\pi)^4 \int_{f_1}^{f_2} f^4 (a \ln f + b) df$$

$$M_4 = (2\pi)^4 \left\{ \frac{f_2^5}{5}\left[a\left(\ln f_2 - \frac{1}{5}\right) + b\right] - \frac{f_1^5}{5}\left[a\left(\ln f_1 - \frac{1}{5}\right) + b\right] \right\} \quad [5.116]$$

where

$$a = \frac{G_2 - G_1}{\ln f_2/f_1} \text{ and } b = \frac{G_2 \ln f_1 - G_1 \ln f_2}{\ln f_1 - \ln f_2}$$

Particular cases

$G_1 = G_2 = G_0 =$ constant, i.e. $a = 0$ and $b = G_0$:

$$M_4 = (2\pi)^4 \frac{G_0}{5}\left(f_2^5 - f_1^5\right) \quad [5.117]$$

If $f_1 = 0$:

$$M_4 = (2\pi)^4 \frac{G_0}{5} f_2^5$$

5.8.4. *Logarithmic-logarithmic scales*

$$G = G_1 \left(\frac{f}{f_1}\right)^b$$

yielding, if $b \neq -5$,

$$M_4 = (2\pi)^4 \frac{G_1}{f_1^b} \frac{f_2^{b+5} - f_1^{b+5}}{b+5}$$

or

$$M_4 = \frac{(2\pi)^4}{b+5} \left(G_2 f_2^5 - G_1 f_1^5\right) \qquad [5.118]$$

If $b = -5$:

$$M_4 = (2\pi)^4 \int_{f_1}^{f_2} G_1 f^4 f_1^5 \frac{df}{f^5}$$

$$M_4 = (2\pi)^4 f_1^5 G_1 \ln \frac{f_2}{f_1} \qquad [5.119]$$

Particular case

If $G_1 = G_2 = G_0 =$ constant and if $b \neq -5$

$$M_4 = \frac{(2\pi)^4}{b+5} G_0 \left(f_2^5 - f_1^5\right) \qquad [5.120]$$

NOTE.– *If the PSD is made up of n horizontal segments, the value of M_4 is obtained by calculating the sum:*

$$M_4 = \sum_{i=1}^{n} M_{4_i} \qquad [5.121]$$

5.9. Generalization: moment of order n

In a more general way, the moment M_n is given, depending on the case, by the following relations.

5.9.1. Linear-linear scales

The order n being positive or zero,

$$M_n = (2\pi)^n \left[\frac{a}{n+2} \left(f_2^{n+2} - f_1^{n+2} \right) + \frac{b}{n+1} \left(f_2^{n+1} - f_1^{n+1} \right) \right] \quad [5.122]$$

5.9.2. Linear-logarithmic scales

If $a \neq 0$

$$M_n = \frac{(2\pi)^n}{a} \left\{ e^{a f_2 + b} \left[f_2^n - \frac{n}{a} f_2^{n-1} + \frac{n(n-1)}{a^2} f_2^{n-2} - \cdots + \frac{n!}{a^n} \right] \right.$$

$$\left. - e^{a f_1 + b} \left[f_1^n - \frac{n}{a} f_1^{n-1} + \frac{n(n-1)}{a^2} f_1^{n-2} - \cdots + \frac{n!}{a^n} \right] \right\} \quad [5.123]$$

If $a = 0$

$$M_n = (2\pi)^n e^b \frac{f_2^{n+1} - f_1^{n+1}}{n+1}$$

5.9.3. Logarithmic-linear scales

$$M_n = (2\pi)^n \left\{ \frac{f_2^{n+1}}{n+1} \left[a \left(\ln f_2 - \frac{1}{n+1} \right) + b \right] - \frac{f_1^{n+1}}{n+1} \left[a \left(\ln f_1 - \frac{1}{n+1} \right) + b \right] \right\} \quad [5.124]$$

$(n \geq 0)$

5.9.4. *Logarithmic-logarithmic scales*

If $b \neq -(n+1)$:

$$M_n = (2\pi)^n \frac{G_2 f_2^{n+1} - G_1 f_1^{n+1}}{b+n+1} = (2\pi)^n \frac{G_1}{f_1^b} \frac{f_2^{b+n+1} - G_1 f_1^{b+n+1}}{b+n+1} \quad [5.125]$$

If $b = -(n+1)$:

$$M_n = (2\pi)^n f_1^{n+1} G_1 \ln \frac{f_2}{f_1} \quad [5.126]$$

Chapter 6

Probability Distribution of Maxima of Random Vibration

6.1. Probability density of maxima

It can be useful, in particular for calculations of damage by fatigue, to know a vibration's average number of peaks per unit time, occurring between two close levels a and a + da, as well as the average total number of peaks per unit time.

NOTE.– *Here we are interested in the maxima of the curve which can be positive or negative (Figure 6.1).*

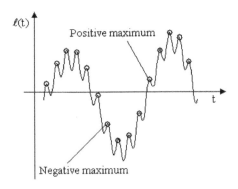

Figure 6.1. *Positive and negative peaks of a random signal*

For a fatigue analysis, it would of course be necessary to also count the minima. We can acknowledge that the average number of minima per unit time of a Gaussian random signal is equal to the average number the maxima per unit time, the distributions of the minima and maxima being symmetric [CAR 68].

A maximum occurs when the velocity (derivative of the signal) cancels out with negative acceleration (second derivative of signal).

This remark leads us to think that the joint probability density between the processes $\ell(t)$, $\dot{\ell}(t)$ and $\ddot{\ell}(t)$ can be used to describe the maxima of $\ell(t)$. This assumes that $\ell(t)$ is derivable twice.

S.O. Rice [RIC 39], [RIC 44] showed that, if $p(a, b, c)$ is the probability density so that $\ell(t)$, $\dot{\ell}(t)$ and $\ddot{\ell}(t)$ respectively lie between a and $a + da$, b and $b + db$, c and $c + dc$, a maximum being defined by a zero derivative and a negative curvature, the average number the maxima located between levels a and $a + da$ in the time interval t, $t + dt$ (window a, $a + da$, t, $t + dt$) is:

$$v_a = -dt \, da \int_{-\infty}^{0} c \, p(a, 0, c) \, dc \qquad [6.1]$$

where, for a Gaussian signal as well as for its first and second derivatives [CRA 67], [KOW 63]:

$$p(a, 0, c) = (2\pi)^{-3/2} |M|^{-1/2} \exp\left[-\frac{\mu_{11} a^2 + \mu_{33} c^2 + 2 \mu_{13} a c}{2 |M|}\right] \qquad [6.2]$$

with

$$\|M\| = \begin{Vmatrix} \ell_{rms}^2 & 0 & -\dot{\ell}_{rms}^2 \\ 0 & \dot{\ell}_{rms}^2 & 0 \\ -\dot{\ell}_{rms}^2 & 0 & \ddot{\ell}_{rms}^2 \end{Vmatrix} \qquad [6.3]$$

Let us recall that:

$$\ell_{rms}^2 = R(0) = M_0$$

$$\dot{\ell}_{rms}^2 = -\ddot{R}(0) = M_2$$

$$\ddot{\ell}_{rms}^2 = R^{(4)}(0) = M_4$$

The determinant $|M|$ is written:

$$|M| = \ell_{rms}^2 \left(\ell_{rms}^2 \ddot{\ell}_{rms}^2 - \dot{\ell}_{rms}^4 \right) = \ell_{rms}^2 \dot{\ell}_{rms}^2 \ddot{\ell}_{rms}^2 \left(1 - r^2\right) \qquad [6.4]$$

$$|M| = M_0 M_2 M_4 \left(1 - r^2\right) \qquad [6.5]$$

if

$$r = \frac{\dot{\ell}_{rms}^2}{\ell_{rms} \ddot{\ell}_{rms}} = \frac{M_2}{\sqrt{M_0 M_4}} = \frac{R^{(2)}(0)}{\sqrt{R(0) R^{(4)}(0)}} \qquad [6.6]$$

r is an important parameter called the *irregularity factor*. $|M|$ is always positive. The cofactors μ_{ij} are respectively equal to:

$$\mu_{11} = \ell_{rms}^2 \ddot{\ell}_{rms}^2 = M_2 M_4 \qquad [6.7]$$

$$\mu_{13} = \dot{\ell}_{rms}^4 = M_2^2 \qquad [6.8]$$

$$\mu_{33} = \ell_{rms}^2 \dot{\ell}_{rms}^2 = M_0 M_2 \qquad [6.9]$$

yielding

$$v_a = -da\, dt \int_{-\infty}^{0} \frac{c\,(2\pi)^{-3/2}}{\sqrt{M_0 M_2 M_4 \left(1 - r^2\right)}}$$

$$\exp\left[-\frac{M_2 M_4 a^2 + M_0 M_2 c^2 + 2 M_2^2 a c}{2 M_0 M_2 M_4 \left(1 - r^2\right)} \right] dc \qquad [6.10]$$

$$v_a = -\frac{da\, dt\, (2\pi)^{-3/2}}{\sqrt{M_0 M_2 M_4 \left(1 - r^2\right)}} e^{-\frac{a^2}{2 M_0 \left(1 - r^2\right)}}$$

$$\int_{-\infty}^{0} c \, \exp\left[-\frac{1}{2\,M_4\left(1-r^2\right)}\left(c^2 + \frac{2\,M_2\,a\,c}{M_0}\right)\right] dc$$

$$v_a = -\frac{da\,dt\,(2\pi)^{-3/2}}{\sqrt{M_0\,M_2\,M_4\left(1-r^2\right)}}\, e^{-\frac{a^2}{2\,M_0\left(1-r^2\right)}}$$

$$\int_{-\infty}^{0} c \, \exp\left\{-\frac{1}{2\,M_4\left(1-r^2\right)}\left[\left(c+\frac{M_2}{M_0}a\right)^2 - \left(\frac{M_2}{M_0}a\right)^2\right]\right\} dc$$

$$v_a = -\frac{da\,dt\,(2\pi)^{-3/2}}{\sqrt{M_0\,M_2\,M_4\left(1-r^2\right)}}\, e^{-\frac{a^2}{2\,M_0\left(1-r^2\right)}}\, e^{\frac{a^2\,r^2}{2\,M_0\left(1-r^2\right)}}$$

$$\left\{\int_{-\infty}^{0}\left(c+\frac{M_2}{M_0}a\right)\exp\left[-\frac{\left(c+\frac{M_2}{M_0}a\right)^2}{2\,M_4\left(1-r^2\right)}\right] dc - \frac{M_2\,a}{M_0}\int_{-\infty}^{0}\exp\left[-\frac{\left(c+\frac{M_2}{M_0}a\right)^2}{2\,M_4\left(1-r^2\right)}\right] dc\right\}$$

Let us set $v = \dfrac{\left(c+\dfrac{M_2}{M_0}a\right)^2}{2\,M_4\left(1-r^2\right)}$ and $w = \sqrt{v}$. It results that:

$$v_a = -\frac{da\,dt\,(2\pi)^{-3/2}}{\sqrt{M_0\,M_2\,M_4\left(1-r^2\right)}}\, e^{-\frac{a^2}{2\,M_0}}\left\{M_4\left(1-r^2\right)\int_{-\infty}^{a^2\,r^2/2\,M_0\,(1-r^2)} e^{-v}\,dv\right.$$

$$-\frac{M_2 \, a}{M_0} \int_{-\infty}^{M_2 \, a/M_0 \sqrt{2 M_4 (1-r^2)}} \sqrt{2 M_4 (1-r^2)} \, e^{-w^2} \, dw \Bigg\}$$

After integration [BEN 58], [RIC 64],

$$v_a = \frac{(2\pi)^{-3/2} \, da \, dt}{M_0 \, M_2} \Bigg\{ \sqrt{M_0 \, M_2 \, M_4 \, (1-r^2)} \, e^{-\frac{a^2}{2 M_0 (1-r^2)}}$$

$$+ \sqrt{\frac{\pi}{2}} \, r \, \frac{M_2^{3/2}}{\sqrt{M_0}} \, e^{-\frac{a^2}{2 M_0}} \left[1 + \text{Erf}\left(\frac{a \, r}{\sqrt{2 M_0 (1-r^2)}} \right) \right] \Bigg\} \quad [6.11]$$

i.e.

$$v_a = n_p^+ \, q(a) \, da \, dt \quad [6.12]$$

where

$$n_p^+ = \frac{1}{2\pi} \sqrt{\frac{M_4}{M_2}} \quad [6.13]$$

(average number of maxima per second). n_p^+ can be also written:

$$n_p^+ = \frac{1}{2\pi} \sqrt{-\frac{R^{(4)}(0)}{R^{(2)}(0)}} \quad [6.14]$$

NOTE.– v_a can be written in the form [RIC 64]:

$$v_a = \frac{-R^{(2)}(0)}{2 \, R(0) \sqrt{R^{(4)}(0) \, R(0)}} \, \exp\left[-\frac{a^2}{2 \, R(0)}\right] \Bigg\{ a \left[1 + \text{erf}\left(-\frac{a \, R^{(2)}(0)}{\sqrt{2} \, k \, R(0)} \right) \right] \Bigg\}$$

$$-\frac{\sqrt{2 k R(0)}}{\sqrt{\pi} R^{(2)}(0)} \exp\left[-\frac{\left(R^{(2)}(0) a\right)^2}{2 k R(0)}\right]\Biggr\}$$ [6.15]

where

$$k = R(0) R^{(4)}(0) - \left[R^{(2)}(0)\right]^2$$ [6.16]

The probability density of maxima per unit time of a Gaussian signal whose amplitude lies between a and a + da is thus [BRO 63], [CAR 56], [LEL 73], [LIN 72]:

$$q(a) = \frac{\sqrt{1-r^2}}{\ell_{rms}\sqrt{2\pi}} e^{-\frac{a^2}{2\ell_{rms}^2(1-r^2)}} + \frac{r a}{2\ell_{rms}^2}\left[1 + \mathrm{erf}\left(\frac{a r}{\ell_{rms}\sqrt{2(1-r^2)}}\right)\right]$$ [6.17]

where $\mathrm{erf}(x) = \frac{2}{\sqrt{\pi}} \int_0^x e^{-\lambda^2} d\lambda$ (Appendix A4.1). The probability so that a maximum taken randomly is, per unit time, contained in the interval a, a + da is $q(a) da$. If we set $u = \frac{a}{\ell_{rms}}$, it becomes:

$$\frac{v_a}{dt} = q(a) da = q(u) du = q\left(\frac{a}{\ell_{rms}}\right)\frac{da}{\ell_{rms}}$$ [6.18]

yielding [BER 77], [CHA 85], [COU 70], [KOW 63], [LEL 73], [LIN 67], [RAV 70], [SCH 63]:

$$q(u) = \frac{\sqrt{1-r^2}}{\sqrt{2\pi}} e^{-\frac{u^2}{2(1-r^2)}} + \frac{r}{2} u e^{-\frac{u^2}{2}}\left[1 + \mathrm{erf}\left(\frac{r u}{\sqrt{2(1-r^2)}}\right)\right]$$ [6.19]

The statistical distribution of the minima follows the same law. The probability density $q(u)$ is thus the weighted sum of a Gaussian law and Rayleigh's law, with coefficient functions of parameter r. This expression can be written in various more or less practical forms according to its application. Since:

$$\int_{-\infty}^{\infty} e^{-\lambda^2} d\lambda = \sqrt{\pi} = 2\int_{0}^{x} e^{-\lambda^2} d\lambda + 2\int_{x}^{\infty} e^{-\lambda^2} d\lambda$$

where

$$\mathrm{erf}(x) = 1 - \frac{2}{\sqrt{\pi}} \int_{x}^{\infty} e^{-\lambda^2} d\lambda \qquad [6.20]$$

it results that:

$$q(u) = \frac{\sqrt{1-r^2}}{\sqrt{2\pi}} e^{-\frac{u^2}{2(1-r^2)}} + r\,u\,e^{-\frac{u^2}{2}} \left[1 - \frac{1}{\sqrt{\pi}} \int_{\frac{r u}{\sqrt{2(1-r^2)}}}^{\infty} e^{-\lambda^2} d\lambda \right] \qquad [6.21]$$

Setting $\lambda = \frac{t}{\sqrt{2}}$ in this relation, we obtain [BEN 61b], [BEN 64], [HIL 70], [HUS 56], [PER 74]:

$$q(u) = \frac{\sqrt{1-r^2}}{\sqrt{2\pi}} e^{-\frac{u^2}{2(1-r^2)}} + r\,u\,e^{-\frac{u^2}{2}} \left[1 - \frac{1}{\sqrt{2\pi}} \int_{\frac{r u}{\sqrt{1-r^2}}}^{\infty} e^{-\frac{t^2}{2}} dt \right] \qquad [6.22]$$

We also find the equivalent expression [BAR 78], [CAR 56], [CLO 03], [CRA 68], [DAV 64], [KAC 76], [KOW 69], [KRE 83], [UDW 73]:

$$q(u) = \frac{\sqrt{1-r^2}}{\sqrt{2\pi}} e^{-\frac{u^2}{2(1-r^2)}} + r\,u\,e^{-\frac{u^2}{2}} \Phi(v) \qquad [6.23]$$

where

$$\Phi(v) = \frac{1}{\sqrt{2\pi}} \int_{-\infty}^{v} e^{-\frac{t^2}{2}} dt$$

and

$$v = \frac{r u}{\sqrt{1-r^2}}$$

$$q(u) = \sqrt{1-r^2}\, e^{-\frac{u^2}{2}} \left[\frac{e^{-\frac{r^2 u^2}{2(1-r^2)}}}{\sqrt{2\pi}} + \frac{r u}{\sqrt{1-r^2}} \Phi(v) \right]$$

$$q(u) = \sqrt{1-r^2}\, e^{-\frac{u^2}{2}} \left[\frac{e^{-\frac{v^2}{2}}}{\sqrt{2\pi}} + v\, \Phi(v) \right] \tag{6.24}$$

or

$$q(u) = \sqrt{1-r^2}\, e^{-\frac{u^2}{2}} \left[\frac{d\Phi(v)}{dv} + v\, \Phi(v) \right]$$

$$q(u) = \sqrt{1-r^2}\, e^{-\frac{u^2}{2}} \frac{d[v\, \Phi(v)]}{dv} \tag{6.25}$$

Particular cases

1. Let us suppose that the parameter r is equal to 1; $q(u)$ then becomes, starting from [6.19], knowing that $\mathrm{erf}(\infty) = 1$,

$$q(u) = u\, e^{-\frac{u^2}{2}} \tag{6.26}$$

which is the probability density of Rayleigh's law of standard deviation equal to 1. Since $u = \dfrac{a}{\ell_{rms}}$ and:

$$q(a) \, da = q(u) \, du = q\left(\frac{a}{\ell_{eff}}\right) \frac{da}{\ell_{eff}} \qquad [6.27]$$

it results that

$$q(a) = \frac{q(u)}{\ell_{rms}} = \frac{a}{\ell_{rms}^2} e^{-\frac{a^2}{2\ell_{rms}^2}} \qquad [6.28]$$

2. If $r = 0$,

$$q(u) = \frac{1}{\sqrt{2\pi}} e^{-\frac{u^2}{2}} \qquad [6.29]$$

(probability density of a normal, i.e. Gaussian law). In this (theoretical) case there are an infinite number of local maxima between two zero crossings with positive slope.

We will reconsider these particular cases

6.2. Moments of the maxima probability distribution

By definition, the nth central moment of the maxima distribution $q(u)$ is:

$$\mu'_n = \int_{-\infty}^{\infty} u^n \, q(u) \, du \qquad [6.30]$$

Even moments ($n = 2p$) [CAR 56]:

$$\mu'_{2p} = 2^p \, p! \left[1 - \frac{(1-r^2)}{2} - \frac{1}{2^2} \frac{1}{2!} (1-r^2)^2 - \ldots - \frac{1.1.3\ldots(2p-3)}{2^p \, p!} \right] \qquad [6.31]$$

Odd moments ($n = 2p+1$):

$$\mu'_{2p+1} = \sqrt{\frac{\pi}{2}} \, r \, \frac{1.3.5\ldots(2p+1)}{(p!)^2} \qquad [6.32]$$

Moments about the origin	Moments about the mean
$\mu'_0 = 1$	$\mu_0 = 1$
$\mu'_1 = 1\sqrt{\dfrac{\pi}{2}}\, r$	$\mu_1 = 0$
$\mu'_2 = 1 + r^2$	$\mu_2 = 1 - \left(\dfrac{\pi}{2} - 1\right) r^2$
$\mu'_3 = 3\sqrt{\dfrac{\pi}{2}}\, r$	$\mu_3 = \sqrt{\dfrac{\pi}{2}}\,(\pi - 3)\, r^3$

Table 6.1. *First moments of the maxima probability distribution*

6.3. Expected number of maxima per unit time

It was seen that the average number of maxima per second (frequency of maxima) can be written [6.13]:

$$n_p^+ = \frac{1}{2\pi}\sqrt{\frac{M_4}{M_2}}$$

Taking into account the preceding definitions, the expected maxima frequency is also equal to [CRA 67], [HUS 56], [LIN 67], [PAP 65], [PRE 56a], [RIC 64], [SJÖ 61]:

$$n_p^+ = \frac{1}{2\pi}\sqrt{-\frac{R^{(4)}(0)}{R^{(2)}(0)}} = \frac{1}{2\pi}\frac{\ddot{\ell}_{rms}}{\dot{\ell}_{rms}} \qquad [6.33]$$

$$n_p^+ = \frac{1}{2\pi}\left[\frac{\int_{-\infty}^{+\infty}\Omega^4\, G(\Omega)\, d\Omega}{\int_{-\infty}^{+\infty}\Omega^2\, G(\Omega)\, d\Omega}\right]^{\frac{1}{2}} = \left[\frac{\int_0^{+\infty} f^4\, G(f)\, df}{\int_0^{+\infty} f^2\, G(f)\, df}\right]^{\frac{1}{2}} \qquad [6.34]$$

In the case of a narrowband noise such as that in Figure 5.3, we have:

$$n_p^+ = \frac{1}{2\pi}\frac{\ddot{\ell}_{rms}}{\dot{\ell}_{rms}} = \frac{1}{2\pi}\sqrt{\frac{G_0\, \omega_0^4\, \Delta\Omega}{G_0\, \omega_0^2\, \Delta\Omega}} \qquad [6.35]$$

i.e.

$$n_p^+ = \frac{\omega_0}{2\pi} \qquad [6.36]$$

n_p^+ is thus approximately equal to n_0^+: there is approximately 1 peak per zero crossing; the signal resembles a sinusoid with modulated amplitude.

NOTE.– *Using the definition of expression [5.81], n_p^+ would be written [BEN 58] [CHA 85]:*

$$n_p^+ = \sqrt{\frac{M_4}{M_2}}$$

Starting from the number of maxima v_a lying between a and a + da in the time interval t, t + dt, we can calculate, by integration between t_1 and t_2 for time, and between $-\infty$ and $+\infty$ for the levels, the average total number of maxima between t_1 and t_2:

$$v_a = \frac{1}{2\pi}\sqrt{\frac{M_4}{M_2}}\, q(a)\, da\, dt \qquad [6.37]$$

Per second,

$$n_p^+ = \frac{1}{2\pi}\sqrt{\frac{M_4}{M_2}} \int_{-\infty}^{+\infty} q(a)\, da \quad \left(= \frac{N_p^+}{dt}\right)$$

$$n_p^+ = \frac{1}{2\pi}\sqrt{\frac{M_4}{M_2}}$$

and, between t_1 and t_2,

$$N_p^+ = \frac{1}{2\pi}\sqrt{\frac{M_4}{M_2}} \int_{t_1}^{t_2} dt$$

$$N_p^+ = \frac{1}{2\pi}\sqrt{\frac{M_4}{M_2}}\,(t_2 - t_1) = n_p^+\,(t_2 - t_1) \qquad [6.38]$$

306 Random Vibration

Application to the case of a noise with constant PSD between two frequencies

Let us consider a vibratory signal $\ell(t)$ whose PSD is constant and equal to G_0 between two frequencies f_1 and f_2 (and zero elsewhere) [COU 70]. We have:

$$M_4 = (2\pi)^4 \frac{G_0}{5} \left(f_2^5 - f_1^5\right)$$

$$M_2 = (2\pi)^2 \frac{G_0}{3} \left(f_2^3 - f_1^3\right)$$

This yields

$$n_p^+ = \left[\frac{3}{5} \frac{f_2^5 - f_1^5}{f_2^3 - f_1^3}\right]^{\frac{1}{2}} \qquad [6.39]$$

If $f_1 \to 0$,

$$n_p^+ \to \sqrt{\frac{3}{5}} \, f_2 = 0.775 \, f_2 \qquad [6.40]$$

If $f_1 = f_0 - \dfrac{\Delta f}{2}$ and $f_2 = f_0 + \dfrac{\Delta f}{2}$ (narrowband noise Δf small).

$$n_p^{+2} = \frac{3}{5} \, \frac{5 f_0^4 + 10 f_0^2 \left(\dfrac{\Delta f}{2}\right)^2 + \left(\dfrac{\Delta f}{2}\right)^4}{3 f_0^2 + \left(\dfrac{\Delta f}{2}\right)^2}$$

$$n_p^{+2} = f_0^2 \, \frac{1 + 2 \left(\dfrac{\Delta f}{2 f_0}\right)^2 + \dfrac{1}{5}\left(\dfrac{\Delta f}{2 f_0}\right)^4}{1 + \dfrac{1}{3}\left(\dfrac{\Delta f}{2 f_0}\right)^2} \qquad [6.41]$$

If $\Delta f \to 0$,

$$n_p^+ \to f_0$$

Figure 6.2 shows the variations of $\dfrac{n_p^+}{f_0}$ versus $\dfrac{\Delta f}{f_0}$.

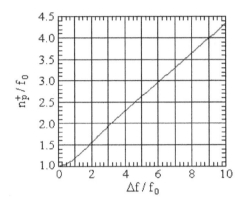

Figure 6.2. *Average number of maxima per second of a narrowband noise versus its width*

6.4. Average time interval between two successive maxima

This average time is calculated directly starting from n_p^+ [COU 70]:

$$\tau_m = \dfrac{1}{n_p^+} \quad [6.42]$$

In the case of a narrowband noise, centered on frequency f_0:

$$\tau_m = \dfrac{1}{f_0} \left[\dfrac{1 + \dfrac{1}{3}\left(\dfrac{\Delta f}{2 f_0}\right)^2}{1 + 2\left(\dfrac{\Delta f}{2 f_0}\right)^2 + \dfrac{1}{5}\left(\dfrac{\Delta f}{2 f_0}\right)^4} \right]^{\frac{1}{2}} \quad [6.43]$$

$$\tau_m \to \frac{1}{f_0} \quad \text{when } \Delta f \to 0.$$

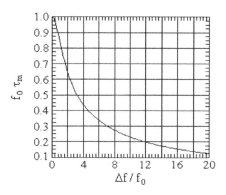

Figure 6.3. *Average time interval between two successive maxima of a narrowband noise versus its width*

6.5. Average correlation between two successive maxima

This correlation coefficient $[\rho(\tau_m)]$ is obtained by replacing τ with τ_m in equation [2.72] previously established [COU 70]. If we set:

$$\delta = \frac{\Delta f}{2 f_0}$$

it becomes:

$$\rho = \frac{1}{2\pi\delta} \left[\frac{1 + 2\delta^2 + \frac{\delta^4}{5}}{1 + \frac{\delta^2}{3}} \right]^{1/2} \cos\left(\frac{1 + \frac{\delta^2}{3}}{1 + 2\delta^2 + \frac{\delta^4}{5}} \right)^{1/2} \sin\left(\frac{1 + \frac{\delta^2}{3}}{1 + 2\delta^2 + \frac{\delta^4}{5}} \right)^{1/2} \quad [6.44]$$

Figure 6.4 shows the variations of $|\rho|$ versus δ.

The correlation coefficient does not exceed 0.2 when δ is greater than 0.4.

We can thus consider the amplitudes of two successive maxima of a wideband process as independent random variables [COU 70].

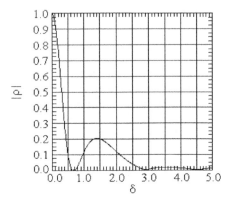

Figure 6.4. *Average correlation between two successive maxima of a narrowband noise versus its bandwidth*

6.6. Properties of the irregularity factor

6.6.1. *Variation interval*

The irregularity factor:

$$r = \frac{M_2}{\sqrt{M_0 M_4}} = \frac{\dot{\ell}_{rms}^2}{\ell_{rms}\,\ddot{\ell}_{rms}} = \frac{-R^{(2)}(0)}{\sqrt{R(0)\,R^{(4)}(0)}}$$

can vary in the interval [0, 1]. Indeed, we obtain [PRE 56b]:

$$r = \frac{M_2}{\sqrt{M_0 M_4}} = \frac{\int_0^\infty \Omega^2\, G(\Omega)\, d\Omega}{\sqrt{\int_0^\infty G(\Omega)\, d\Omega \int_0^\infty \Omega^4\, G(\Omega)\, d\Omega}} \qquad [6.45]$$

According to Cauchy-Schwarz's inequality,

$$\left[\int_0^\infty u(x)\,v(x)\,dx\right]^2 \le \int_0^\infty u^2(x)\,dx \int_0^\infty v^2(x)\,dx\;)$$

we obtain

$$\int_0^\infty \Omega^2\, G(\Omega)\, d\Omega \le \sqrt{\int_0^\infty G(\Omega)\, d\Omega}\, \sqrt{\int_0^\infty \Omega^4\, G(\Omega)\, d\Omega}$$

310 Random Vibration

i.e.

$$M_2 \leq \sqrt{M_0 M_4} \qquad [6.46]$$

Since $M_2 \geq 0$, it results that:

$$0 \leq \frac{M_2}{\sqrt{M_0 M_4}} \leq 1 \qquad [6.47]$$

Another definition

The irregularity factor r can also be defined like the ratio of the average number of zero crossings per unit time with positive slope to the average number of positive and negative maxima (or minima) per unit time. Indeed,

$$r = \frac{M_2}{\sqrt{M_0 M_4}} = \frac{1}{2\pi}\sqrt{\frac{M_2}{M_0}} 2\pi \sqrt{\frac{M_2}{M_4}} = \frac{n_0^+}{n_p^+} = \frac{n_0}{2 n_p^+} \qquad [6.48]$$

Example 6.1.

Let us consider the sample of acceleration signal as a function of time represented in Figure 6.5 (with not many peaks to facilitate calculations).

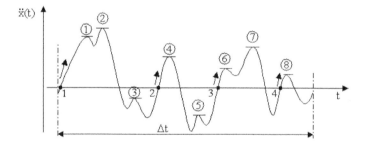

Figure 6.5. *Example of peaks of a random signal*

The number of maxima in the considered time interval Δt is equal to 8, the number of zero-crossing with positive slope to 4 yielding:

$$r = \frac{4}{8} = 0.5$$

The parameter r is a measure of the width of the noise:

– for a broadband process, the number of maxima is much higher than the number of zeros. This case corresponds to the limiting case where $r = 0$. The maxima occur above or below the zero line with an equal probability [CAR 68]. We saw that the probability density of the peaks then tends towards that of a Gaussian law [6.29]:

$$q(u) = \frac{1}{\sqrt{2\pi}} e^{-\frac{u^2}{2}}$$

– when the number of passages through zero is equal to the number of peaks, r is equal to 1 and the signal appears as a sinusoidal wave, of about constant frequency and slowly modulated amplitude passing successively through a zero, one peak (positive or negative), a zero, etc. We are dealing with what is called a *narrowband signal*, obtained in response to a narrow rectangular filter or in response of a one-degree-of-freedom system of a rather high Q factor (higher than 10 for example).

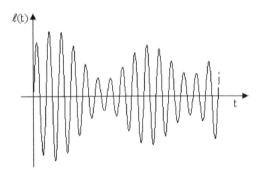

Figure 6.6. *Narrowband signal*

All the maxima are positive and the minima negative. For this value of r, $q(u)$ tends towards Rayleigh's law [6.26]:

$$q(u) = u\, e^{-\frac{u^2}{2}}$$

The value of the parameter r depends on the PSD of the noise via n_0 and n_p (or the moments M_0, M_2 and M_4). Figure 6.7 shows the variations of $q(u)$ for r varying from 0 to 1 per step $\Delta r = 0.15$.

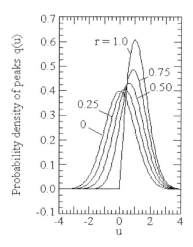

Figure 6.7. *Probability density function of peaks for various values of r*

Example 6.2.

The probability that $u_0 \leq u \leq u_0 + \Delta u$ is defined by:

$$\frac{n_u}{n_R} = \int_{u_0}^{u_0 + \Delta u} q(u)\, du$$

where n_R is the total number of occurrences.

For example, the probability that a peak exceeds the rms value is approximately 60.65%. The probability of exceeding 3 times the rms value is only approximately 1.11% [CLY 64].

NOTES.–

1. *Some authors prefer to use the parameter $k = 1/r$ [SCH 63] instead of r. More commonly, others prefer the quantity [CAR 56], [KRE 83]:*

$$q = \sqrt{1 - r^2} \qquad [6.49]$$

(sometimes noted ε) called the bandwidth parameter[NAG 06] or the effective bandwidth, whose properties are similar:

– since r varies between 0 and 1, q lies between 0 and 1,

– q is close to 0 for a narrowband process and close to 1 for a wideband process,

– q = 0 for a pure sinusoid with random phase [UDW 73].

We should not confuse this parameter q with the quantity

$$q = \frac{\text{rms value of the slope of the envelope of the process}}{\text{rms value of the slope of the process}}, \text{ often noted using the}$$

same letter; this spectral parameter also varies between 0 and 1 (according to the Schwartz inequality) and is a function of the form of the PSD [VAN 70], [VAN 72], [VAN 75], [VAN 79]. It is shown that it is equal to the ratio of the rms value of the envelope of the signal to the rms value of the slope of the signal. To avoid any confusion, it will hereafter be noted q_E (Chapter 10).

2. The parameter r depends on the form of the PSD and there is only one probability density of maxima for a given r. However, the PSD of different forms can have the same r.

3. A measuring instrument for the parameter r ("R meter") has been developed by the Brüel and Kjaer Company [CAR 68].

6.6.2. Calculation of irregularity factor for band-limited white noise

The following definition can be used:

$$r^2 = \frac{M_2^2}{M_0 M_4}$$

$$M_0 = G(f_2 - f_1) \quad [6.50]$$

$$M_2 = (2\pi)^2 \int_{f_1}^{f_2} G f^2 \, df = (2\pi)^2 G \frac{f_2^3 - f_1^3}{3} \quad [6.51]$$

$$M_4 = (2\pi)^4 \int_{f_1}^{f_2} G f^4 \, df = (2\pi)^4 G \frac{f_2^5 - f_1^5}{5} \quad [6.52]$$

yielding

$$r^2 = \frac{5}{9} \frac{\left(f_2^3 - f_1^3\right)^2}{\left(f_2 - f_1\right)\left(f_2^5 - f_1^5\right)} \quad [6.53]$$

i.e., if $h = \dfrac{f_2}{f_1}$,

$$r^2 = \frac{5}{9} \frac{\left(h^3 - 1\right)^2}{\left(h-1\right)\left(h^5 - 1\right)} \quad [6.54]$$

$$r^2 = \frac{5}{9} \frac{\left(h^2 + h + 1\right)^2}{h^4 + h^3 + h^2 + h + 1}$$

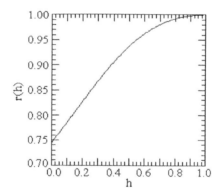

Figure 6.8. *Irregularity factor of band-limited white noise with respect to h*

If $f_2 \to f_1$, $h \to 1$ and $r \to 1$. If $f_2 \to \infty$, $h \to \infty$ and $r \to \dfrac{\sqrt{5}}{3}$.

When the bandwidth tends towards the infinite, the parameter r tends towards $\dfrac{\sqrt{5}}{3} = 0.7454$. This is also true if $f_1 \to 0$, whatever the value of f_2 [PRE 56b].

The limiting case r = 0 can be obtained only if the number of peaks between two zero crossings is very large, infinite at the limit. This is, for example, the case for a composite signal made up of the sum of a harmonic process of low frequency f_2 and of a band-limited process at very high frequency and of low amplitude compared with the harmonic movement.

L.P. Pook [POO 76] uses the rectangular filter as an analogy– a one-degree-of-freedom mechanical filter in which $\Delta f = \dfrac{f_0}{Q} = 2\,\xi\,f_0$ to demonstrate, by considering that the band-limited PSD is the response of the system (f_0, Q) to a white noise, that:

$$r^2 = \frac{4}{9}\xi^2 \frac{15+\xi^2}{5+10\xi^2+\xi^4}$$

This expression is obtained while setting $f_1 = f_0 - \dfrac{\Delta f}{2}$ and $f_2 = f_0 + \dfrac{\Delta f}{2}$ in [6.53]:

$$r = \frac{1+\dfrac{\xi^2}{3}}{\sqrt{1+2\,\xi^2+\dfrac{\xi^4}{5}}} \qquad [6.55]$$

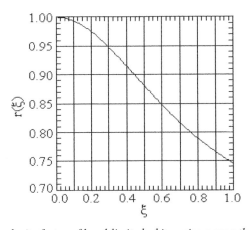

Figure 6.9. *Irregularity factor of band-limited white noise versus the damping factor*

It should be noted that $r \to 1$ if $\xi \to 0$.

NOTE.– *The parameter r of a narrowband noise centered on frequency f_0, whose PSD has a width Δf, is written, from the above expressions [COU 70], [RUD 75]:*

$$r = \frac{n_0^+}{n_p^+} = \frac{1 + \frac{1}{3}\left(\frac{\Delta f}{2 f_0}\right)^2}{\sqrt{1 + 2\left(\frac{\Delta f}{2 f_0}\right)^2 + \frac{1}{5}\left(\frac{\Delta f}{2 f_0}\right)^4}} \quad [6.56]$$

6.6.3. Calculation of irregularity factor for noise of form $G = \text{Const.} \, f^b$

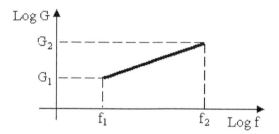

Figure 6.10. *PSD of a noise defined by a straight line segment in logarithmic scales*

$$r = \frac{M_2}{\sqrt{M_0 \, M_4}}$$

The moments are expressed

$$\left. \begin{aligned} M_0 &= \frac{G_1}{f_1^b} \frac{f_2^{b+1} - f_1^{b+1}}{b+1} & \text{if } b \neq -1 \\ M_0 &= f_1 G_1 \ln\frac{f_2}{f_1} & \text{if } b = -1 \end{aligned} \right\} \quad [6.57]$$

$$\left. \begin{aligned} M_2 &= (2\pi)^2 \frac{G_1}{f_1^b (b+3)} \left(f_2^{b+3} - f_1^{b+3}\right) & \text{if } b \neq -3 \\ M_2 &= (2\pi)^2 f_1 G_1 \ln\frac{f_2}{f_1} & \text{if } b = -3 \end{aligned} \right\} \quad [6.58]$$

$$M_4 = (2\pi)^4 \frac{G_1}{f_1^b(b+5)}\left(f_2^{b+5} - f_1^{b+5}\right) \quad \text{if } b \neq -5$$

$$M_4 = (2\pi)^4 f_1^5 G_1 \ln\frac{f_2}{f_1} \quad \text{if } b = -5$$

[6.59]

Case study: $b \neq -1$, $b \neq -3$, $b \neq -5$

Let us set $h = \dfrac{f_2}{f_1}$. Then:

$$r^2 = \frac{(b+1)(b+5)}{(b+3)^2} \frac{\left(h^{b+3}-1\right)^2}{\left(h^{b+1}-1\right)\left(h^{b+5}-1\right)} \qquad [6.60]$$

The curves of Figures 6.11 and 6.12 show the variations of $r(h)$ for various values of b ($b \leq 0$ and $b \geq 0$).

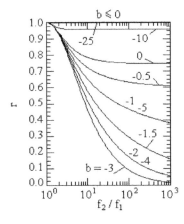
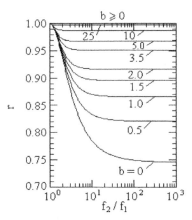

Figure 6.11. *Irregularity factor versus h, for various values of the negative exponent b*

Figure 6.12. *Irregularity factor versus h, for various values of the positive exponent b*

For $b < 0$, we note (Figure 6.11) that, when b varies from 0 to -25, the curve, always issuing from the point $r = 1$ for $h = 1$, goes down to $b = -3$, then rises; the curves for $b = -2$ and $b = -4$ are thus superimposed, just like those for $b = -1$ and $b = -5$. This behavior can be highlighted in a more detailed way while plotting, for a given h, the variations of r with respect to b (Figure 6.13) [BRO 63].

318 Random Vibration

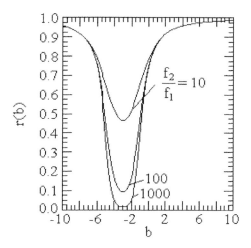

Figure 6.13. *Irregularity factor versus the exponent b*

Moreover, we observe that for $b = 0$, the curve $r(h)$ tends, for a large h, towards $r_0 = \frac{\sqrt{5}}{3} = 0.7454$. This is similar to the case where f_1 is zero (signal filtered by a low-pass filter).

Case study: $b = -1$

$$M_0 = f_1 G_1 \ln \frac{f_2}{f_1}$$

$$M_2 = (2\pi)^2 \frac{f_1 G_1}{2} \left(f_2^2 - f_1^2\right)$$

$$M_4 = (2\pi)^4 \frac{f_1 G_1}{4} \left(f_2^4 - f_1^4\right)$$

yielding

$$r = \frac{h^2 - 1}{\sqrt{\left(h^4 - 1\right) \ln h}} \qquad [6.61]$$

Case study: b = –3

$$M_0 = -\frac{f_1^3 G_1}{2}\left(\frac{1}{f_2^2} - \frac{1}{f_1^2}\right)$$

$$M_2 = (2\pi)^2 f_1^3 G_1 \ln \frac{f_2}{f_1}$$

$$M_4 = (2\pi)^4 \frac{f_1^3 G_1}{2}\left(f_2^2 - f_1^2\right)$$

$$r = \frac{2h \ln h}{\left|h^2 - 1\right|} \qquad [6.62]$$

This curve gives, for given h, the lowest value of r.

Case study: b = –5

$$M_0 = \frac{f_1^5 G_1}{4}\left(\frac{1}{f_1^4} - \frac{1}{f_2^4}\right)$$

$$M_2 = -(2\pi)^2 \frac{f_1^5 G_1}{2}\left(\frac{1}{f_2^2} - \frac{1}{f_1^2}\right)$$

$$M_4 = (2\pi)^4 f_1^5 G_1 \ln \frac{f_2}{f_1}$$

$$r = \frac{\left|h^2 - 1\right|}{\sqrt{\left(h^4 - 1\right)\ln h}} \qquad [6.63]$$

6.6.4. *Case study: variations of irregularity factor for two narrowband signals*

Let us set $\Delta f = f_2 - f_1$ in the case of a single narrowband noise. Expressions [6.50], [6.51] and [6.52] can be approximated by assuming that, Δf being small, the frequencies f_1 and f_2 are close to the central frequency of the band $f_0 = \dfrac{f_1 + f_2}{2}$. We then obtain:

$$M_0 = G\,\Delta f$$

$$M_2 \approx (2\pi)^2\,G\,\Delta f\,f_0^2$$

and

$$M_4 \approx (2\pi)^4\,G\,\Delta f\,f_0^4$$

Now let us apply the same process to two narrowband noises whose central frequencies and widths are respectively equal to f_0, Δf_0 and f_1, Δf_1.

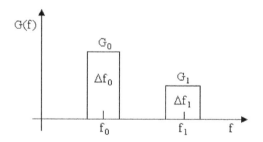

Figure 6.14. *Random noise composed of two narrowbands*

With the same procedure, the factor r obtained is roughly given by [BRO 63]:

$$r^2 = \frac{(2\pi)^4 \left(\Delta f_0\,f_0^2\,G_0 + \Delta f_1\,f_1^2\,G_1\right)^2}{\left(\Delta f_0\,G_0 + \Delta f_1\,G_1\right)(2\pi)^4\left(\Delta f_0\,f_0^4\,G_0 + \Delta f_1\,f_1^4\,G_1\right)}$$

$$r = \frac{1 + \dfrac{\Delta f_1}{\Delta f_0} \dfrac{f_1^2}{f_0^2} \dfrac{G_1}{G_0}}{\sqrt{\left(1 + \dfrac{\Delta f_1}{\Delta f_0} \dfrac{G_1}{G_0}\right)\left(1 + \dfrac{\Delta f_1}{\Delta f_0} \dfrac{f_1^4}{f_0^4} \dfrac{G_1}{G_0}\right)}}$$ [6.64]

Figures 6.15 and 6.16 show the variations of r with $\dfrac{f_1}{f_0}$ and of $\dfrac{\Delta f_1}{\Delta f_0} \dfrac{G_1}{G_0}$. It is observed that if $\dfrac{f_1}{f_0} = 1$, r is equal to 1, whatever the value of $\dfrac{\Delta f_1}{\Delta f_0} \dfrac{G_1}{G_0}$.

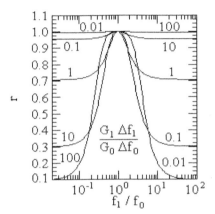

Figure 6.15. *Irregularity factor of a two narrowband noise versus f_1/f_0*

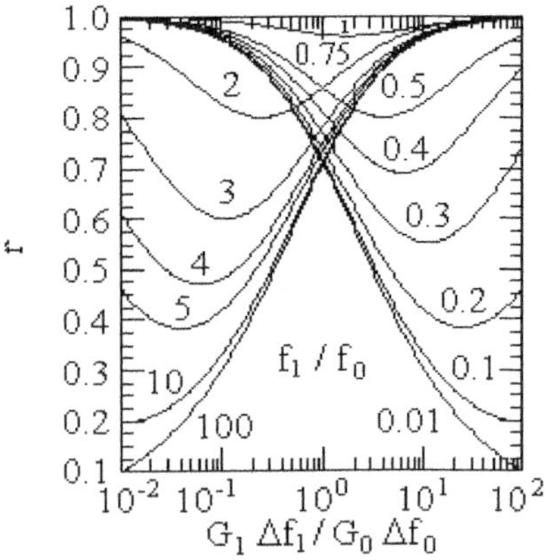

Figure 6.16. *Irregularity factor of a two narrowband noise versus $G_1 \Delta f_1 / G_0 / \Delta f_0$*

These results can be useful to interpret the response of a two-degree-of-freedom linear system to a white noise, each of the two peaks of the PSD response being able to be compared to a rectangle of amplitude equal to Q_i^2 times the PSD of the excitation, and of width $\Delta f_i = \dfrac{\pi}{2} \dfrac{f_0}{Q}$ [BRO 63].

6.7. Error related to the use of Rayleigh's law instead of a complete probability density function

This error can be evaluated by plotting, for various values of r, variations of the ratio [BRO 63]:

$$\frac{q(u)}{p_r(u)}$$

where $q(u)$ is given by [6.19] and where $p_r(u)$ is the probability density from Rayleigh's law (Figure 6.17):

$$p_r(u) = u \, e^{-\frac{u^2}{2}}$$

When u becomes large, these curves tend towards a limit equal to r. This result can be easily shown from the above ratio, which can be written:

$$\frac{q(u)}{p_r(u)} = \frac{\sqrt{1-r^2}}{\sqrt{2\pi}} \frac{e^{\frac{r^2 u^2}{2(1-r^2)}}}{u} + \frac{r}{2}\left\{1 + \text{erf}\left[\frac{r\,u}{\sqrt{2(1-r^2)}}\right]\right\} \quad [6.65]$$

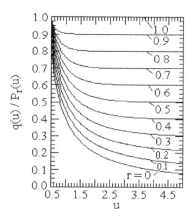

Figure 6.17. *Error related to the approximation of the peak distribution by Rayleigh's law*

It is verified that, when u becomes large, $\frac{q(u)}{p_r(u)} \to r$. In addition, we note from these curves that:

– this ratio is closer to 1 the larger r is;

– the greatest maxima tend to obey a law close to Rayleigh's law, the difference being related to the value of r (which characterizes the number of maxima which occur in an alternation between two zero-crossings).

6.8. Peak distribution function

6.8.1. General case

From the probability density $q(u)$, we can calculate by integration the probability that a peak (maximum) randomly selected among all the maxima of a random process is higher than a given value (per unit time) [CAR 56], [LEY 65]:

$$Q_p(u) = \int_u^\infty q(u) \, du = P\left(\frac{u}{\sqrt{1-r^2}}\right) + r\, e^{-\frac{u^2}{2}}\left[1 - P\left(\frac{r\, u}{\sqrt{1-r^2}}\right)\right] \qquad [6.66]$$

where

$$P(x_0) = \frac{1}{\sqrt{2\pi}} \int_{x_0}^\infty e^{-\frac{\lambda^2}{2}} \, d\lambda$$

$P(x_0)$ is the probability that the normal random variable x exceeds a given threshold x_0. If $u \to \infty$, $P(x_0) \to 1$ and $Q_p(u) \to 0$. Figure 6.18 shows the variations of $Q_p(u)$ for $r = 0$; 0.25; 0.5; 0.75 and 1.

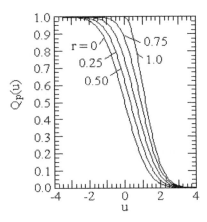

Figure 6.18. *Probability that a peak is higher than a given value u*

NOTES.–

1. The distribution function of the peaks is obtained by calculating $1-Q_p(u)$.

2. The function $Q_p(u)$ can also be written in several forms.

Knowing that:

$$A = \frac{1}{\sqrt{2\pi}} \int_{x_0}^{\infty} e^{-\frac{\lambda^2}{2}} d\lambda = \frac{1}{2} - \frac{1}{\sqrt{2\pi}} \int_{0}^{x_0} e^{-\frac{\lambda^2}{2}} d\lambda$$

$$A = \frac{1}{2} - \frac{1}{\sqrt{\pi}} \int_{0}^{x_0/\sqrt{2}} e^{-\lambda^2} d\lambda = \frac{1}{2}\left[1 - erf\left(\frac{x_0}{\sqrt{2}}\right)\right]$$

it results that [HEA 56], [KOW 63]:

$$Q_p(u) = \frac{1}{2}\left\{1 - erf\left[\frac{u}{\sqrt{2(1-r^2)}}\right]\right\} + \frac{r}{2} e^{-u^2/2}\left\{1 + erf\left[\frac{ru}{\sqrt{2(1-r^2)}}\right]\right\} \quad [6.67]$$

or [HEA 56]:

$$Q_p(u) = \frac{1}{2}\left\{erfc\left[\frac{u}{\sqrt{2(1-r^2)}}\right]\right\} + \frac{r}{2} e^{-u^2/2}\left\{1 + erf\left[\frac{ru}{\sqrt{2(1-r^2)}}\right]\right\} \quad [6.68]$$

This form is the most convenient to use, the error function erf being able to be approximated by a series expansion with very high precision (see Appendix A4.1). We also sometimes encounter the following expression:

$$Q_p(u) = 1 - \frac{\sqrt{1-r^2}}{\sqrt{2\pi}} \int_{-\infty}^{u} e^{-\frac{\lambda^2}{2(1-r^2)}} d\lambda$$

$$-\frac{r^2}{\sqrt{2\pi}}\int_{-\infty}^{u/\sqrt{1-r^2}} e^{-\frac{\lambda^2}{2}}\,d\lambda + \frac{r}{\sqrt{2\pi}} e^{-\frac{u^2}{2}} \int_{-\infty}^{ru/\sqrt{1-r^2}} e^{-\frac{\lambda^2}{2}}\,d\lambda \qquad [6.69]$$

3. *For large u [HEA 56],*

$$Q_p(u) \approx r\, e^{-\frac{u^2}{2}}.$$

yielding the average amplitude of the maximum (or minimum):

$$\overline{u_{max}} = r\sqrt{\frac{\pi}{2}} \qquad [6.70]$$

6.8.2. *Particular case of narrowband Gaussian process*

For a narrowband Gaussian process ($r = 1$), we saw that [6.28]:

$$q(a) = \frac{a}{\ell_{rms}^2}\, e^{-\frac{a^2}{2\ell_{rms}^2}}$$

The probability so that a maximum is greater than a given threshold a is then:

$$Q_p(a) = e^{-\frac{a^2}{2\ell_{rms}^2}} \qquad [6.71]$$

It is observed that, in this case [5.38],

$$Q_p(a) = \frac{n_a^+}{n_0^+} = \frac{p(a)}{p(0)}$$

yielding

$$q(a) = -\frac{d[p(a)]/da}{p(0)} \qquad [6.72]$$

These two last relationships assume that the functions $\ell(t)$ and $\dot\ell(t)$ are independent. If this is not the case, in particular if $p(\ell)$ is not Gaussian, J.S. Bendat [BEN 64] notes that these relationships nonetheless give acceptable results in the majority of practical cases.

NOTE.– *Relationship [6.28] can also be established as follows [CRA 63], [FUL 61], [POW 58]. We showed that the number of threshold level crossings with positive slope, per unit time, n_a^+ is, for a Gaussian stationary noise [5.44]:*

$$n_a^+ = n_0^+ \; e^{-\frac{a^2}{2\,\ell_{rms}^2}}$$

where

$$n_0^+ = \frac{1}{2\pi} \frac{\dot\ell_{rms}}{\ell_{rms}}$$

The average number of maxima per unit time between two neighboring levels a and a + da must be equal, for a narrowband process, to:

$$n_a^+ - n_{a+da}^+ = -\frac{dn_a^+}{da} da$$

yielding, by definition of $q(a)$,

$$n_p^+ \, q(a) \, da = -\frac{dn_a^+}{da} da$$

The signal being assumed narrowband, $n_p^+ = n_0^+$. This yields

$$q(a) = -\frac{1}{n_0^+} \frac{dn_a^+}{da}$$

and

$$q(a) = \frac{a}{\ell_{rms}^2} \; e^{-\frac{a^2}{2\,\ell_{rms}^2}}$$

It is shown that the calculation of the number of peaks from the number of threshold crossings using the difference $n_a^+ - n_{a+da}^+$ is correct only for one perfectly narrowband process [LAL 92]. In general, this method can lead to errors.

Probability Distribution of Maxima of Random Vibration   327

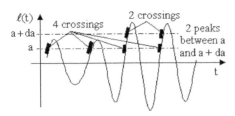

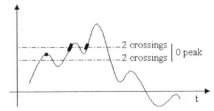

Figure 6.19. *Threshold crossings of a narrowband noise*

Figure 6.20. *Threshold crossings of a wideband noise*

Particular case where $f_1 \to 0$

We saw that, for a band-limited noise, $r \to \dfrac{\sqrt{5}}{3}$ when $f_1 \to 0$. Figures 6.21 and 6.22 respectively show the variations of the density $q(u)$ and of $P(a < u\, \ell_{rms}) = 1 - Q_p(u)$ versus u, for $r = \dfrac{\sqrt{5}}{3}$.

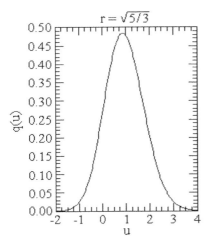

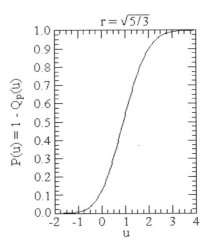

Figure 6.21. *Peak probability density of a band-limited noise with zero initial frequency*

Figure 6.22. *Peak distribution function of a band-limited noise with zero initial frequency*

6.9. Mean number of maxima greater than the given threshold (by unit time)

The mean number of maxima which, per unit time, exceeds a given level $a = u\, \ell_{rms}$ is equal to:

$$M_a = n_p^+ \, Q_p(u) \qquad [6.73]$$

If a is large and positive, the functions $P\left(\dfrac{u}{\sqrt{1-r^2}}\right)$ and $P\left(\dfrac{u\,r}{\sqrt{1-r^2}}\right)$ tend towards zero; yielding:

$$Q_p \approx r\, e^{-u^2/2} \qquad [6.74]$$

and

$$M_a \approx n_p^+ \, r\, e^{-\dfrac{u^2}{2}} \qquad [6.75]$$

i.e. [RAC 69], since $r = \dfrac{n_0^+}{n_p^+}$,

$$M_a \approx n_0^+ \, e^{-\dfrac{u^2}{2}} \qquad [6.76]$$

This expression gives acceptable results for $u \geq 2$ [PRE 56b]. For $u < 2$, it results in underestimating the number of maxima. To evaluate this error, we have plotted in Figure 6.23 variations of the ratio $\dfrac{\text{exact value}}{\text{approximate value}}$ of M_a:

$$\dfrac{n_p^+ \, Q_p(u)}{n_0^+ \, e^{-u^2/2}} = \dfrac{Q_p(u)\, e^{u^2/2}}{r}$$

with respect to u, for various values of r. This ratio is equal to 1 when $r = 1$ (narrowband process).

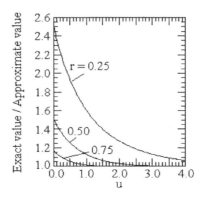

Figure 6.23. *Error related to the use of the approximate expression of the average number of maxima greater than a given threshold*

This yields $Q_p(u) \approx r\, e^{-u^2/2}$ and $M_a \approx n_0^+ e^{-u^2/2}$ (the same result as for large a). In these two particular cases, the average number per second of the maxima located above a threshold a is thus equal to the average number of times per second which $\ell(t)$ crosses the threshold a with a positive slope; this is equivalent to saying that there is only one maximum between two successive threshold crossings (with positive slope). For a narrowband noise, we thus obtain:

$$M_a = n_p^+ \, Q_p(a)$$

$$M_a = \frac{1}{2\pi}\sqrt{\frac{M_4}{M_2}}\, e^{-\frac{a^2}{2\ell_{rms}^2}} = \frac{1}{2\pi}\sqrt{\frac{M_2}{M_0}}\, e^{-\frac{a^2}{2M_0}} \qquad [6.77]$$

NOTE.–

Expression [5.44] ($n_a^+ = n_0^+\, e^{-\frac{a^2}{2\ell_{rms}^2}}$) *is an asymptotic expression for large a*
[PRE 56b]. The average frequency $n_0^+ = \dfrac{1}{2\pi}\left(\dfrac{\int_0^\infty \Omega^2\, G(\Omega)\, d\Omega}{\int_0^\infty G(\Omega)\, d\Omega}\right)^{1/2}$ *is independent of noise intensity and depends only on the form of the PSD. In logarithmic scales, [5.44] becomes:*

$$\ln n_a^+ = \ln n_0^+ - \frac{a^2}{2\,\ell_{rms}^2}$$

$\ln n_a^+$ is thus a linear function of a^2, the corresponding straight line having a slope $-\dfrac{1}{2\,\ell_{rms}^2}$. We often observe this property in practice. Sometimes, however, the curve $\left(\ln n_a^+, a^2\right)$ resembles that in Figure 6.24. This is particularly the case for turbulence phenomena. We then carry out a combination of Gaussian processes [PRE 56b] when calculating:

$$M(a) = \sum_{i=1}^{k} P_i\, n_{a\,i}^+(a) \qquad [6.78]$$

where P_i is a coefficient characterizing the contribution brought by the ith component and $n_{a\,i}^+$ is the number of crossings per second for this ith component. If it is assumed that the shape of the atmospheric turbulence spectrum is invariant and that only the intensity varies, n_0^+ is constant. A few components then often suffice to represent the curve correctly.

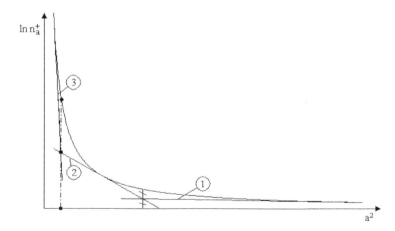

Figure 6.24. *Decomposition of the number of threshold crossings into Gaussian components*

We can for example proceed according to the following (arbitrary) steps:

– plot the tangent at the tail of the observed distribution ①;

– plot the straight line ② starting from the point of the straight line 1 which underestimates the distribution observed by a factor 2, and tangent to the higher part of the distribution;

– plot straight line ③ from ② in the same way.

The sum of these three lines gives a good enough approximation of the initial curve. The slopes of these lines allow the calculation of the squares of the rms values of each component. The coefficients P_i are obtained from:

$$M_i(a) = P_i \, n_0^+ \, e^{-\frac{a^2}{2 \ell_{rms}^2}} \qquad [6.79]$$

for each component. Each term M_i can be evaluated directly by reading the ordinate at the beginning of each line (for $a = 0$), yielding

$$P_i = \frac{M_i}{n_0^+ \, e^{-\frac{a^2}{2 \ell_{rms}^2}}} \qquad [6.80]$$

6.10. Mean number of maxima above given threshold between two times

If a is the threshold, and t_1 and t_2 the two times, this number is given by [CRA 67], [PAP 65]:

$$E(a) = N_a^+ = \frac{1}{2\pi}(t_2 - t_1)\sqrt{\frac{M_4}{M_2}} \, e^{-\frac{a^2}{2 \ell_{rms}^2}} \qquad [6.81]$$

6.11. Mean time interval between two successive maxima

Let T be the duration of the sample. The average number of positive maxima which exceeds the level a in time T is:

$$M_a T = n_p^+ \, Q(a) \, T \qquad [6.82]$$

and the average time between positive peaks above a is:

$$T_a = \frac{1}{M_a} = \frac{1}{n_p^+ \, Q(a)} \qquad [6.83]$$

For a narrowband noise,

$$T_a = \frac{1}{M_a} = \frac{1}{n_p^+ \, Q_p(a)} = \frac{1}{n_0^+ \, Q_p(a)}$$

$$T_a = \frac{2\pi \, e^{\frac{a^2}{2\ell_{rms}^2}}}{\sqrt{\dfrac{M_4}{M_2}}} = 2\pi \, \frac{M_4}{M_2} \, e^{\frac{a^2}{2\ell_{rms}^2}} \qquad [6.84]$$

or

$$T_a = 2\pi \, \frac{M_0}{M_2} \, e^{\frac{a^2}{2 M_0}} \qquad [6.85]$$

6.12. Mean number of maxima above given level reached by signal excursion above this threshold

The parameter $r = \dfrac{n_0}{2 \, n_p^+}$ makes it possible to compare the number of zero-crossings and the number of peaks of the signal. Another interesting parameter can be the ratio N_m of the mean number, per unit time, of maxima which occur above a level a_0 to the mean number, per unit time, of crossings of the same level a_0 with a positive slope [CRA 68].

The mean number, per unit of time, of maxima which occur above a level a_0 is equal to:

$$M_{a_0} = n_p \int_{u_0}^{\infty} q(u) \, du \qquad [6.86]$$

where $u_0 = \dfrac{a_0}{\ell_{rms}}$ and $q(u)$ are given by [6.19]. The mean number, per unit of time, of crossings of the level a_0 with a positive slope is [5.44]:

$$n_a^+ = n_0^+ \, e^{-\dfrac{u_0^2}{2}}$$

This yields

$$N_m = \dfrac{M_{a_0}}{n_a^+} \qquad [6.87]$$

$$N_m = \dfrac{1}{r} \, Q(u_0) \, e^{\dfrac{u_0^2}{2}} \qquad [6.88]$$

Figure 6.25 shows the variations of N_m versus u_0, for various values of r.

It should be noted that N_m is large for small u_0 and r: there are several peaks of amplitude greater than u_0 for only one crossing of this u_0 threshold.

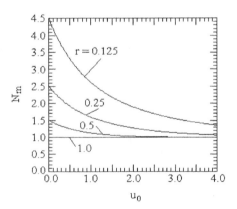

Figure 6.25. *Average number of maxima above a given level through excursion of the signal above this threshold*

For large u_0, N_m decreases quickly and tends towards unity whatever the value of r. In this case, there is on average only one peak per level crossing. During a time interval $t_1 - t_0$, the average number of maxima which exceed level a is:

334   Random Vibration

$$M_a(t_1 - t_0) = n_0^+ (t_1 - t_0) Q_p\left(\frac{a}{\ell_{rms}}\right) \quad [6.89]$$

Let us replace the rms value ℓ_{rms} with ℓ_{rms_1} and seek the rms value ℓ_{rms2} of another random vibration which has the same number n_p^+ of peaks so that, over time $t_3 - t_2 = t_1 - t_0$, we have [BEN 61b], [BEN 64]:

$$M_a(t_3 - t_2) = n_p^+ (t_1 - t_0) Q_p\left(\frac{a}{\ell_{rms_1}}\right) = n_p^+ (t_3 - t_2) Q\left(\frac{a}{\ell_{rms_2}}\right) \quad [6.90]$$

It is thus necessary that:

$$\frac{t_3 - t_2}{t_1 - t_0} = \frac{Q_p\left(\frac{a}{\ell_{rms_1}}\right)}{Q_p\left(\frac{a}{\ell_{rms_2}}\right)} \quad [6.91]$$

If the two vibrations follow each other, applied successively over $t_1 - t_0$ and $t_2 - t_1$, the *equivalent* stationary noise of rms value $\ell_{rms\,eq}$ applied over:

$$T = (t_1 - t_0) + (t_2 - t_1)$$

which has the same number of maxima n_p^+ exceeding the threshold a as the two vibrations ℓ_{rms_1} and ℓ_{rms_2}, is such that:

$$M_a T = M_a(t_1 - t_0) + M_a(t_2 - t_1)$$

$$M_a T = n_p^+ (t_1 - t_0) Q_p\left(\frac{a}{\ell_{rms_1}}\right) + n_p^+ (t_2 - t_1) Q_p\left(\frac{a}{\ell_{rms_2}}\right) \quad [6.92]$$

and

$$M_a T = n_p^+ T Q_p\left(\frac{a}{\ell_{rms\,eq}}\right) \quad [6.93]$$

This yields

$$T = \left[\frac{Q_p\left(\frac{a}{\ell_{rms_1}}\right)}{Q_p\left(\frac{a}{\ell_{rms\,eq}}\right)}\right](t_1 - t_0) + \left[\frac{Q_p\left(\frac{a}{\ell_{rms_2}}\right)}{Q_p\left(\frac{a}{\ell_{rms\,eq}}\right)}\right](t_2 - t_1) \qquad [6.94]$$

and

$$Q_p\left(\frac{a}{\ell_{rms eq}}\right) = \frac{t_1 - t_0}{T} Q_p\left(\frac{a}{\ell_{rms_1}}\right) + \frac{t_2 - t_1}{T} Q_p\left(\frac{a}{\ell_{rms_2}}\right) \qquad [6.95]$$

This expression makes it possible to calculate the value of $\ell_{rms eq}$ (for $a \neq 0$).

6.13. Time during which the signal is above a given value

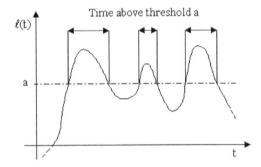

Figure 6.26. *Time during which the signal is above a given value*

Let a be the selected threshold; the time during which $\ell(t)$ is greater than a is a random variable [RAC 69]. The problem of research of the statistical distribution of this time is not yet solved.

We can however consider the average value of this time for a stationary random process. The average time during which we obtain $a \leq \ell(t) \leq b$ is equal to:

$$\overline{T}_{ab} = \int_a^b \frac{1}{\ell_{rms} \sqrt{2\pi}} e^{-\frac{\ell^2}{2\ell_{rms}^2}} d\ell \qquad [6.96]$$

and, if $b \to \infty$, the time for which $\ell(t) \geq a$ is given by:

$$\overline{T}_{a\infty} = \int_a^\infty \frac{1}{\ell_{rms} \sqrt{2\pi}} e^{-\frac{\ell^2}{2\ell_{rms}^2}} d\ell \qquad [6.97]$$

(ℓ_{rms} = rms value of $\ell(t)$). This result is a consequence of the theorem of ergodicity. It should be noted that this average time does not describe in any way how time is spent above the selected threshold. For high frequency vibrations, the response of the structure can have many excursions above the threshold with a relatively small average time between two excursions. For low frequency vibrations, having the same probability density p as for the preceding high frequencies, there would be fewer excursions above the threshold, but these would be longer, with the excursions being more spaced.

Proportion of time during which $\ell(t) > a$

Given a process $\ell(t)$ defined in [0, T] and a threshold a, let us set [CRA 67]:

$$\left.\begin{array}{ll} \eta(t) = 1 & \text{if } \ell(t) > a \\ \eta(t) = 0 & \text{elsewhere} \end{array}\right\} \qquad [6.98]$$

and

$$Z_0(t) = \frac{1}{T} \int_0^T \eta(t) \, dt = m_\eta \qquad [6.99]$$

the proportion of time during which $\ell(t) > a$, the average of Z_0 is:

$$m_Z = \frac{1}{T} \int_0^T m_\eta \, dt = m_\eta$$

$$m_Z = P[\ell(t) > a]$$

$$m_Z = 1 - \phi\left(\frac{a}{\ell_{rms}}\right) \quad [6.100]$$

where $\ell_{rms}^2 = M_0 = R(0)$ and $\phi(\ )$ is the Gaussian law. The variance of Z_0 is of the form $\dfrac{A}{\pi} e^{-\frac{a^2}{\ell_{rm}^2}} \dfrac{\ln T}{T}$ when $T \to \infty$.

6.14. Probability that a maximum is positive or negative

These probabilities, respectively q_{max}^+ and q_{max}^-, are obtained directly from the expression of $Q_p(u)$. If we set $u = 0$, it results that [CAR 56], [COU 70], [KRE 83]:

$$q_{max}^+ = \frac{1+r}{2} \quad [6.101]$$

$q_{max}^- = 1 - q_{max}^+$ since, for u equal to $-\infty$, $Q_p(u) = 1$ [POO 76], [POO 78], yielding

$$q_{max}^- = \frac{1-r}{2} \quad [6.102]$$

q_{max}^+ is the percentage of positive maxima (number of positive maxima divided by the total number of maxima) and q_{max}^- is the percentage of negative maxima [CAR 56]. These relations can be used to estimate r by simply counting the number of positive and negative maxima over a fairly long time.

For a wid band process, $r = 0$ and $q_{max}^+ = q_{max}^- = \dfrac{1}{2}$.

For a narrowband process, $r = 1$ and $q_{max}^+ = 1$, $q_{max}^- = 0$.

6.15. Probability density of the positive maxima

This density has the expression [BAR 78], [COU 70]:

$$q^+(u) = \frac{2}{1+r} q(u) \qquad [6.103]$$

6.16. Probability that the positive maxima is lower than a given threshold

Let u be this threshold. This probability is given by [COU 70]:

$$P(u) = 1 - \frac{2}{1+r} Q_p(u) \qquad [6.104]$$

yielding

$$P(u) = \frac{1}{1+r} \operatorname{erf}\left(\frac{u}{\sqrt{2(1-r^2)}}\right) + \frac{r}{1+r}\left\{1 - e^{-\frac{u^2}{2}}\left[1 + \operatorname{erf}\left(\frac{u\,r}{\sqrt{2(1-r^2)}}\right)\right]\right\} \qquad [6.105]$$

6.17. Average number of positive maxima per unit of time

The average number of maxima per unit time is equal to [BAR 78]:

$$n_p^+ = \int_{-\infty}^{0} \int_{-\infty}^{+\infty} |\ddot{\ell}|\, p(\ell, 0, \ddot{\ell})\, d\ell\, d\ddot{\ell} \qquad [6.106]$$

i.e. [6.13]

$$n_p^+ = \frac{1}{2\pi} \sqrt{\frac{M_4}{M_2}}$$

(the notation + means that it is a maximum, which is not necessarily positive). The average number of positive maxima per unit time is written:

$$n_{p>0}^+ = \int_{-\infty}^{0} \int_{0}^{+\infty} |\ddot{\ell}|\, p(\ell,0,\ddot{\ell})\, d\ell\, d\ddot{\ell}$$

$$n_{p>0}^+ = \frac{1}{4\pi}\left(\sqrt{\frac{M_2}{M_0}} + \sqrt{\frac{M_4}{M_2}}\right) \qquad [6.107]$$

6.18. Average amplitude jump between two successive extrema

Being given a random signal $\ell(t)$, the total height swept in a time interval $(-T, T)$ is [RIC 64]:

$$\int_{-T}^{T} \left| \frac{d\ell(t)}{dt} \right| dt$$

Let $dn(t)$ be the random function which has the value 1 when an extremum occurs and 0 at all the other times. The number of extrema in $(-T, T)$ is $\int_{-T}^{T} dn(t)$.

Figure 6.27. *Amplitude jump between two successive extrema*

The average height $\overline{h}_T$ between two successive extrema (maximum – minimum) in $(-T, T)$ is the total distance divided by the number of extrema:

$$\overline{h}_T = \frac{\int_{-T}^{T} \left| \frac{d\ell}{dt} \right| dt}{\int_{-T}^{T} dn(t)} = \frac{\frac{1}{2T}\int_{-T}^{T} \left| \frac{d\ell}{dt} \right| dt}{\frac{1}{2T}\int_{-T}^{T} dn(t)} \qquad [6.108]$$

340 Random Vibration

If the temporal averages are identical to the ensemble averages, the average height $\bar{h}$ is:

$$\bar{h} = \lim_{T \to \infty} \overline{h_T} = \frac{\lim_{T \to \infty} \frac{1}{2T} \int_{-T}^{T} \left|\frac{d\ell}{dt}\right| dt}{\lim_{T \to \infty} \frac{1}{2T} \int_{-T}^{T} dn(t)} = \frac{E\left(\left|\frac{d\ell}{dt}\right|\right)}{n_p} \qquad [6.109]$$

where n_p is the number of extrema per unit time.

For a Gaussian process, the average height $\bar{h}$ of the rises or falls is equal to [KOW 69], [LEL 73], [RIC 65], [SWA 68]:

$$E(h) = \bar{h} = \sqrt{2\pi} \frac{n_0^+}{n_p^+} \ell_{rms} = \sqrt{2\pi}\, r\, \ell_{rms} \qquad [6.110]$$

or

$$\bar{h} = \sqrt{2\pi}\, \frac{\dot{\ell}_{rms}}{\ddot{\ell}_{rms}} = -R^{(2)}(0)\sqrt{\frac{2\pi}{R^{(4)}(0)}} = M_2 \sqrt{\frac{2\pi}{M_4}} \qquad [6.111]$$

For a narrowband process, $r = 1$ and:

$$\bar{h} = \ell_{rms} \sqrt{2\pi} \qquad [6.112]$$

This value constitutes an upper limit when r varies [RIC 64].

NOTE.– *The calculation of $\bar{h}$ can be also carried out starting from the average number of crossings per second of the threshold [KOW 69]. For a Gaussian signal, this number is equal to [5.47]:*

$$n_a = n_0\, \exp\left(-\frac{a^2}{2\,\ell_{rms}^2}\right)$$

The total rise or fall (per second) is written:

$$R = \int_{-\infty}^{+\infty} n_a\, da = n_0 \int_{-\infty}^{+\infty} e^{-\frac{a^2}{2\,\ell_{rms}^2}}\, da = \sqrt{2\pi}\, n_0\, \ell_{rms} \qquad [6.113]$$

This yields the average rise or fall [PAR 62]:

$$\bar{h} = \frac{R}{n_p} = \frac{\sqrt{2\pi}\, n_0}{n_p} \ell_{rms} = \sqrt{2\pi}\, r\, \ell_{rms} \qquad [6.114]$$

Example 6.3.

Let us consider a stationary random process defined by [RIC 65]:

$$\left.\begin{array}{l} G(\Omega) = \dfrac{\ell_{rms}^2}{(1-\beta)\omega_0} \quad \text{for } \beta\omega_0 < \Omega < \omega_0 \\ (0 \leq \beta \leq 1) \\ G(\Omega) = 0 \qquad\qquad \text{elsewhere} \end{array}\right\} \qquad [6.115]$$

J.R. Rice and F.P. Beer [RIC 65] show that:

$$\frac{\bar{h}}{\ell_{rms}} = \sqrt{\frac{10\pi}{(1-\beta)(1-\beta^5)}\, \frac{(1-\beta^3)}{3}} \qquad [6.116]$$

For $\beta = 0$ (perfect low-pass filter),

$$\frac{\bar{h}}{\ell_{rms}} = \sqrt{\frac{10\pi}{3}} \qquad [6.117]$$

If $\beta \to 1$ (narrowband process),

$$\frac{\bar{h}}{\ell_{rms}} \to \sqrt{2\pi} \qquad [6.118]$$

6.19. Average number of inflection points per unit of time

K.A Sweitzer *et al.* show that [SWE 04]

$$n_{IP} = \left(\frac{M_6}{M_4}\right)^{\frac{1}{2}} \qquad [6.119]$$

Chapter 7

Statistics of Extreme Values

7.1. Probability density of maxima greater than a given value

Let us consider a signal $\ell(t)$ having a distribution of instantaneous values of probability density $p(\ell)$ and distribution function $P(\ell)$:

$$\text{prob}\left[\ell < \ell_{peak} < \ell + d\ell\right] = p(\ell)\, d\ell$$

$$P(\ell) = \text{prob}\left[\ell_{peak} < \ell\right] = \int_{-\infty}^{\ell} p(\ell)\, d\ell$$

Let λ_N be a new random variable such that $\lambda_N = \max_{i=1,n} \ell_{peak_i}$. λ_N is the largest peak obtained among the N_p peaks of the signal $\ell(t)$ over a given duration. The distribution function of λ_N is equal to:

$$P(\lambda_N < \ell) = P_N(\ell) = [P(\ell)]^{N_p} \qquad [7.1]$$

and the probability density function to:

$$p_N(\ell) = \frac{dP_N}{d\ell}$$

$$p_N(\ell) = N_p \left[P(\ell)\right]^{N_p-1} p(\ell) \qquad [7.2]$$

If the probability Q that a maximum is higher than a given value is used,

$$Q = 1 - P$$

we have

$$p_N(\ell) = N_p \left[1 - Q(\ell)\right]^{N_p-1} p(\ell) \qquad [7.3]$$

where $N_p \left[1 - Q(\ell)\right]^{N_p-1}$ is the probability of having $\left(N_p - 1\right)$ peaks less than a value ℓ among the N_p peaks.

7.2. Return period

The *return period* $T(X)$ is the number of peaks necessary such that, on average, there is a peak equal to or higher than X. $T(X)$ is a monotonous increasing function of X:

$$T(X) = \frac{1}{1 - P(X)} \qquad [7.4]$$

where $P(X)$ is related to the distribution of ℓ. It becomes:

$$T(X)\left[1 - P(X)\right] = T(X)\,\text{Prob}(x > X) = 1 \qquad [7.5]$$

7.3. Peak ℓ_p expected among N_p peaks

ℓ_p is the value exceeded once on average in a sample containing N_p peaks. We have:

$$P(\ell_p) = 1 - \frac{1}{N_p} \qquad [7.6]$$

and

$$N_p \left[1 - P(\ell_p)\right] = N_p \, \text{Prob}\left(\ell > \ell_p\right)$$

The return period of ℓ_p is equal to:

$$T(\ell_p) = N_p \qquad [7.7]$$

7.4. Logarithmic rise

The logarithmic rise α_N characterizes the increase in the expected maximum ℓ_p in accordance with the Napierian logarithm of the sample size:

$$\frac{1}{\alpha_N} = \frac{d\ell_p}{d(\ln N_p)} \qquad [7.8]$$

From [7.6], we obtain

$$\frac{dP(\ell_p)}{\ell_p} d\ell_p = p(\ell_p) d\ell_p = \frac{dN_p}{N_p^2}$$

yielding

$$N_p \, p(\ell_p) \, d\ell_p = \frac{dN_p}{N_p} = d(\ln N_p)$$

and

$$N_p \, p(\ell_p) = \frac{d(\ln N_p)}{d\ell_p} = \alpha_N$$

i.e.

$$\alpha_N = N_p \, p(\ell_p) \qquad [7.9]$$

7.5. Average maximum of N_p peaks

$$\overline{\ell_N} = \int_{-\infty}^{+\infty} \ell \, p_N(\ell) \, d\ell \qquad [7.10]$$

7.6. Variance of maximum

$$s_n^2 = \int_{-\infty}^{+\infty} (x - \overline{\ell_N}) \, p_N(\ell) \, d\ell \qquad [7.11]$$

7.7. Mode (most probable maximum value)

Let us set ℓ_M such that $p_N(\ell_M)$ is maximum. The calculation of $\dfrac{dp_N(\ell)}{d\ell} = 0$ gives:

$$(N_p - 1) \frac{p(\ell_M)}{P(\ell_M)} + \frac{p'(\ell_M)}{p(\ell_M)} = 0 \qquad [7.12]$$

7.8. Maximum value exceeded with risk α

This value, noted $\ell_{N\alpha}$, is defined by:

$$P_N(\ell_{N\alpha}) = 1 - \alpha \qquad [7.13]$$

α is the probability of recording a maximum value higher than $\ell_{N\alpha}$ among N_p peaks.

7.9. Application to the case of a centered narrowband normal process

7.9.1. *Distribution function of largest peaks over duration T*

If it is considered that the maxima are distributed according to a Rayleigh density law

$$p(\ell) = \frac{\ell}{s_\ell^2} \exp\left(-\frac{\ell^2}{2 s_\ell^2}\right)$$

and if it is assumed that the peaks of the narrowband random signal are themselves randomly distributed (a broad assumption in a strict sense, because such a signal may have a correlation between consecutive peaks), the probability that an arbitrary peak ℓ_{peak} is lower than a given value ℓ is equal to:

$$P(\ell_{peak} \leq \ell) = \int_0^\ell \frac{\ell}{s_\ell^2} \exp\left(-\frac{\ell^2}{2 s_\ell^2}\right) d\ell$$

i.e.

$$P(\ell) = 1 - \exp\left(-\frac{\ell^2}{2 s_\ell^2}\right)$$

We obtain, from the above relationships, the distribution function of the largest peaks

$$P_N = P(\ell_{peak_i} \leq \ell) = \left[1 - \exp\left(-\frac{\ell^2}{2 s_\ell^2}\right)\right]^{N_p} \quad [7.14]$$

$(1 \leq i \leq N_p)$. P_N is the probability that each of the N_p peaks is lower than ℓ, if the peaks are independent [KOW 69]. Figure 7.1 shows this probability for some values of $n_0^+ T$ (equal to N_p since, for a narrowband noise, $n_p^+ = n_0^+$), plotted versus $u = \dfrac{\ell}{\ell_{rms}}$.

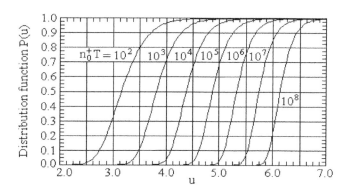

Figure 7.1. *Distribution function of largest peaks of a narrowband noise*

Figure 7.2 presents the variations of the function $Q_N = 1 - P_N$, Q_N being the probability so that the largest peak is higher than a given value U during a length of time T.

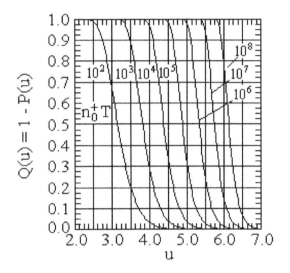

Figure 7.2. *Probability that the largest peak is higher than a given value*

NOTES.–

1. For N_p large (i.e., in practice, for $s_\ell/\ell \leq 0.2$) [KOW 69], we have

$$P_N \approx e^{-N_p \exp(-\ell^2/2 s_\ell^2)} \qquad [7.15]$$

2. *This relation can be written in the form*:

$$\frac{\ell}{s_\ell} = \sqrt{-2 \ln\{1 - \exp[(\ln P_N)/N_p]\}} \qquad [7.16]$$

Figure 7.3 shows the variations of ℓ/s_ℓ versus N_p, for various values of P_N.

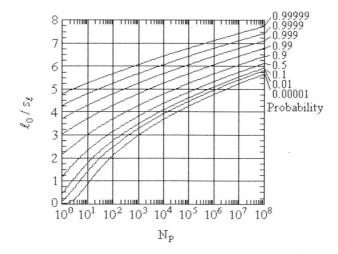

Figure 7.3. *Amplitude of the largest peak against number of peaks, for a given probability*

7.9.2. Probability that one peak at least exceeds a given threshold

The probability that one peak at least exceeds the threshold ℓ is equal to:

$$P(\ell_i \geq \ell) = 1 - \left(1 - e^{-\frac{\ell^2}{2 s_\ell^2}}\right)^{N_p} \qquad [7.17]$$

where $\left(1 \leq i \leq N_p\right)$, yielding the probability so that a maximum ℓ_{peak_i} lies between ℓ and $\ell + d\ell$:

$$P\!\left(\ell \leq \ell_{peak_i} \leq \ell + d\ell\right) = P\!\left(\ell_{peak_i} \geq \ell\right) - P\!\left(\ell_{peakc_i} \geq \ell + d\ell\right)$$

i.e.

$$P\!\left(\ell \leq \ell_{peak_i} \leq \ell + d\ell\right) = -d\left[1 - \left(1 - e^{-\frac{\ell^2}{2 s_\ell^2}}\right)^{N_p}\right] \qquad [7.18]$$

7.9.3. *Probability density of the largest maxima over duration T*

The probability density of the largest maxima is thus

$$p_N(\ell) = N_p \left[1 - \exp\left(-\frac{\ell^2}{2 s_\ell^2}\right) \right]^{N_p - 1} \frac{\ell}{s_\ell^2} \exp\left(-\frac{\ell^2}{2 s_\ell^2}\right) \qquad [7.19]$$

or, while noting $v = \left(\dfrac{\ell}{\sqrt{2}\, s_\ell}\right)^2$:

$$p_N(v) = N_p \left(1 - e^{-v}\right)^{N_p - 1} v^{1/2} e^{-v} \qquad [7.20]$$

Over time T, the number of maxima higher than $u = \dfrac{\ell}{\ell_{rms}}$ is

$$v = Q(u)\, N_p \qquad [7.21]$$

where v is such that $0 \leq v \leq N_p$.

$$p_N(u)\, du = -\left\{ N_p \left[1 - Q(u)\right] \right\}^{N_p - 1} dQ$$

$$p_N(u)\, du = \left\{ N_p \left[1 - Q(u)\right] \right\}^{N_p - 1} d[1 - Q(u)]$$

$$p_N(u)\, du = d\left[\left(1 - \frac{v}{N_p}\right)^{N_p} \right]$$

$$p_N(u)\, du = d\left(e^{-v}\right) = -e^{-v}\, dv \qquad [7.22]$$

For large u, we can accept that $Q(u)$ can be approximated by [CAR 56]:

$$Q(u) \approx r\, e^{-\frac{u^2}{2}} \qquad [7.23]$$

Statistics of Extreme Values   351

(Rayleigh's law). For large N_p, we have, on average, for a given duration T,

$$N_p = n_p^+ T$$

In addition, we still have $v = N_p Q(u)$, yielding, since $r = 1$,

$$v \approx n_0^+ T e^{-\frac{u^2}{2}} \qquad [7.24]$$

and

$$p_N(u)du = d\left[\exp\left(-n_0^+ T e^{-u^2/2}\right)\right] \qquad [7.25]$$

From relation [7.25], we can express, by integration, this density in the form:

$$p_N(u) = n_0^+ T u \exp\left\{-\left[\frac{u^2}{2} + n_0^+ T \exp\left(-\frac{u^2}{2}\right)\right]\right\} \qquad [7.26]$$

Figure 7.4 shows the variations of $p_N(u)$ for various values of $n_0^+ T$ between 10^2 and 10^8.

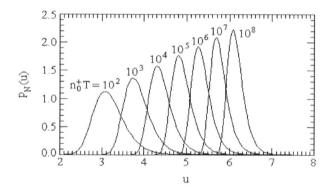

Figure 7.4. *Probability density of the largest maximum over duration T*

Each of these curves gives the distribution law of the largest maximum over duration T of n signal samples to be studied (Figure 7.5).

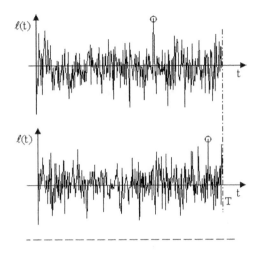

Figure 7.5. *Largest peak of a sample of given duration*

Figure 7.6 shows this same probability density for $n_0^+ T = 3.6\,10^4$ to $3.6\,10^6$, superimposed over the probability density curve of the instantaneous values of the random signal (Gauss's law).

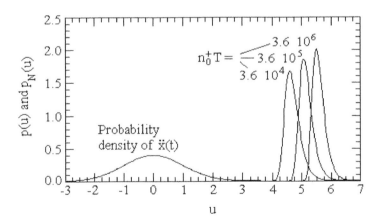

Figure 7.6. *Probability densities of peaks and highest maxima*

7.9.4. *Average of highest peaks*

$$\overline{u_0} = \int_{-\infty}^{+\infty} u \, p_N(u) \, du = \int_{-\infty}^{+\infty} u \, e^{-v} \, dv$$

Relation [7.24] makes it possible to express u according to v:

$$u = 2 \ln\left(n_0^+ T\right) - 2 \ln(v)$$

On the assumption that $\ln\left(n_0^+ T\right)$ is large compared to $\ln(v)$, A.G. Davenport [DAV 64] deduces the average value of ℓ_0:

$$\overline{\ell_N} = E(\ell) = -\int_0^\infty \ell \, d\left[1 - \left(1 - e^{-\frac{\ell^2}{2 s_\ell^2}}\right)^{N_p}\right] \qquad [7.27]$$

i.e.

$$\overline{\ell_N} = \int_0^\infty \ell \, N_p \left[1 - \exp\left(-\frac{\ell^2}{2 s_\ell^2}\right)\right]^{N_p - 1} \frac{\ell}{s_\ell^2} \exp\left(-\frac{\ell^2}{2 s_\ell^2}\right) d\ell \qquad [7.28]$$

After a MacLaurin series development and an integration by parts [KOW 69], [LON 52]:

$$\overline{u_0} = \frac{E(\ell)}{s_\ell} = \sqrt{\frac{\pi}{2}} \left[\frac{N_p}{1! \sqrt{1}} - \frac{N_p (N_p - 1)}{2! \sqrt{2}} \right.$$

$$\left. + \frac{N_p (N_p - 1)(N_p - 2)}{3! \sqrt{3}} - \cdots + (-1)^{N_p + 1} \frac{1}{\sqrt{N_p}} \right] \qquad [7.29]$$

For large values of N_p, M.S. Longuet-Higgins [KRE 83], [LON 52] shows that we can use the asymptotic expression

$$\overline{u_0} \approx \sqrt{2 \ln\left(n_0^+ T\right)} + \frac{\varepsilon}{\sqrt{2 \ln\left(n_0^+ T\right)}} \qquad [7.30]$$

where ε is the Euler's constant equal to 0.57721566490 ... (cf. Appendix A4.3), the difference with the whole expression being about $\left(\ln N_p\right)^{-3/2}$ [UDW 73].

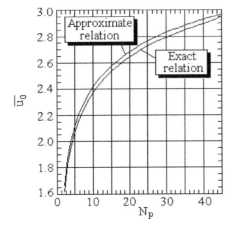

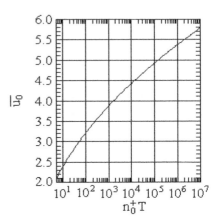

Figure 7.7. *Comparison of the approximate average value of the distribution of the highest peaks to the exact value*

Figure 7.8. *Average of the highest peaks*

The approximation is very good [CAR 56], even for small N_p (error less than 3% for all $N_p \geq 2$ and less than 1% if $N_p > 50$).

NOTE.–

Let us take the assumption $\ln\left(n_0^+ T\right) \gg \ln(v)$.

The ratio

$$\frac{\ln v}{\ln n_0^+ T} = \frac{-\frac{u^2}{2} + \ln n_0^+ T}{\ln n_0^+ T} = -\frac{u^2}{2 \ln n_0^+ T} + 1 \qquad [7.31]$$

is small with regard to the unit if $u^2 \approx 2 \ln n_0^+ T$. Approximation [7.30] is very acceptable for a narrowband process, i.e. for r close to 1 [CAR 56], [POO 76].

7.9.5. *Mean value probability*

If we consider expression [7.30] of the mean value of the law of largest peaks, we obtain, by carrying $\overline{u_0} = \sqrt{2\ln(N_p)} + \dfrac{\varepsilon}{\sqrt{2\ln(N_p)}}$ into [7.14]:

$$P(u_i < \overline{u_0}) = \left[1 - \exp\left(-\frac{\overline{u_0}^2}{2}\right)\right]^{N_p} \tag{7.32}$$

$$P = \left\{1 - \exp\left[-\frac{1}{2}\left(\sqrt{2\ln(N_p)} + \frac{\varepsilon}{\sqrt{2\ln(N_p)}}\right)^2\right]\right\}^{N_p} \tag{7.33}$$

$$P = \left\{1 - \exp\left[-\ln(N_p) - \varepsilon - \frac{\varepsilon^2}{4\ln(N_p)}\right]\right\}^{N_p} \tag{7.34}$$

Figure 7.9 shows the variations of probability of mean [7.34] according to the N_p variable between 10^3 and 10^{11}.

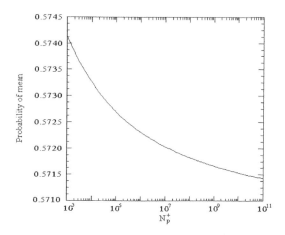

Figure 7.9. *Probability of the highest peak mean*

When N_p is great,

$$\ln P \approx N_p \ln\left(1 - \frac{e^{-\varepsilon}}{N_p}\right) \approx N_p \left(-\frac{e^{-\varepsilon}}{N_p}\right) \quad [7.35]$$

and

$$P \to \exp(-e^{-\varepsilon}) \approx 0.570376... \quad [7.36]$$

The probability of a larger peak than mean [7.30] is approximately 0.43.

7.9.6. Standard deviation of highest peaks

On the same assumptions, the standard deviation of the largest peak distribution is calculated from

$$s_{u_0} = \sqrt{\overline{u_0^2} - (\overline{u_0})^2}$$

$$s_{u_0} \approx \frac{\pi}{\sqrt{6}} \frac{1}{\sqrt{2 \ln(n_0^+ T)}} \quad [7.37]$$

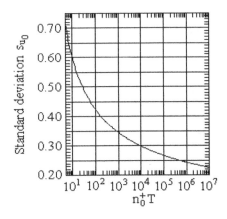

Figure 7.10. *Standard deviation of the distribution law of the highest peaks*

Figures 7.8 and 7.10 respectively show the average $\overline{u_0}$ and the standard deviation s_u as a function of $n_0^+ T$. We note on these curves that, when $n_0^+ T$ increases, the average increases and the standard deviation decreases very quickly.

We notice in Figure 7.11 that the slope of the curve $P_N(u)$ increases with $n_0^+ T$, result in conformity with the decrease of s_u.

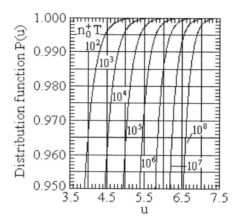

Figure 7.11. *Probability density of the largest peaks close to 1*

7.9.7. *Variation coefficient*

The variation coefficient (CoV) is equal to the ratio of the standard deviation s_{u_0} [7.37] and the mean $\overline{u_0}$ [7.30]:

$$\text{CoV} = \frac{\dfrac{\pi}{\sqrt{6}} \dfrac{1}{\sqrt{2 \ln(N_p)}}}{\sqrt{2 \ln(N_p)} + \dfrac{\varepsilon}{\sqrt{2 \ln(N_p)}}} \qquad [7.38]$$

If N_p is large

$$\text{CoV} \approx \frac{\dfrac{\pi}{\sqrt{6}} \dfrac{1}{\sqrt{2 \ln(N_p)}}}{\sqrt{2 \ln(N_p)}} = \frac{\pi}{\sqrt{6}} \frac{1}{2 \ln(N_p)} \qquad [7.39]$$

7.9.8. *Most probable value*

The most probable value of ℓ corresponds to the peak of the probability density curve defined by [7.19], i.e. to the mode ℓ_m (or to the reduced mode $m = \dfrac{\ell_m}{s_\ell}$). If we let $v = \left(\dfrac{\ell}{\sqrt{2}\, s_\ell}\right)^2$, it occurs when

$$\frac{d}{dv}\left[\left(1-e^{-v}\right)^{N_p-1} v^{1/2}\, e^{-v}\right] = 0$$

i.e. when [PRA 70], [UDW 73]

$$v = \ln N_p - \ln\left[1 - \frac{1}{2v}\left(1 - e^{-v}\right)\right] \qquad [7.40]$$

If N_p is large

$$v \approx \ln N_p \qquad [7.41]$$

yielding the most probable value

$$m = \frac{\ell_m}{s_\ell} = \sqrt{2}\,\sqrt{v} \approx \sqrt{2\ln N_p} \qquad [7.42]$$

$$m = \sqrt{2\ln\left(n_0^+\, T\right)} \qquad [7.43]$$

7.9.9. *Median*

The median is defined by u_m such that $P = \dfrac{1}{2} = \left[1 - \exp\left(-\dfrac{u_m^2}{2}\right)\right]^{N_p}$. Thus,

$$N_p \ln\left[1 - \exp\left(-\frac{u_m^2}{2}\right)\right] = -\ln 2$$

and

$$u_m = \sqrt{-2\ln\left[1 - \exp\left(-\frac{\ln 2}{N_p}\right)\right]} \qquad [7.44]$$

For N_p large

$$u_m \approx \sqrt{-2\ln\left(\frac{\ln 2}{N_p}\right)} = \sqrt{-2\left[\ln(\ln 2) - \ln(N_p)\right]} \qquad [7.45]$$

Figures 7.12 and 7.13 show the position of the mode, median and mean of the law of highest peaks for $N_p = 100$ on the distribution function and on probability density.

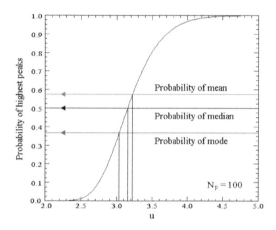

Figure 7.12. *Distribution function of the largest peaks ($N_p = 100$)*

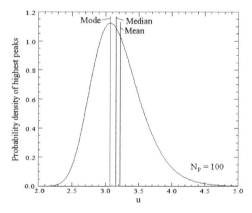

Figure 7.13. *Probability density of largest peaks ($N_p = 100$)*

Median/mean ratio

The ratio R of median [7.45] and mean [7.30] is equal to

$$R = \frac{\text{Median}}{\text{Mean}} \approx \frac{\sqrt{-2\left[\ln(\ln 2) - \ln(N_p)\right]}}{\sqrt{2\ln(N_p)} + \dfrac{\varepsilon}{\sqrt{2\ln(N_p)}}} \qquad [7.46]$$

When N_p becomes very large, ratio R tends toward 1 very slowly. However, we cannot ignore the second term of the mean ($\dfrac{\varepsilon}{\sqrt{2\ln(N_p)}}$), because its influence is still felt for the very large number of cycles N_p.

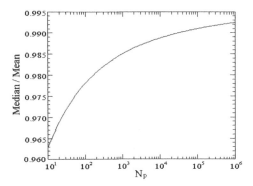

Figure 7.14. *Variations of the median/mean ratio according to N_p*

7.9.10. *Value of density at mode*

$$p_{N_m} = \frac{1}{e}\sqrt{2\ln\left(n_0^+ T\right)} \qquad [7.47]$$

A typical example of the use of the preceding relations relates to the study of the distribution of the wave heights, starting from an empirical relationship of the acceleration spectral density [PIE 63].

7.9.11. Value of distribution function at mode

From [7.14] and [7.43], setting $N_p = n_0^+ T$, we obtain:

$$P = \left[1 - \exp\left(-\ln(N_p)\right)\right]^{N_p} \quad [7.48]$$

$$P = \left(1 - \frac{1}{N_p}\right)^{N_p} \quad [7.49]$$

Still with the hypothesis where N_p is large,

$$\ln(N_p) = N_p \ln(1 - \frac{1}{N_p}) \approx -1.$$

Hence

$$P \to \frac{1}{e} = 0.36787944\ldots \quad [7.50]$$

The probability of finding a higher peak than the mode then is $1 - P = 0.6329\ldots$ [OCH 81]

7.9.12. Expected maximum

The expected maximum ℓ_p is such that [7.6]

$$P(\ell_p) = 1 - \frac{1}{N_p} = 1 - \exp\left(-\frac{\ell_p^2}{2 s_\ell^2}\right) \quad [7.51]$$

$$\ell_p = 2 s_\ell \ln N_p \quad [7.52]$$

7.9.13. Maximum exceeded with given risk α

$$P_N(\ell_{N\alpha}) = 1 - \alpha = \left[1 - \exp\left(-\frac{\ell_{N\alpha}^2}{2 s_\ell^2}\right)\right]^{N_p} \quad [7.53]$$

$$\ell_{N\alpha} = \sqrt{2 \ln \frac{1}{\left[1 - (1-\alpha)^{1/N_p}\right]}} \, s_\ell \quad [7.54]$$

i.e., for $\alpha \ll 1$,

$$\ell_{N\alpha} \approx s_\ell \sqrt{2 \ln \frac{N_p}{\alpha}} \qquad [7.55]$$

We find in Table 7.1 the value of the parameters defined above from relationships [7.30], [7.37], [7.43] and [7.47] for some values of $n_0^+ T$.

$n_0^+ T$	$\overline{u_0}$	s_{u_0}	$s_{u_0}/\overline{u_0}$	m	p_{Nm}
$3.6 \cdot 10^2$	3.5993	0.3738	$10.386 \cdot 10^{-2}$	3.4311	1.2622
$3.6 \cdot 10^3$	4.1895	0.3169	$7.565 \cdot 10^{-2}$	4.0469	1.4888
$3.6 \cdot 10^4$	4.7067	0.2800	$5.949 \cdot 10^{-2}$	4.5807	1.6851
$3.6 \cdot 10^5$	5.1725	0.2535	$4.901 \cdot 10^{-2}$	5.0584	1.8609
$3.6 \cdot 10^6$	5.5999	0.2334	$4.168 \cdot 10^{-2}$	5.4948	2.0214

Table 7.1. *Examples of values of parameters from the distribution law of highest peaks*

It should be noted that $s_{u_0}/\overline{u_0}$ is always very small and tends to decrease when $n_0^+ T$ increases. Table 7.2 gives, with respect to $n_0^+ T$, the values of $Q = 1 - P$ for $u = \overline{u_0}$ and $u = m$.

$n_0^+ T$	$\overline{u_0}$	$Q(\overline{u_0})$	m	$Q(m)$
10^2	3.2250	0.4239	3.0349	0.6321
10^3	3.8722	0.4258	3.7169	0.6321
10^4	4.4264	0.4267	4.2919	0.6321
$3.6 \cdot 10^4$	4.7067	0.4271	4.5807	0.6321
10^5	4.9188	0.4273	4.7985	0.6321
$3.6 \cdot 10^5$	5.1725	0.4275	5.0584	0.6321
10^6	5.3663	0.4277	5.2565	0.6321
$3.6 \cdot 10^6$	5.5999	0.4279	5.4948	0.6321
10^7	5.7794	0.4280	5.6777	0.6321
10^8	6.1648	0.4282	6.0697	0.6321

Table 7.2. *Examples of values of probability Q from the distribution law of highest peaks*

It should be noted that, for any value of $n_0^+ T$, $Q(\overline{u_0})$ and $Q(m)$ are almost constant.

In many problems, we can assume that with slight error the highest value is equal to the average value $\overline{u_0}$. It should also be noted that the average is higher than the mode, but the deviation decreases when $n_0^+ T$ increases.

Over one hour of vibrations and for an average frequency n_0^+ of the signal varying between 10 Hz and 1,000 Hz, we note that the average $\overline{u_0}$ varies between 4.7 and 5.6 times the rms value ℓ_{rms} (Figure 7.8 and Table 7.2). The amplitude of the largest peak therefore remains lower than 5.6 ℓ_{rms}.

The amplitude of the probability density to the mode increases with respect to $n_0^+ T$ (Figure 7.15).

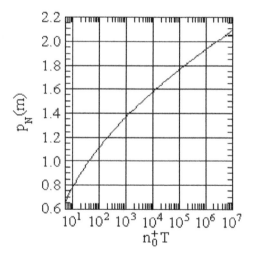

Figure 7.15. *Value of density of largest peaks at the mode*

7.10. Wideband centered normal process

7.10.1. *Average of largest peaks*

The preceding calculations were carried out on the assumption of a narrowband noise $(r \approx 1)$. For a wideband noise $(r \neq 1)$, D.E. Cartwright and M.S. Longuet-

Higgins [CAR 56] showed that the average value of the largest peak in a sample of N_p peaks is equal to:

$$\overline{u}_0 \approx \sqrt{2\ln(r\,N_p)} + \frac{\varepsilon}{\sqrt{2\ln(r\,N_p)}} \qquad [7.56]$$

($\varepsilon = 0.57721566490... =$ Euler's constant). We obtain relation [7.30] for $r = 1$, N_p then being equal to $n_0^+\,T$. Let us set $\sqrt{m_2}$ as the rms value of the peak distribution, where

$$m_2 = 1 + r^2 \qquad [7.57]$$

Figure 7.16 shows the variations of $\dfrac{\overline{u}_0}{\sqrt{m_2}}$ with respect to r, for various values of N_p.

For large N_p, $\dfrac{\overline{u}_0}{\sqrt{m_2}}$ is a decreasing function of r. When the spectrum widens, the average value of the highest peak decreases. When $r \to 0$ (Gaussian case), expression [7.56] can no longer be used, the quantity $r\,N_p$ becoming small compared to 1. The general expression is complicated and without much interest. R.A. Fisher and L.H.C. Tippett [CAR 56], [FIS 28], [TIP 25] propose an asymptotic expression of the form:

$$\overline{u}_0 = m + \frac{\varepsilon\,m}{1 + m^2} \qquad [7.58]$$

where m is the mode of the distribution of maxima, given in this case by

$$m\,e^{\frac{m^2}{2}} = \frac{N_p}{\sqrt{2\pi}} \qquad [7.59]$$

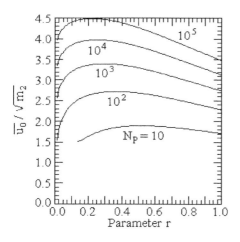 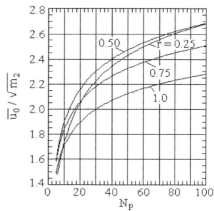

Figure 7.16. *Average value of the highest peak of a wideband process versus the irregularity factor*

Figure 7.17. *Average value of the highest peak of a wideband process versus the number of peaks*

Distribution [7.19] is thus centered around this mode for large N_p.

From [7.59], it results that:

$$m^2 = \ln\left(\frac{N_p^2}{2\pi}\right) - \ln(m^2)$$

yielding

$$m \approx \left\{\ln\left(\frac{N_p^2}{2\pi}\right) - \ln\left[\ln\left(\frac{N_p^2}{2\pi}\right)\right]\right\}^{1/2}$$

and

$$\overline{u_0} \approx \sqrt{2\ln\left(\frac{N_p}{\sqrt{2\pi}}\right)} \qquad [7.60]$$

We can show that $\overline{u_0}$ converges only very slowly towards this limit.

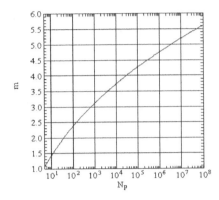

Figure 7.18. *Mode of the distribution law of the highest peaks of a wideband noise*

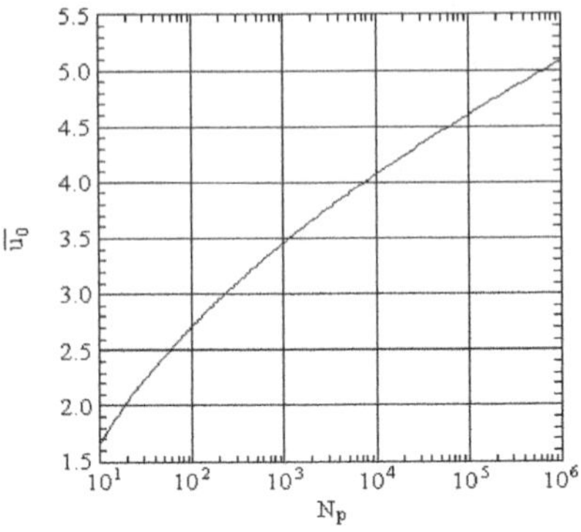

Figure 7.19. *Average value of the highest peaks of a wideband noise over duration T*

7.10.2. *Variance of the largest peaks*

The variance is given by [FIS 28]

$$s^2 = \frac{\pi^2}{6} \frac{m^2}{\left(m^2+1\right)^2} \qquad [7.61]$$

The standard deviation is plotted against N_p in Figure 7.20.

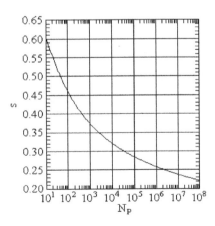

Figure 7.20. *Standard deviation of the distribution law of the highest peaks of a wideband noise*

Table 7.3 makes it possible to compare the values of $\dfrac{E(\ell)}{s_\ell}$ calculated from [7.29] with those given exactly by L.H.C. Tippet for some values of N_p [TIP 25].

		N_p	10	20	100	200	500	1000
		m	1.43165	1.74393	2.37533	2.61467	2.90799	3.11528
$E(\ell)$		Relation [7.29]	1.70263	1.99302	2.58173	2.80726	3.08549	3.28326
s_ℓ		L.H.C. Tippett	1.53875	1.86747	2.50758	2.74604	3.03670	3.24138

Table 7.3. *Comparison of exact and approximate values of* $\dfrac{E(\ell)}{s_\ell}$

7.10.3. *Variation coefficient*

The CoV, ratio of standard deviation calculated from [7.61] and the mean [7.58] of the largest peaks of a wideband noise (for N_p large), is equal to:

$$\text{CoV} = \frac{\pi}{\sqrt{6}} \frac{m}{1+\varepsilon+m^2} \qquad [7.62]$$

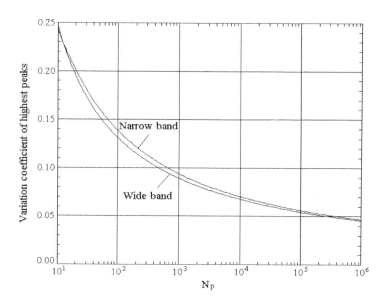

Figure 7.21. *CoV as a function of* N_p

In both narrowband and wideband hypotheses, the CoV of laws of probability of largest peaks tends toward zero when N_p tends toward the infinite (Figure 7.21).

7.11. Asymptotic laws

The use of exact laws of probability for extreme values, established from the initial distribution law of the instantaneous values or from the distribution law of the maxima, leads to calculations which quickly become very complicated.

They can be simplified by treating only the tail of the initial law, but with many precautions because, as we can well imagine, several asymptotic laws can be used in this domain. Moreover, the values contained therein, of weak probability, appear only occasionally and the real law is not well known.

7.11.1. Gumbel asymptote

This approximation is used for distribution functions, which tend towards 1 at least as quickly as exponential for the great values of the variable [GUM 54]. This asymptotic law applies in particular to the normal and lognormal laws. Let us consider a distribution function which, for x large, is of the form

$$P(x) = 1 - a\, \exp(-b\, x) \qquad [7.63]$$

The constants a and b are selected according to the law being simulated. If, for example, we want to respect the values of the expected maximum x_p and the logarithmic increase α_N:

$$P(x_p) = 1 - \frac{1}{N_p} \qquad [7.64]$$

$$\alpha_N = N_p\, p(x_p) \qquad [7.65]$$

In comparing these expressions with those derived from the P(x) law, it becomes:

$$\frac{1}{N_p} = a\, e^{-b\, x_p} \qquad [7.66]$$

$$\alpha_N = N_p \, a \, b \, e^{-b x_p} \qquad [7.67]$$

yielding

$$\begin{cases} b = \alpha_N \\ a = \dfrac{1}{N_p} e^{\alpha_N x_p} \end{cases}$$

The adjusted distribution function around x_p is thus

$$P(x) = 1 - \frac{1}{N_p} \exp\left[-\alpha_N (x - x_p)\right] \qquad [7.68]$$

For large N_p, we obtain an approximate value of the distribution function of the extreme values making use of the relationship

$$\left(1 - \frac{x}{N_p}\right)^{N_p} \approx e^{-x}$$

which yields

$$P_N(x) \approx \exp\left\{-\exp\left[-\alpha_N (x - x_p)\right]\right\} \qquad [7.69]$$

7.11.2. Case study: Rayleigh peak distribution

We have

$$x_p = s_x \sqrt{2 \ln N_p}$$

$$\alpha_N = \frac{1}{s_x} \sqrt{2 \ln N_p}$$

If we set $x = u \, s_x$ and if the reduced variable $\eta = \alpha_N (x - x_p)$ is considered, we have

$$\eta = \sqrt{2 \ln N_p}\left(u - \sqrt{2 \ln N_p}\right) \qquad [7.70]$$

The distribution function is expressed as

$$P_N(x) = \exp[-\exp(-\eta)] \qquad [7.71]$$

while the probability density is written:

$$p(\eta) = \exp[-\eta - \exp(-\eta)] \qquad [7.72]$$

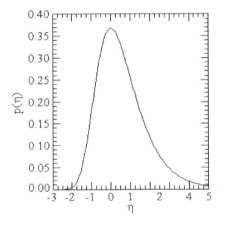

Figure 7.22. *Probability density of extreme values for a Rayleigh peak distribution*

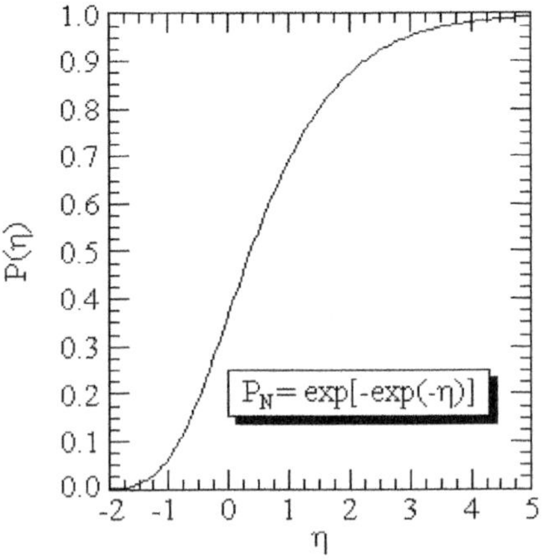

Figure 7.23. *Distribution function of extreme values for a Rayleigh peak distribution*

7.11.3. *Expressions for large values of* N_p

Average maximum

$$\overline{x_{N_p}} = x_p + \frac{\varepsilon}{\alpha_N} \qquad [7.73]$$

where $\varepsilon = 0.57722\ldots$ (Euler's constant).

Standard deviation of maxima

$$s_N = \frac{\pi}{\sqrt{6}} \frac{1}{\alpha_N} \qquad [7.74]$$

Probability of an extreme value less than x_p

$$P_N(x_p) \approx \frac{1}{e} \approx 0.36788 \qquad [7.75]$$

7.12. Choice of type of analysis

The prime objective is to simplify the analysis by reducing the number and duration of the signals studied. The starting datum is in general composed of one or more records of an acceleration time history. If there are several records, the first step is to carry out a check of the stationarity of the process and, if it is the case, its ergodicity. If we only have one record, we check the autostationarity of the signal and its ergodicity. These properties make it possible to reduce the analysis of the whole of the process to that of only one signal sample of short duration (a few tens of seconds for example).

This procedure is not always followed and we often prefer to plot the rms value of the record with respect to time (sliding average on a few tens of points). In a complementary way, we can add the time variations of skewness and kurtosis. This work makes it possible to identify the various events characteristic of the phenomenon, to isolate the shocks, the transitional phases and the time intervals when, the rms value varying little, the signal can be analyzed from a sample of short duration. It also makes it possible to make sure that the signal is Gaussian.

The *rms value* ℓ_{rms} of the signal gives an overall picture of the excitation intensity. It can be useful to calculate the average $E(\ell) = m$. If it differs from zero, we can either center the signal, if it is estimated that the physical phenomenon has indeed a zero average and that the DC component is due to an imperfection of measurement, or calculate the rms value of the total signal and the standard deviation $s = \sqrt{\ell_{rms}^2 - m^2}$.

In order to have a precise idea of the frequency content of the vibration, it is also important to calculate the *power spectral density (PSD)* of the signal in a sufficiently broad range not to truncate its frequency contents. If we have measurements carried out at several points of a structure, the PSDs can be used to calculate the transfer functions between these various points. The PSDs are in addition very often used as source data for other more specific analyses.

The test facilities are controlled starting from the PSD and it is still from the PSD that we can most easily evaluate the test feasibility on a given facility: calculation of the rms value of acceleration (on the whole frequency band or a given band), of the velocity and displacement, average frequency, etc.

The autocorrelation function is a rather more specific mode of analysis. We saw that this function is the inverse Fourier transform of the PSD. Strictly speaking, there is no more information in the autocorrelation than in the PSD. However, these two functions underline different properties of the signal. The autocorrelation makes it possible in particular to identify more easily the periodic signals which can be superimposed on the random vibration (measurement of the periods of the periodic components, measurement of coherence time, etc.) [VIN 72].

The identification of the nature of *the probability density law of the instantaneous values* of the signal is seldom carried out, for two essential reasons [BEN 61b]:

– this analysis is very long if we require points representative of the density around 3 to 4 times that of the rms value ℓ_{rms} (a recording lasting 18.5 minutes is necessary to estimate the probability density to $4\ \ell_{rms}$ of a normal law with an error of 30%);

– the tendency is generally, and sometimes wrongly, to consider *a priori* that the signal studied is Gaussian. However, skewness and kurtosis are simple indicators to use.

Peak value distribution

It is especially useful to know the distribution of the peak values when we wish to make a study of the fatigue damage. The parameter as a function of time to study must be, in this case, not acceleration at the input or in a point of the specimen, but rather the relative displacement between two given points (or, even better, directly strains or stresses in the part). The maxima of this displacement are proportional to the maximum stresses in the part on the assumption of linearity. We saw that if the signal is Gaussian, the probability density of the distribution of the peak values follows a law made up of the sum of a Gaussian law and Rayleigh law.

Extreme values analysis

This type of analysis can also be interesting either for studies of fatigue damage or for studies of damage due to crossing a threshold stress, while working under the same conditions as above.

It can also be useful to determine these values directly on the acceleration signal to anticipate possible disjunctions of the test facility as a result of going beyond its possibilities.

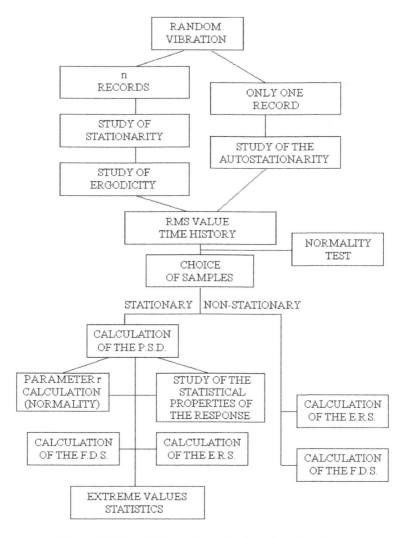

Figure 7.24. *Possibilities of analysis of random vibrations*

Threshold level crossings

The study of threshold crossings of a random signal can be of interest in certain cases:

– to reduce the test duration by preserving the shape of the PSD and that of the threshold level crossings curve (by rotation of this last curve) [HOR 75], [LAL 81]. This method is not often used;

– to predict collisions between parts of a structure or to choose the dimension of the clearance between parts (the signal being a relative displacement);

– to anticipate disjunctions of the test facility.

7.13. Study of the envelope of a narrowband process

7.13.1. *Probability density of the maxima of the envelope*

It was previously shown how we can estimate the maxima distribution of a random vibration.

Another method of analyzing the properties of the maxima can consist of studying the smoothed curve connecting all the peaks of the signal [BEN 58], [BEN 64], [CRA 63], [CRA 67], [RIC 44].

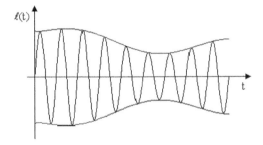

Figure 7.25. *Narrowband vibration and its envelope*

Given a random vibration $\ell(t)$, we can use a diagram giving $\dot{\ell}(t)$ with respect to $\ell(t)$. For a sinusoidal movement, we would have:

$$\ell(t) = A \sin \omega_0 t \qquad [7.76]$$

$$\dot{\ell}(t) = A \omega_0 \cos \omega_0 t \qquad [7.77]$$

and the diagram $\dfrac{\dot{\ell}(t)}{\omega_0}$ according to $\ell(t)$ would be a circle of radius A, since:

$$\ell^2(t) + \frac{\dot{\ell}^2(t)}{\omega_0^2} = A^2 \qquad [7.78]$$

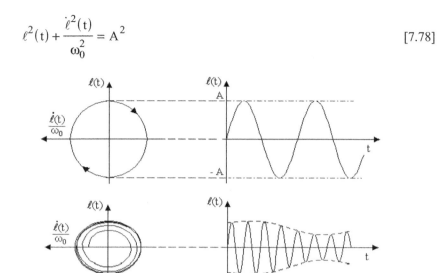

Figure 7.26. *Study of the envelope of a sinusoidal signal and of a narrowband signal*

The envelope of this sinusoid is made up of two straight lines: ± A. In the case of a narrowband random signal, envelope A is a time function and can be regarded as the amplitude of a function of the form [DEE 71]:

$$u(t) = A(t) \sin[\omega_0 t + \theta(t)] \qquad [7.79]$$

in which $A(t)$ and the phase $\theta(t)$ are random functions that are assumed to be slowly variable with ω_0. There are in reality two symmetric curves with respect to the time axis which are envelopes of the curve $\ell(t)$.

By analogy with the case of a pure sinusoid, $A(t)$ can be considered the radius of the image point in the diagram $\ell(t)$, $\dot{\ell}(t)$:

$$A^2(t) = \ell^2(t) + \frac{\dot{\ell}^2(t)}{\omega_0^2}$$

$(A \geq 0)$, where

$$\ell(t) = A(t) \sin[\theta(t)]$$

$$\dot{\ell}(t) = A(t)\,\omega_0\,\cos[\theta(t)]$$

The probability that the envelope lies between A and A + dA is equal to the joint probability that the curves ℓ and $\dfrac{\dot{\ell}}{\omega_0}$ are located in the hatched field ranging between the two circles of radius A and A + dA (Figure 7.27).

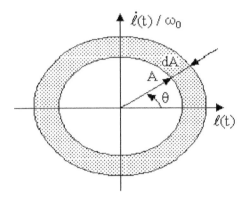

Figure 7.27. *Probability that the envelope lies between A and A + dA*

Consider the corresponding two dimensional probability density $p\!\left(\ell, \dfrac{\dot{\ell}}{\omega_0}\right)$. We have:

$$p\!\left(\ell, \frac{\dot{\ell}}{\omega_0}\right) dx\, d\!\left(\frac{\dot{\ell}}{\omega_0}\right) = \omega_0\, p(\ell, \dot{\ell})\, d\ell\, d\!\left(\frac{\dot{\ell}}{\omega_0}\right)$$

$$p\!\left(\ell, \frac{\dot{\ell}}{\omega_0}\right) d\ell\, d\!\left(\frac{\dot{\ell}}{\omega_0}\right) = p(A\sin\theta, A\,\omega_0\cos\theta)\, A\, dA\, d\theta$$

$$p\!\left(\ell, \frac{\dot{\ell}}{\omega_0}\right) d\ell\, d\!\left(\frac{\dot{\ell}}{\omega_0}\right) = q(A, \theta)\, dA\, d\theta \qquad [7.80]$$

where

$$q(A, \theta) = A\; p(A\sin\theta,\, A\,\omega_0\cos\theta) \qquad [7.81]$$

Statistics of Extreme Values

The probability density function $q(A)$ of the envelope $A(t)$ is obtained by making the sum of all the angles θ:

$$q(\theta) = \int_0^{2\pi} q(A,\theta)\, d\theta \qquad [7.82]$$

Let us now assume that the random vibration $\ell(t)$ and its derivative $\dot{\ell}(t)$ are statistically independent, with zero averages and equal variances $s_\ell^2 = s_{\dot{\ell}}^2$ ($= \ell_{rms}^2$), according to a two-dimensional Gaussian law:

$$p\left(\ell, \frac{\dot{\ell}}{\omega_0}\right) = \frac{1}{2\pi s_\ell^2} \exp\left[-\frac{\ell^2 + \frac{\dot{\ell}^2}{\omega_0^2}}{2 s_\ell^2}\right] = \frac{1}{2\pi s_\ell^2} \exp\left[-\frac{A^2}{2 s_\ell^2}\right] \qquad [7.83]$$

$$q(A, \theta) = \frac{A}{2\pi s_\ell^2} \exp\left[-\frac{A^2}{2 s_\ell^2}\right] \qquad [7.84]$$

and

$$q(A) = \frac{A}{s_\ell^2} \exp\left[-\frac{A^2}{2 s_\ell^2}\right] \qquad [7.85]$$

($A \geq 0$). The probability density of the envelope $A(t)$ follows Rayleigh's law.

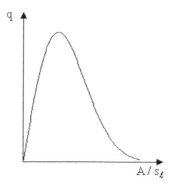

Figure 7.28. *Probability density of envelope A(t)*

NOTES.–

1. *The probability density $q(A)$, calculated at a given time t, is independent of t, the process being assumed to be stationary.*

We could calculate this density from an arbitrary signal $^{i}\ell(t)$. The result would be independent of the sample chosen in $^{i}\ell(t)$ if the process is ergodic [CRA 63].

2. *The density $q(A)$ has the same form as the probability density $q(a)$ of maxima [CRA 63]. It is a consequence of the assumption of a Gaussian law for $\ell(t)$ and $\dot{\ell}(t)$. In the case of a narrowband noise for which this assumption would not be observed, or if the system were non-linear, the densities $q(A)$ and $q(a)$ would have different forms [BEN 64], [CRA 61].*

When the process has only one maximum per cycle, the maxima have the same distribution as its envelope (this remark is strictly true when r = 1).

When the number of maxima per second increases and tends towards infinity, we have seen that the distribution of maxima becomes identical to that of the instantaneous values of the signal (Gaussian law) [CRA 68].

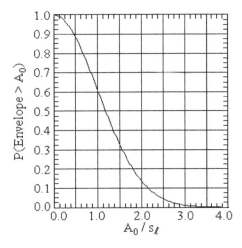

Figure 7.29. *Probability that the envelope exceeds a given threshold A_0*

The probability that the envelope exceeds a certain given value A_0 is obtained by integrating $q(A)$ between A_0 and infinity:

$$P(\text{Envelope} > A_0) = \int_{A_0}^{\infty} q(A)\, dA$$

$$P(\text{Envelope} > A_0) = \exp\left[-\frac{A_0^2}{2\,s_\ell^2}\right]$$

$\dfrac{A_0}{s_\ell}$	P
0.5	0.8825
1	0.6065
2	0.1353
3	0.0111

Table 7.4. *Examples of probabilities of threshold crossings*

7.13.2. *Distribution of maxima of envelope*

S.O. Rice [RIC 44] showed that the average number of maxima (per second) of the envelope of a white noise between two frequencies f_a and f_b is:

$$N \approx 0.64110\,(f_b - f_a) \qquad [7.86]$$

Let us set $v = \dfrac{A_{max}}{s_\ell}$. If v is large (greater than 2.5), the probability density $q(v)$ can be approximated by:

$$q(v) \approx \frac{\sqrt{\dfrac{\pi}{6}}}{0.64110}\,(v^2 - 1)\,e^{-\dfrac{v^2}{2}} \qquad [7.87]$$

and the corresponding distribution function by:

380   Random Vibration

$$Q_A = Q(A_{max} < v\, s_\ell) \approx 1 - \frac{\sqrt{\frac{\pi}{6}}}{0.64110} v\, e^{-\frac{v^2}{2}} \qquad [7.88]$$

Q_A is the probability that a maximum of the envelope chosen randomly is lower than a given value $A = v\, s_\ell$. The functions $q(v)$ and Q_A are respectively plotted in the general case (arbitrary v) in Figures 7.30 and 7.31.

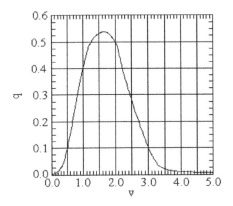 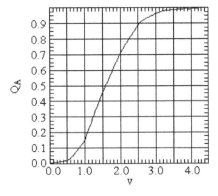

Figure 7.30. *Probability density of envelope maxima*   **Figure 7.31.** *Distribution function of envelope maxima*

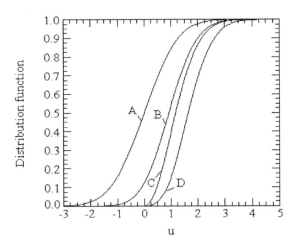

Figure 7.32. *Comparison of distribution functions*

Figure 7.32 shows, by way of comparison, the distribution functions of:

– the instantaneous values of the signal (normal law) (A): $P = \dfrac{1}{2}\left[1 + \mathrm{erf}\left(\dfrac{u}{\sqrt{2}}\right)\right]$;

– the maxima of signal [6.67] (B);

– the instantaneous values of the envelope (Rayleigh's law) (C): $P = 1 - e^{-\dfrac{u^2}{2}}$;

– the maxima of the envelope (curve given by S.O. Rice [RIC 44]) (D).

7.13.3. *Average frequency of envelope of narrowband noise*

It is shown that [BOL 84]:

$$\varphi_M^2 = \dfrac{\int_0^\infty (\Omega - \theta)^2\, G(\Omega)\, d\Omega}{\ell_{rms}} \qquad [7.89]$$

where

ℓ_{rms} = rms value of the noise $\ell(t)$;

θ = average pulsation of the noise $(2\pi f_0)$;

f_0 = average frequency of $\ell(t)$.

For a signal $\ell(t)$ whose PSD $G(f)$ is constant between frequencies f_1 and f_2 and centered on f_0, this relationship leads to:

$$\varphi_M^2 = \dfrac{\int_{f_1}^{f_2} (f - f_0)^2\, df}{f_2 - f_1}$$

i.e. to:

$$\varphi_M = \left[\dfrac{f_1^2 + f_1 f_2 + f_2^2}{3} - f_0(f_1 + f_2) + f_0^2\right]^{1/2} \qquad [7.90]$$

Summary tables of the main results

Parameter	Relation	Expression
Number of crossings of a threshold a with positive slope per unit time	[5.44]	$n_a = n_0 \, e^{-\frac{a^2}{2\ell_{rms}^2}}$
Average frequency	[5.53] [5.79]	$n_0^+ = \left[\dfrac{\int_0^\infty f^2 \, G(f) \, df}{\int_0^\infty G(f) \, df}\right]^{\frac{1}{2}} = \dfrac{1}{2\pi}\sqrt{\dfrac{M_2}{M_0}}$
Moments	[5.77]	$M_n = (2\pi)^n \int_0^\infty f^n \, G(f) \, df$
Irregularity factor	[6.6]	$r = \dfrac{\dot{\ell}_{rms}^2}{\ell_{rms}^2 \, \ddot{\ell}_{rms}^2} = \dfrac{M_2}{\sqrt{M_0 \, M_4}} = \dfrac{R^{(2)}(0)}{\sqrt{R(0)\,R^{(4)}(0)}}$
Probability density of the maxima	[6.19]	$q(u) = \dfrac{\sqrt{1-r^2}}{\sqrt{2\pi}} \, e^{-\frac{u^2}{2(1-r^2)}} + \dfrac{r}{2} \, u \, e^{-\frac{u^2}{2}}\left[1+\text{Erf}\left(\dfrac{r\,u}{\sqrt{2(1-r^2)}}\right)\right]$
Average number of maxima per second	[6.13] [6.34]	$n_p^+ = \left[\dfrac{\int_0^{+\infty} f^4 \, G(f) \, df}{\int_0^{+\infty} f^2 \, G(f) \, df}\right]^{\frac{1}{2}} = \dfrac{1}{2\pi}\sqrt{\dfrac{M_4}{M_2}}$
Average time between two successive maxima (narrowband noise)	[6.43]	$\tau_m = \dfrac{1}{f_0}\left[\dfrac{1+\dfrac{1}{3}\left(\dfrac{\Delta f}{2 f_0}\right)^2}{1+2\left(\dfrac{\Delta f}{2 f_0}\right)^2 + \dfrac{1}{5}\left(\dfrac{\Delta f}{2 f_0}\right)^4}\right]^{\frac{1}{2}}$
Average correlation between two successive maxima	[6.44]	$\rho = \dfrac{1}{2\pi\delta}\left[\dfrac{1+2\delta^2+\dfrac{\delta^4}{5}}{1+\dfrac{\delta^2}{3}}\right]^{1/2}\left\{\cos\left[\dfrac{1+\dfrac{\delta^2}{3}}{1+2\delta^2+\dfrac{\delta^4}{5}}\right]^{1/2} - \sin\left[\dfrac{1+\dfrac{\delta^2}{3}}{1+2\delta^2+\dfrac{\delta^4}{5}}\right]^{1/2}\right\}$

Table 7.5(a). *Main results*

Statistics of Extreme Values 383

Parameter	Relation	Expression
Distribution function of the peaks	[6.67]	$Q_p(u) = \dfrac{1}{2}\left\{1-\text{Erf}\left[\dfrac{u}{\sqrt{2(1-r^2)}}\right]\right\} + \dfrac{r}{2}e^{-\dfrac{u^2}{2}}\left\{1+\text{Erf}\left[\dfrac{r\,u}{\sqrt{2(1-r^2)}}\right]\right\}$
Average number of maxima greater than a threshold a per unit time	[6.77]	$M_a = \dfrac{1}{2\pi}\sqrt{\dfrac{M_4}{M_2}}\, e^{-\dfrac{a^2}{2\ell_{rms}^2}} = \dfrac{1}{2\pi}\sqrt{\dfrac{M_2}{M_0}}\, e^{-\dfrac{a^2}{2M_0}}$
Average number of positive maxima per second	[6.107]	$n_p^+ = \dfrac{1}{4\pi}\left(\sqrt{\dfrac{M_2}{M_0}} + \sqrt{\dfrac{M_4}{M_2}}\right)$
Average time interval between the maxima	[6.83]	$T_a = \dfrac{1}{M_a} = \dfrac{1}{n_p^+\, Q(a)}$
Average time interval between the maxima (narrowband noise)	[6.85]	$T_a = 2\pi\,\dfrac{M_0}{M_2}\, e^{\dfrac{a^2}{2M_0}}$
Probability so that a maximum is positive or negative	[6.101] [6.102]	$q_{max}^+ = \dfrac{1+r}{2} \qquad q_{max}^- = \dfrac{1-r}{2}$
Time during which the signal is above a given value	[6.96]	$\overline{T}_{ab} = \displaystyle\int_a^b \dfrac{1}{\ell_{rms}\sqrt{2\pi}}\, e^{-\dfrac{\ell^2}{2\ell_{rms}^2}}\, d\ell$
Average amplitude jump between two successive maxima	[6.110]	$\overline{h} = \sqrt{2\pi}\, r\, \ell_{rms}$
Probability density of the largest peaks	[7.19]	$p_N(\ell) = N_p\left[1 - \exp\left(-\dfrac{\ell^2}{2s_\ell^2}\right)\right]^{N_p-1} \dfrac{\ell}{s_\ell^2}\exp\left(-\dfrac{\ell^2}{2s_\ell^2}\right)$

Table 7.5(b). *Main results*

Parameter	Relation	Expression
Probability density of the largest maximum over duration T	[7.26]	$p_N(u) = n_0^+ \, T \, u \, \exp\left\{-\left[\dfrac{u^2}{2} + n_0^+ \, T \, \exp\left(-\dfrac{u^2}{2}\right)\right]\right\}$
Average for the large values of the number of peaks (narrowband noise)	[7.30]	$\overline{u}_0 \approx \sqrt{2 \ln\left(n_0^+ \, T\right)} + \dfrac{\varepsilon}{\sqrt{2 \ln\left(n_0^+ \, T\right)}}$
Standard deviation (narrowband noise)	[7.37]	$s_{u_0} \approx \dfrac{\pi}{\sqrt{6}} \dfrac{1}{\sqrt{2 \ln\left(n_0^+ \, T\right)}}$
Most probable value (mode)	[7.43]	$m = \sqrt{2 \ln\left(n_0^+ \, T\right)}$
Maximum exceeded with a risk α	[7.54]	$\ell_{N\alpha} = \sqrt{2 \ln \dfrac{1}{[1-(1-\alpha)]^{1/N_p}}} \, s_\ell$
Average for the great values of the number of peaks (wideband noise)	[7.56]	$\overline{u}_0 \approx \sqrt{2 \ln\left(r \, N_p\right)} + \dfrac{\varepsilon}{\sqrt{2 \ln\left(r \, N_p\right)}}$
Standard deviation (wideband noise)	[7.61]	$s^2 = \dfrac{\pi^2}{6} \dfrac{m^2}{\left(m^2+1\right)^2}$
Probability density of the envelope of a narrowband Gaussian process	[7.85]	$q(A) = \dfrac{A}{s_\ell^2} \exp\left[-\dfrac{A^2}{2 \, s_\ell^2}\right]$
Distribution of maxima of the envelope of a narrowband process	[7.87]	$q(v) \approx \dfrac{\sqrt{\dfrac{\pi}{6}}}{0.64110} \left(v^2 - 1\right) e^{-\dfrac{v^2}{2}}$
Average frequency of the envelope of a narrowband noise of a constant PSD	[7.90]	$\varphi_M = \left[\dfrac{f_1^2 + f_1 \, f_2 + f_2^2}{3} - f_0\left(f_1 + f_2\right) + f_0^2\right]^{1/2}$

Table 7.5(c). *Main results*

Chapter 8

Response of a One-Degree-of-Freedom Linear System to Random Vibration

8.1. Average value of the response of a linear system

The response of a linear system to an excitation assumed to be random stationary is given by the general relation

$$q(\theta) = \int_{-\infty}^{+\infty} \lambda(\theta - \alpha) \; h(\alpha) \; d\alpha \qquad [8.1]$$

where

$\lambda(\theta)$ is the excitation;

α is a variable of integration;

and $h(\theta)$ is the impulse response of the system.

The ensemble average of $q(\theta)$ (*through the process*) is

$$E[q(\theta)] = E\left[\int_{-\infty}^{+\infty} \lambda(\theta - \alpha) \; h(\alpha) \; d\alpha\right]$$

The operators E and sum being linear, we can, by permuting them, write

$$E[q(\theta)] = \int_{-\infty}^{+\infty} E[\lambda(\theta-\alpha)] \, h(\alpha) \, d\alpha$$

Since, by assumption, the process $\lambda(\theta)$ is stationary, the ensemble averages are independent of the time at which they are calculated:

$$E[q(\theta)] = E[\lambda(\theta)] \int_{-\infty}^{+\infty} h(\theta) \, d\theta$$

This relation links the average value of the response to that of the excitation. We saw that the Fourier transform of $h(\theta)$ is equal to:

$$H(\Omega) = \int_{-\infty}^{+\infty} h(\theta) \, e^{-i\theta\Omega} \, d\theta$$

and

$$H(0) = \int_{-\infty}^{+\infty} h(\theta) \, d\theta$$

$$E[q(\theta)] = E[\lambda(\theta)] \, H(0) \qquad [8.2]$$

If the process $\lambda(\theta)$ is zero average, the response is also zero average [CRA 63].

8.2. Response of perfect bandpass filter to random vibration

Let us consider a perfect bandpass filter defined by:

$H(f) = 0 \qquad$ for $0 \le f \le f_a$

$H(f) = H_0 \qquad$ for $f_a \le f \le f_b$

$H(f) = 0 \qquad$ for $f > f_b$

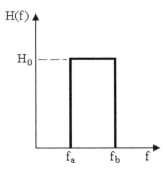

Figure 8.1. *Bandpass filter*

The response of this filter to the excitation $\ell(t)$ is given by the convolution integral

$$u(t) = \int_0^\infty h(\lambda)\, \ell(t-\lambda)\, d\lambda = h(t) * \ell(t) \qquad [8.3]$$

where $h(t)$ is the impulse response of the system. In the frequency domain, we saw that, if $G_\ell(f)$ is the input PSD of the signal $\ell(t)$ and $G_u(f)$ that of the response $u(t)$:

$$G_u(f) = |H(f)|^2\, G_\ell(f) \qquad [8.4]$$

yielding the rms value u_{rms} of the response

$$u_{rms}^2 = \int_0^\infty G_u(f)\, df$$

$$u_{rms}^2 = \int_0^\infty |H(f)|^2\, G_\ell(f)\, df$$

$$u_{rms}^2 = H_0^2 \int_{f_a}^{f_b} G_\ell(f)\, df \qquad [8.5]$$

If the PSD of the input is a white noise, $G_\ell(f) = G_{\ell 0}$ = constant:

$$u_{rms}^2 = H_0^2\, G_{\ell 0}\, (f_b - f_a) \qquad [8.6]$$

If the transfer function of the filter is such that $H_0 = 1$:

$$u_{rms} = \sqrt{G_{\ell 0}\, (f_b - f_a)} \qquad [8.7]$$

$$u_{rms} = \ell_{rms}(f_a, f_b)$$

The rms value of the response signal is equal to the rms value of the input in the band f_a, f_b.

NOTE.– *The PSD symbol is modified by an index indicating on which signal it is calculated ($\ell(t)$, $\ddot{x}(t)$,....). The symbol G is followed by Ω or f between brackets to* *specify the nature of the variable, pulsation or frequency. If the PSD is constant, we add the index 0 (e.g. $G_{\ell 0}(\Omega)$).*

8.3. The PSD of the response of a one-degree-of-freedom linear system

Consider a mass-spring-damping linear system of natural frequency $f_0 = \omega_0/2\pi$ and quality factor $Q = 1/2\xi$. As above, the response $u(t)$ of the system to the input $\ell(t)$ has as a general expression (Volume 1):

$$u(t) = \int_0^\infty h(\alpha)\,\ell(t-\alpha)\,d\alpha = h(t) * \ell(t) \tag{8.8}$$

where the impulse response of the system $h(t)$ is equal to

$$h(t) = \omega_0 \frac{e^{-\omega_0 \xi t}}{\sqrt{1-\xi^2}} \sin\left(\omega_0 \sqrt{1-\xi^2}\,t\right) \tag{8.9}$$

For a physical system, $h(t) = 0$ for $t < 0$ (causal system). In the frequency domain, the PSDs are related by

$$S_u(\Omega) = |H(\Omega)|^2\, S_\ell(\Omega) \tag{8.10}$$

For a relative response, the function $H(\Omega)$ is given by

$$H(\Omega) = \frac{1}{\sqrt{\left[1-\left(\dfrac{\Omega}{\omega_0}\right)^2\right]^2 + \left(\dfrac{\Omega}{Q\,\omega_0}\right)^2}} \tag{8.11}$$

and for an absolute response by

$$H(\Omega) = \frac{\sqrt{1+\dfrac{1}{Q}\left(\dfrac{\Omega}{\omega_0}\right)^2}}{\sqrt{\left[1-\left(\dfrac{\Omega}{\omega_0}\right)^2\right]^2 + \left(\dfrac{\Omega}{Q\,\omega_0}\right)^2}} \tag{8.12}$$

Response of a One-Degree-of-Freedom Linear System 389

In a more general way, we can write the complex transfer in the form

$$H(i\Omega) = \frac{B_0 + i\Omega B_1}{A_0 - \Omega^2 A_2 + i\Omega A_1} \quad [8.13]$$

where $B_1 = 0$ or 1 in terms of the case selected.

8.4. Rms value of response to white noise

We will initially assume that the PSDs $S(\Omega)$ are defined in the most general case for Ω between $-\infty$ and $+\infty$. The rms value of the response, u_{rms}, is given by

$$u_{rms}^2 = \int_{-\infty}^{+\infty} S_u(\Omega) \, d\Omega$$

$$u_{rms}^2 = \int_{-\infty}^{+\infty} |H(\Omega)|^2 S_\ell(\Omega) \, d\Omega$$

Let us examine the theoretical particular case of a white noise where $S_\ell(\Omega) = \text{constant} = S_{\ell 0}$ from $-\infty$ to $+\infty$. Then

$$u_{rms}^2 = S_{\ell 0} \int_{-\infty}^{+\infty} |H(\Omega)|^2 \, d\Omega$$

It can be shown [LAL 94] that the integral

$$I = \int_{-\infty}^{+\infty} |H(i\Omega)|^2 \, d\Omega = \int_{-\infty}^{+\infty} \left| \frac{B_0 + i\Omega B_1}{A_0 - \Omega^2 A_2 + i\Omega A_1} \right|^2 d\Omega$$

has as a value

$$I = \pi \frac{\left(\dfrac{B_0^2}{A_0}\right) A_2 + B_1^2}{A_1 A_2}$$

Assumption	n° 1 (relative response)	n° 2 (absolute response)
$H(\Omega)$	$H(\Omega) = \dfrac{1}{\sqrt{\left[1-\left(\dfrac{\Omega}{\omega_0}\right)^2\right]^2 + \left(\dfrac{\Omega}{Q\,\omega_0}\right)^2}}$	$H(\Omega) = \dfrac{\sqrt{1+\dfrac{1}{Q}\left(\dfrac{\Omega}{\omega_0}\right)^2}}{\sqrt{\left[1-\left(\dfrac{\Omega}{\omega_0}\right)^2\right]^2 + \left(\dfrac{\Omega}{Q\,\omega_0}\right)^2}}$
B_0	1	1
B_1	0	$1/Q\,\omega_0$
A_0	1	1
A_1	$1/Q\,\omega_0$	$1/Q\,\omega_0$
A_2	$1/\omega_0^2$	$1/\omega_0^2$
I	$\pi\,Q\,\omega_0$	$\dfrac{\pi}{Q}\left(Q^2+1\right)\omega_0$

Table 8.1. *Parameters for the calculation of the integral I*

This yields

assumption 1 (relative response)

$$u_{rms} = \sqrt{\pi\,Q\,\omega_0\,S_{\ell_0}}$$ [8.14]

assumption 2 (absolute response)

$$u_{rms} = \sqrt{\dfrac{\pi\,\omega_0}{Q}\left(1+Q^2\right)S_{\ell_0}}$$ [8.15]

Particular cases

Case 1. $\ell(t) = -\dfrac{\ddot{x}(t)}{\omega_0^2}$

$$S_{\ell_0} = \dfrac{S_{\ddot{x}_0}}{\omega_0^4}$$

Response of a One-Degree-of-Freedom Linear System   391

and [CRA 63], [CRA 66]

$$u_{rms}^2 = z_{rms}^2 = \frac{\pi Q S_{\ddot{x}0}}{\omega_0^3} \qquad [8.16]$$

Case 2. The PSD is defined, in terms of Ω, in $(0, \infty)$ [GAL 57]:

$$G_\ell(\Omega) = 2 S_\ell(\Omega)$$

$$u_{rms}^2 = \frac{\pi \omega_0 Q}{2} G_{\ell 0} \qquad [8.17]$$

Case 3. From case 2, $\ell(t) = -\dfrac{\ddot{x}(t)}{\omega_0^2}$ [CRA 68], [CRA 70], [MIL 61]:

$$u_{rms}^2 = z_{rms}^2 = \frac{\pi}{2} \frac{Q G_{\ddot{x}0}}{\omega_0^3} \qquad [8.18]$$

Case 4. The PSD is defined in terms of f in $(-\infty, +\infty)$.

From [8.14], since $S_{\ell 0}(f) = 2\pi S_{\ell 0}(\Omega)$

$$u_{rms}^2 = \frac{\omega_0 Q}{2} S_{\ell 0} \qquad [8.19]$$

Case 5. From case 4, if $\ell(t) = -\dfrac{\ddot{x}(t)}{\omega_0^2}$

$$u_{rms}^2 = z_{rms}^2 = \frac{Q}{2\omega_0^3} S_{\ddot{x}0} \qquad [8.20]$$

Case 6. The PSD is defined in terms of f in $(0, \infty)$:

$$G_\ell(f) = 2 S_\ell(f)$$

and [8.19] [BEN 61b], [MOR 63], [MOR 75], [PIE 64]:

$$u_{rms}^2 = \frac{\omega_0 Q}{4} G_{\ell 0} \qquad [8.21]$$

Case 7. From case 6, if $\ell(t) = -\dfrac{\ddot{x}(t)}{\omega_0^2}$ [BEN 62], [BEN 64], [CRA 63], [PIE 64],

$$u_{rms}^2 = z_{rms}^2 = \frac{\omega_0 \, Q}{4} \frac{G_{\ddot{x}0}(f)}{\omega_0^4}$$

$$u_{rms}^2 = \frac{Q \, G_{\ddot{x}0}}{4 \, \omega_0^3} \qquad [8.22]$$

NOTES.–

1. *We could find, in the same way, the results obtained on assumption 2 while replacing, in the relations above, Q with* $\frac{1+Q^2}{Q}$ *or* $\frac{1+Q^2}{Q} \omega_0^4$ *(in terms of whether it concerns* $G_{\ell 0}$ *or* $G_{\ddot{x}0}$*).*

2. *The response of a linear one-degree-of-freedom system subjected to a sine wave excitation is proportional to Q. It is proportional to* $\sqrt{Q}$ *if the excitation is random [CRE 56a].*

Table 8.2 lists various expressions of the rms value in terms of the definition of the PSD. If the PSD is defined in $(0, \infty)$, we now consider various inputs (acceleration, velocity, displacement or force). The transfer function of a linear one-degree-of-freedom system can be written:

$$H(\Omega) = \frac{1}{\sqrt{\left(1 - \Omega^2/\omega_0^2\right)^2 + 4\, \xi^2 \, \Omega^2/\omega_0^2}}$$

The rms value of the response is calculated, for an excitation defined by a displacement, a velocity or an acceleration, respectively using the following integrals:

$$\int_0^\infty H^2(h) \, G_\ell(h) \, dh = G_{\ell 0} \int_0^\infty H^2(h) \, dh = \frac{\pi}{4\,\xi} \omega_0 \, G_{\ell 0}(\Omega) \qquad [8.23]$$

$$\int_0^\infty h^2 \, H^2(h) \, G_\ell(h) \, dh = G_{\ell 0} \int_0^\infty h^2 \, H^2(h) \, dh = \frac{\pi}{4\,\xi} \omega_0^3 \, G_{\ell 0}(\Omega) \qquad [8.24]$$

$$\int_0^\infty h^4 \, H^2(h) \, G_\ell(h) \, dh = G_{\ell 0} \int_0^\infty h^4 \, H^2(h) \, dh = \frac{\pi}{4\,\xi} \left(1 + 4\,\xi^2\right) \omega_0^5 \, G_{\ell 0}(\Omega)$$

$$[8.25]$$

Table 8.3 lists the rms responses of a linear one-degree-of-freedom system for these excitations (of which the PSD is constant) [PIE 64].

Summary chart

	Type of transfer function	
Definition of PSD	$H = \dfrac{1}{\sqrt{\left[1-\left(\dfrac{f}{f_0}\right)^2\right]^2 + \left(\dfrac{f}{Q f_0}\right)^2}}$	$H = \dfrac{\sqrt{1+\left(\dfrac{f}{Q f_0}\right)^2}}{\sqrt{\left[1-\left(\dfrac{f}{f_0}\right)^2\right]^2 + \left(\dfrac{f}{Q f_0}\right)^2}}$
PSD of the reduced variable in terms of Ω in $(-\infty, \infty)$	$\sqrt{\pi\, \omega_0\, S_{\ell_0}\, Q}$	$\sqrt{\pi\, \omega_0\, \dfrac{1+Q^2}{Q}\, S_{\ell_0}}$
PSD of $\ddot{x}(t)$ in terms of Ω in $(-\infty, \infty)$	$\sqrt{\dfrac{\pi\, Q}{\omega_0^3}\, S_{\ddot{x}_0}}$	$\sqrt{\pi\, \dfrac{1+Q^2}{Q}\, S_{\ddot{x}_0}\, \omega_0}$
PSD of $\ell(t)$ in terms of Ω in $(0, \infty)$	$\sqrt{\dfrac{\pi\, \omega_0\, Q}{2}\, G_{\ell_0}}$	$\sqrt{\dfrac{\pi\, \omega_0}{2}\, \dfrac{1+Q^2}{Q}\, G_{\ell_0}}$
PSD of $\ddot{x}(t)$ in terms of Ω in $(0, \infty)$	$\sqrt{\dfrac{\pi\, Q}{2\, \omega_0^3}\, G_{\ddot{x}_0}}$	$\sqrt{\dfrac{\pi}{2}\, \dfrac{1+Q^2}{Q}\, G_{\ddot{x}_0}\, \omega_0}$
PSD of $\ell(t)$ in terms of f in $(-\infty, \infty)$	$\sqrt{\dfrac{\omega_0\, Q}{2}\, S_{\ell_0}}$	$\sqrt{\dfrac{\omega_0}{2}\, \dfrac{1+Q^2}{Q}\, S_{\ell_0}}$
PSD of $\ddot{x}(t)$ in terms of f in $(-\infty, \infty)$	$\sqrt{\dfrac{Q}{2\, \omega_0^3}\, S_{\ddot{x}_0}}$	$\sqrt{\dfrac{1+Q^2}{2\, Q}\, S_{\ddot{x}_0}\, \omega_0}$
PSD of $\ell(t)$ in terms of f in $(0, \infty)$	$\sqrt{\dfrac{\omega_0\, Q}{4\, \omega_0^3}\, G_{\ell_0}}$	$\sqrt{\dfrac{\omega_0}{4}\, \dfrac{1+Q^2}{Q}\, G_{\ell_0}}$
PSD of $\ddot{x}(t)$ in terms of f in $(0, \infty)$	$\sqrt{\dfrac{Q\, G_{\ddot{x}_0}}{4\, \omega_0^3}}$	$\sqrt{\dfrac{1+Q^2}{4\, Q}\, G_{\ddot{x}_0}\, \omega_0}$

Table 8.2. *Expressions of the rms values of the response in terms of the definition of the PSD*

Excitation	Parameter response	Rms value
Absolute displacement G_x in $(0, \infty)$	Absolute displacement	$y_{rms} = \left[\dfrac{\pi f_0 \left(1 + 4\xi^2\right) G_x}{4\xi} \right]^{1/2}$
Absolute velocity $G_{\dot{x}}$	Absolute velocity	$\dot{y}_{rms} = \left[\dfrac{\pi f_0 \left(1 + 4\xi^2\right) G_{\dot{x}}}{4\xi} \right]^{1/2}$
Absolute acceleration $G_{\ddot{x}}$	Absolute acceleration	$\ddot{y}_{rms} = \left[\dfrac{\pi f_0 \left(1 + 4\xi^2\right) G_{\ddot{x}}}{4\xi} \right]^{1/2}$
	Relative displacement	$z_{rms} = \left[\dfrac{G_{\ddot{x}}}{64 \pi^3 f_0^3 \xi} \right]^{1/2}$
Force $G_x = \dfrac{G_F}{k^2}$	Relative or absolute displacement	$y_{rms} = z_{rms} = \left[\dfrac{\pi f_0 G_F}{4\xi k^2} \right]^{1/2}$

Table 8.3. *Rms values of the response in terms of the definition of excitation*

If the noise is not white, we obtain the relationships by taking the value of G for $\Omega = \omega_0$ and by supposing that the value of G varies little with Ω between the half-power points. The approximation is much better when ξ is smaller ($0.005 \leq \xi \leq 0.05$). For a weak hysteretic damping, we would set $\eta = 2\xi$ in these relations.

These relations are thus applicable only:

– if the spectrum has a constant amplitude;

– if not, with slightly damped systems (for which the transfer function presents a narrow peak), with the condition that the PSD varies little around the natural frequency of the system.

In all cases, the natural frequency must be in the frequency range of the PSD.

8.5. Rms value of response of a linear one-degree-of-freedom system subjected to bands of random noise

8.5.1. *Case where the excitation is a PSD defined by a straight line segment in logarithmic scales*

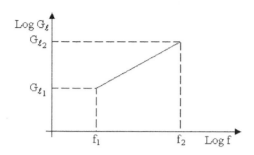

Figure 8.2. *PSD defined by a segment of straight line in logarithmic scales*

We propose to examine the case of signals whose spectrum can be broken up into several bands, each one having a constant or linearly variable amplitude in logarithmic scales; the PSD will be defined in a band f_1, f_2 (or Ω_1, Ω_2) by:

$$G_\ell(\Omega) = G_{\ell_1} \left(\frac{\Omega}{\Omega_1}\right)^b = \frac{G_{\ell_1}}{h_1^b} h^b \qquad [8.26]$$

where ω_0 is the natural pulsation of the one-degree-of-freedom system and b is a constant whose value characterizes the slope of the segment defined in logarithmic scales ($h = \Omega/\omega_0$).

NOTE.– *We saw that the number of octaves n which separate two frequencies f_1 and f_2 is equal to [LAL 82]:*

$$n = \frac{\log \dfrac{f_2}{f_1}}{\log 2} \qquad [8.27]$$

In addition, we define the number of decibels m between two levels of the PSD by

$$m(dB) = 10 \log_{10} \frac{G_\ell(\Omega_2)}{G_\ell(\Omega_1)} \qquad [8.28]$$

yielding, from

$$m(dB) = 10 b \log_{10} \frac{f_2}{f_1} \qquad [8.29]$$

the number of dB per octave,

$$N = \frac{m}{n} = \frac{10 b \log \frac{\Omega_2}{\Omega_1}}{\log \frac{\Omega_2}{\Omega_1}} \log 2$$

$$N \approx 3.01 b \qquad [8.30]$$

The mean square value of the response is

$$u_{rms}^2 = \int_{\Omega_1}^{\Omega_2} G_\ell(\Omega) H^2(\Omega) d\Omega \qquad [8.31]$$

where $G_\ell(\Omega)$ is the PSD of the input $\ell(t)$, defined from 0 to infinity, and $H(\Omega)$ is the transfer function of the system. With the notation $h = \dfrac{\Omega}{\omega_0}$, the function H can be written as follows (Table 8.4).

In general,

$$u_{rms}^2 = \int_{h_1}^{h_2} \omega_0 \frac{G_{\ell_1}}{h_1^b} h^b H^2 dh \qquad [8.32]$$

As in case 1,

$$u_{rms}^2 = \omega_0 \frac{G_{\ell_1}}{h_1^b} \int_{h_1}^{h_2} \frac{h^b}{\left(1 - h^2\right)^2 + \dfrac{h^2}{Q^2}} dh$$

$$u_{rms}^2 = \frac{\pi \omega_0 G_{\ell_1}(\Omega)}{4 \xi h_1^b} \left[I_b(h_2) - I_b(h_1)\right] \qquad [8.33]$$

where $h_1 = \dfrac{f_1}{f_0} = \dfrac{\Omega_1}{\omega_0}$, $h_2 = \dfrac{f_2}{f_0} = \dfrac{\Omega_2}{\omega_0}$ and

$$I_b(h) = \frac{4\xi}{\pi} \int \frac{h^b}{\left(1-h^2\right)^2 + 4\xi^2 h^2} \, dh \qquad [8.34]$$

Case	Transfer function	Excitation and response
1	$H(h) = \dfrac{1}{\left[\left(1-h^2\right)^2 + \dfrac{h^2}{Q^2}\right]^{1/2}}$	$\ell(t) = -\dfrac{\ddot{x}(t)}{\omega_0^2}$ and $u(t) = z(t)$ or $\ell(t) = \dfrac{F(t)}{k}$ and $u(t) = y(t) = z(t)$
2	$H(h) = \dfrac{\sqrt{1 + \dfrac{h^2}{Q^2}}}{\left[\left(1-h^2\right)^2 + \dfrac{h^2}{Q^2}\right]^{1/2}}$	$\ell(t) = x(t)$ and $u(t) = y(t)$ or $\ell(t) = \ddot{x}(t)$ and $u(t) = \ddot{y}(t)$ or $\ell(t) = \dot{x}(t)$ and $u(t) = \dot{y}(t)$
3	$H(h) = \dfrac{h^2}{\left[\left(1-h^2\right)^2 + \dfrac{h^2}{Q^2}\right]^{1/2}}$	$\ell(t) = \dfrac{F(t)}{k}$ and $u(t) = \ddot{y}(t)$ or $\ell(t) = x(t)$ and $u(t) = z(t)$ or $\ell(t) = \dot{x}(t)$ and $u(t) = \dot{z}(t)$ or $\ell(t) = \ddot{x}(t)$ and $u(t) = \ddot{z}(t)$

Table 8.4. *Transfer functions relating to the various values of the generalized variables*

For case 2,

$$u_{rms}^2 = \omega_0 \frac{G_{\ell_1}}{h_1^b} \int_{h_1}^{h_2} h^b \frac{1 + \frac{h^2}{Q^2}}{(1-h^2)^2 + \frac{h^2}{Q^2}} dh$$

$$u_{rms}^2 = \frac{\pi}{4\xi} \omega_0 \frac{G_{\ell_1}(\Omega)}{h_1^b} [I_b(h) + 4\xi I_{b+2}(h)]_{h_1}^{h_2} \qquad [8.35]$$

For case 3,

$$u_{rms}^2 = \omega_0 \frac{G_{\ell_1}(\Omega)}{h_1^b} \int_{h_1}^{h_2} h^b \frac{h^4}{(1-h^2)^2 + \frac{h^2}{Q^2}} dh$$

$$u_{rms}^2 = \frac{\pi}{4\xi} \omega_0 \frac{G_{\ell_1}(\Omega)}{h_1^b} [I_{b+4}(h_2) - I_{b+4}(h_1)] \qquad [8.36]$$

If we set

$$\beta = 2(1 - 2\xi^2) \qquad [8.37]$$

it is shown (Appendix A6.1) that, for $b \neq 3$,

$$I_b = \frac{4\xi}{\pi} \frac{h^{b-3}}{b-3} + \beta I_{b-2} - I_{b-4} \qquad [8.38]$$

and for $b = 3$,

$$I_3 = \frac{4\xi}{\pi} \ln h + \beta I_1 - I_{-1} \qquad [8.39]$$

This recurrence formula makes it possible to calculate I_b for arbitrary b [PUL 68]. Table A6.1 in Appendix A6 lists expressions for I_b.

Particular cases

It can be useful to use a series expansion in terms of h around zero to improve the precision of the calculation of the rms values for larger values of f_0, or an asymptotic development in terms of 1/h for the small values of f_0.

1. For small h, lower than 0.02, a development in limited series shows that, for the fifth order,

$$I_0 \approx \frac{4\xi}{\pi} h + \frac{8\xi}{3\pi}\left(1 - 2\xi^2\right) h^3 + \frac{4\xi}{5\pi}\left(3 - 16\xi^2 + 16\xi^4\right) h^5 \qquad [8.40]$$

$$I_1 \approx -\frac{4}{\pi\alpha} \arctan\left(\frac{\alpha}{2\xi}\right) + \frac{2\xi}{\pi}\left[1 + \left(1 - 2\xi^2\right)h^2\right]h^2 \qquad [8.41]$$

$$(\alpha = 2\sqrt{1-\xi^2})$$

$$I_2 \approx \frac{4\xi}{3\pi} h^3 + \frac{8\xi}{5\pi}\left(1 - 2\xi^2\right) h^5 \qquad [8.42]$$

$$I_3 \approx -\frac{2\beta}{\pi\alpha} \arctan\left(\frac{\alpha}{2\xi}\right) + \frac{\xi}{\pi} h^4 \qquad [8.43]$$

2. For large h, greater than 40, we have, for the fifth order,

$$I_0 \approx 1 - \frac{4\xi}{3\pi h^3} - \frac{8\xi}{5\pi h^5}\left(1 - 2\xi^2\right) \qquad [8.44]$$

$$I_1 \approx -\frac{2\xi}{\pi h^2} - \frac{2\xi}{\pi h^4}\left(1 - 2\xi^2\right) \qquad [8.45]$$

$$I_2 \approx 1 - \frac{4\xi}{\pi h} - \frac{8\xi}{3\pi h^3}\left(1 - 2\xi^2\right) - \frac{4\xi}{5\pi h^5}\left(3 - 16\xi^2 + 16\xi^4\right) \qquad [8.46]$$

$$I_3 \approx \frac{4\xi}{\pi} \ln(h) - \frac{4\xi}{\pi h^2}\left(1 - 2\xi^2\right) - \frac{\xi}{\pi h^4}\left(3 - 16\xi^2 + 16\xi^4\right) \qquad [8.47]$$

NOTE.– *For large values of h (>100) and more particularly with $I_0(h)$, it can be interesting to directly evaluate $\Delta I = I(h_1 + \Delta h) - I(h_1)$ starting from relationships [8.44] to [8.47] when the width $\Delta h = h_2 - h_1$ is small. We then have*

$$\Delta I_0 \approx \frac{4\xi}{3\pi h_1^3}\left[1 - \frac{1}{(1+\Delta h/h_1)^3}\right] + \frac{8\xi(1-2\xi^2)}{5\pi h_1^5}\left[1 - \frac{1}{(1+\Delta h/h_1)^5}\right] \quad [8.48]$$

$$\Delta I_1 \approx \frac{2\xi}{\pi h_1^2}\left[1 - \frac{1}{(1+\Delta h/h_1)^2}\right] + \frac{2\xi(1-2\xi^2)}{\pi h_1^4}\left[1 - \frac{1}{(1+\Delta h/h_1)^4}\right] \quad [8.49]$$

$$\Delta I_2 \approx \frac{4\xi}{\pi h_1}\left[1 - \frac{1}{1+\Delta h/h_1}\right] + \frac{8\xi(1-2\xi^2)}{3\pi h_1^3}\left[1 - \frac{1}{(1+\Delta h/h_1)^3}\right]$$

$$+ \frac{4\xi(3-16\xi^2+16\xi^4)}{5\pi h_1^5}\left[1 - \frac{1}{(1+\Delta h/h_1)^5}\right] \quad [8.50]$$

$$\Delta I_3 \approx \frac{4\xi}{\pi}\ln\left(1+\frac{\Delta h}{h_1}\right) + \frac{4\xi(1-2\xi^2)}{\pi h_1^2}\left[1 - \frac{1}{(1+\Delta h/h_1)^2}\right]$$

$$+ \frac{\xi(3-16\xi^2+16\xi^4)}{\pi h_1^4}\left[1 - \frac{1}{(1+\Delta h/h_1)^4}\right] \quad [8.51]$$

where the terms $\dfrac{1}{(1+\Delta h/h_1)^n}$ can be replaced by series developments.

On the most current assumption where the excitation is an acceleration $[\ell(t) = -\ddot{x}(t)/\omega_0^2]$ and the response is relative displacement $z(t)$, relation [8.33], transformed while replacing G_ℓ with $G_\ell(\Omega) = \dfrac{G_{\ddot{x}}(\Omega)}{\omega_0^4} = \dfrac{G_{\ddot{x}_0}(\Omega)}{\omega_0^4}$, and $G_{\ddot{x}}(\Omega)$ with $\dfrac{G_{\ddot{x}}(f)}{2\pi}$, then becomes:

$$z_{rms}^2 = \dfrac{G_{\ddot{x}_0}(f)}{2\pi\,\omega_0^3\,h_1^b}\,\dfrac{\pi}{4\xi}\left[I_b(h_2) - I_b(h_1)\right] \qquad [8.52]$$

Since $G_{\dot{z}} = (2\pi f)^2 G_z$ and $G_{\ddot{z}} = (2\pi f)^2 G_{\dot{z}}$, the rms velocity $\dot{z}_{rms}$ is given in the same way by:

$$\dot{z}_{rms}^2 = \dfrac{G_{\ddot{x}_0}(f)}{2\pi\,\omega_0\,h_1^b}\,\dfrac{\pi}{4\xi}\left[I_{b+2}(h_2) - I_{b+2}(h_1)\right] \qquad [8.53]$$

and rms acceleration by:

$$\ddot{z}_{rms}^2 = \dfrac{\omega_0\,G_{\ddot{x}_0}(f)}{2\pi\,h_1^b}\,\dfrac{\pi}{4\xi}\left[I_{b+4}(h_2) - I_{b+4}(h_1)\right] \qquad [8.54]$$

8.5.2. *Case where the vibration has a PSD defined by a straight line segment of arbitrary slope in linear scales*

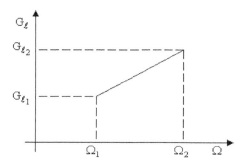

Figure 8.3. *PSD defined by a straight line segment in linear scales*

Let us suppose that the vibration is characterized by a signal $\ell(t)$ of which the PSD can be represented by an expression of the form

$$G_\ell(\Omega) = A\,\Omega + B \qquad [8.55]$$

where

$$A = \frac{G_{\ell_2} - G_{\ell_1}}{\Omega_2 - \Omega_1}$$

$$B = G_{\ell_1} - \Omega_1 \frac{G_{\ell_2} - G_{\ell_1}}{\Omega_2 - \Omega_1}$$

$$B = \frac{G_{\ell_1} \Omega_2 - G_{\ell_2} \Omega_1}{\Omega_2 - \Omega_1}$$

$$u_{rms}^2 = \int_{\Omega_1}^{\Omega_2} H^2(\Omega) \, G_\ell(\Omega) \, d\Omega$$

In case 1,

$$u_{rms}^2 = \omega_0 \int_{h_1}^{h_2} \frac{A\Omega + B}{\left(1 - h^2\right)^2 + 4\xi^2 h^2} \, dh$$

This expression can be written, with the notations of the above sections:

$$u_{rms}^2 = \frac{\pi A \omega_0^2}{4\xi} [I_1(h_2) - I_1(h_1)] + \frac{\pi B \omega_0}{4\xi} [I_0(h_2) - I_0(h_1)]$$

i.e., by replacing A and B with their expressions,

$$u_{rms}^2 = \frac{\pi \omega_0}{4\xi} \left\{ \frac{G_{\ell_2} - G_{\ell_1}}{h_2 - h_1} [I_1(h_2) - I_1(h_1)] \right.$$

$$\left. + \frac{G_{\ell_1} h_2 - G_{\ell_2} h_1}{h_2 - h_1} [I_0(h_2) - I_0(h_1)] \right\} \qquad [8.56]$$

Response of a One-Degree-of-Freedom Linear System   403

If the vibration is defined by an acceleration $[\ell(t) = -\ddot{x}(t)/\omega_0^2]$ and if the PSD is defined in terms of f, we have $G_{\ddot{x}}(\Omega) = \dfrac{G_{\ddot{x}}(f)}{2\pi}$ and

$$z_{rms}^2 = \dfrac{\pi}{4\xi(2\pi)^4 f_0^3} \left\{ \dfrac{G_{\ddot{x}2} - G_{\ddot{x}1}}{h_2 - h_1}[I_1(h_2) - I_1(h_1)] \right.$$

$$\left. + \dfrac{G_{\ddot{x}1} h_2 - G_{\ddot{x}2} h_1}{h_2 - h_1}[I_0(h_2) - I_0(h_1)] \right\} \qquad [8.57]$$

Likewise, it can be shown that

$$\dot{u}_{rms}^2 = \dfrac{\pi A \omega_0^4}{4\xi}[I_3(h_2) - I_3(h_1)] + \dfrac{\pi B \omega_0^3}{4\xi}[I_2(h_2) - I_2(h_1)]$$

$$\dot{z}_{rms}^2 = \dfrac{\pi}{4\xi(2\pi)^2 f_0} \left\{ \dfrac{G_{\ddot{x}2} - G_{\ddot{x}1}}{h_2 - h_1}[I_3(h_2) - I_3(h_1)] \right.$$

$$\left. + \dfrac{G_{\ddot{x}1} h_2 - G_{\ddot{x}2} h_1}{h_2 - h_1}[I_2(h_2) - I_2(h_1)] \right\} \qquad [8.58]$$

and that

$$\ddot{u}_{rms}^2 = \dfrac{\pi A \omega_0^6}{4\xi}[I_5(h_2) - I_5(h_1)] + \dfrac{\pi B \omega_0^5}{4\xi}[I_4(h_2) - I_4(h_1)]$$

$$\ddot{z}_{rms}^2 = \dfrac{\pi f_0}{4\xi} \left\{ \dfrac{G_{\ddot{x}2} - G_{\ddot{x}1}}{h_2 - h_1}[I_5(h_2) - I_5(h_1)] \right.$$

$$\left. + \dfrac{G_{\ddot{x}1} h_2 - G_{\ddot{x}2} h_1}{h_2 - h_1}[I_4(h_2) - I_4(h_1)] \right\} \qquad [8.59]$$

The functions $I_0(h)$ to $I_5(h)$ are defined by expressions [A6.20] to [A6.25] in Appendix A6.

8.5.3. *Case where the vibration has a constant PSD between two frequencies*

8.5.3.1. *PSD defined in a frequency interval of arbitrary width*

For a white noise of constant level $G_{\ell_0}(\Omega)$ between two frequencies Ω_1 and Ω_2, [CRA 63], by making $G_{\ell_1} = G_{\ell_2} = G_{\ell_0}$ in [8.56]:

$$u_{rms}^2 = \frac{\pi \omega_0 \, G_{\ell_0}}{4 \xi} \left[I_0(h_2) - I_0(h_1) \right]$$

$$u_{rms}^2 = \frac{\omega_0 \, G_{\ell_0}(\Omega)}{4 \xi} \left[\frac{\xi}{\alpha} \ln \frac{h^2 + \alpha h + 1}{h^2 - \alpha h + 1} + \arctan \frac{2h + \alpha}{2\xi} + \arctan \frac{2h - \alpha}{2\xi} \right]_{h_1}^{h_2}$$

[8.60]

NOTE.–

If $h_1 \to 0$, the term between square brackets tends towards zero. It tends towards 1 when $h_2 \to \infty$. This term is thus the corrective factor to apply to the corresponding expression of section 8.4 to take account of the limits f_1 and f_2, which are respectively non-zero and finite.

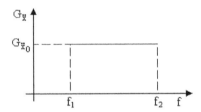

Figure 8.4. *Constant PSD between two frequencies*

More simply, if the excitation $\ddot{x}(t)$ is a noise of constant PSD $G_{\ddot{x}_0}$ between two arbitrary frequencies f_1 and f_2, we have, since $\ell(t) = -\ddot{x}(t)/\omega_0^2$, $G_\ell(\Omega) = G_{\ddot{x}}(\Omega)/\omega_0^4$ and $G_{\ddot{x}}(f) = 2\pi \, G_{\ddot{x}}(\omega)$,

$$z_{rms}^2 = \frac{G_{\ddot{x}_0}}{(2\pi)^4 \, f_0^3} \frac{\pi}{4\xi} \left[I_0(h_2) - I_0(h_1) \right]$$

[8.61]

Response of a One-Degree-of-Freedom Linear System   405

i.e.

$$z_{rms}^2 = \frac{G_{\ddot{x}0}(f)}{4\,\xi\,(2\,\pi)^4\,f_0^3}\left[\frac{\xi}{\alpha}\ln\frac{h^2+\alpha\,h+1}{h^2-\alpha\,h+1}+\text{arc tan}\frac{2\,h+\alpha}{2\,\xi}+\text{arc tan}\frac{2\,h-\alpha}{2\,\xi}\right]_{h_1}^{h_2}$$

[8.62]

The rms values of the velocity and acceleration have the following expressions:

$$\dot{z}_{rms}^2 = \frac{G_{\ddot{x}0}}{(2\,\pi)^2\,f_0}\,\frac{\pi}{4\,\xi}\,[I_2(h_2)-I_2(h_1)]$$

[8.63]

$$\ddot{z}_{rms}^2 = G_{\ddot{x}0}\,f_0\,\frac{\pi}{4\,\xi}\,[I_4(h_2)-I_4(h_2)]$$

[8.64]

Example 8.1.

$$\left.\begin{array}{l}f_1 = 1\text{ Hz}\\ f_2 = 2{,}000\text{ Hz}\end{array}\right\} \qquad G_{\ddot{x}0} = 1\ (m/s^2)^2/\text{Hz} \qquad f_0 = 100\text{ Hz}$$

$$(\ddot{x}_{rms} = 44.71\text{ m/s}^2) \qquad \xi = 0.05$$

Rms values of the response

$$z_{rms} = 1.0036\ 10^{-4}\text{ m}$$

$$\dot{z}_{rms} = 0.063\text{ m/s}$$

$$\ddot{z}_{rms} = 59.54\text{ m/s}^2$$

$$\omega_0^2\,z_{rms} = 39.62\text{ m/s}^2$$

Average frequency

$$n_0^+ = \frac{1}{2\,\pi}\frac{\dot{z}_{rms}}{z_{rms}} = 99.87\text{ Hz}$$

Number of maxima per second

$$n_p^+ = \frac{1}{2\,\pi}\frac{\ddot{z}_{rms}}{\dot{z}_{rms}} = 150.5$$

Parameter r

$$r = \frac{n_0^+}{n_p^+} = 0.6637$$

Particular case where $\xi = 0$

In this case

$$z_{rms}^2 = \int_{\Omega_1}^{\Omega_2} \frac{G_{\ddot{x}0}}{\omega_0^4 (1-h^2)} \, d\Omega$$

It is assumed here that $G_{\ddot{x}0}$ is expressed in terms of Ω. If $G_{\ddot{x}0}$ is defined in terms of f, we have, as previously,

$$z_{rms}^2 = \frac{G_{\ddot{x}0}}{2\pi \omega_0^3} \int_{h_1}^{h_2} \frac{dh}{(1-h^2)^2}$$

Let us set

$$h = \text{th } x$$

$$dh = \frac{1}{\text{ch}^2 x} dx$$

$$z_{rms}^2 = \frac{G_{\ddot{x}0}}{2\pi \omega_0^3} \int_{x_1}^{x_2} \frac{1}{\text{ch}^2 x} \frac{dx}{(1-\text{th}^2 x)^2} = \frac{G_{\ddot{x}0}}{2\pi \omega_0^3} \int_{x_1}^{x_2} \text{ch}^2 x \, dx$$

$$z_{rms}^2 = \frac{G_{\ddot{x}0}}{2\pi \omega_0^3} \int_{x_1}^{x_2} \frac{1 + \text{ch } 2x}{2} dx = \frac{G_{\ddot{x}0}}{2\pi \omega_0^3} \left[\frac{x}{2} + \frac{1}{2} \frac{\text{sh } 2x}{2} \right]_{x_1}^{x_2}$$

$$x_1 = \text{arg th } h_1$$

and

$$z_{rms}^2 = \frac{G_{\ddot{x}0}}{4\pi \omega_0^3} \left\{ \text{arg th}(h_2) - \text{arg th}(h_1) + \frac{1}{2} \text{sh}[2\,\text{arg th}(h_2)] - \frac{1}{2} \text{sh}[2\,\text{arg th}(h_1)] \right\}$$

[8.65]

8.5.3.2. *Case of a narrowband noise of width* $\Delta f = \lambda f_0/Q$

It is assumed that

$$\Delta f = \lambda \frac{f_0}{Q} \qquad [8.66]$$

where λ is a constant. From [8.61], let us calculate I_0 for

$$h_2 = \frac{f_0 + \lambda \frac{\Delta f}{2}}{f_0}$$

and

$$h_1 = \frac{f_0 - \lambda \frac{\Delta f}{2}}{f_0}$$

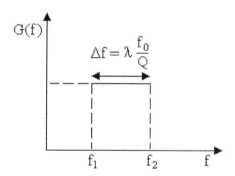

Figure 8.5. *Narrowband noise of width* $\Delta f = \lambda f_0 / Q$

If G is the value of the PSD and assuming that Q sufficiently large so that $\sqrt{4Q^2 - 1} \approx 2Q$,

$$z_{rms}^2 \approx \frac{GQ}{2(2\pi)^4 f_0^3} \left[\frac{1}{4Q} \ln \frac{1+\lambda/2Q}{1-\lambda/2Q} + \arctan \frac{2\lambda}{16Q^2 + 1 - \lambda^2} + 2 \arctan \lambda \right]$$

i.e., at first approximation

$$z_{rms}^2 \approx \frac{G}{(2\pi)^4 f_0^3} \left\{ \frac{\lambda}{8Q} + Q \left[\frac{\lambda}{16Q^2 + 1 - \lambda^2} + \arctan \lambda \right] \right\} \qquad [8.67]$$

and, if $\lambda \ll 16\, Q$,

$$z_{rms}^2 \approx \frac{G\,Q}{(2\pi)^4 f_0^3}\, \text{arc tan}(\lambda) \qquad [8.68]$$

Equivalence with white noise

There will be equivalence (same rms response) with white noise $G_{\ddot{x}_0}$, if

$$z_{rms}^2 = \frac{\pi}{2}\frac{f_0\, G_{\ddot{x}_0}\, Q}{(2\pi f_0)^4} = \frac{G\,Q\,f_0}{(2\pi f_0)^4}\, \text{arc tan}(\lambda)$$

i.e., if

$$G \approx \frac{\pi}{2}\frac{G_{\ddot{x}_0}}{\text{arc tan}(\lambda)} \qquad [8.69]$$

Particular cases

If $\lambda = 1$,

$$\Delta f = \frac{f_0}{Q} \qquad [8.70]$$

$$z_{rms}^2 \approx \frac{\pi}{4}\frac{G\,Q}{(2\pi)^4 f_0^3} \qquad [8.71]$$

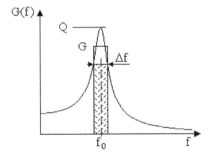

Figure 8.6. *Determination of an equivalence with white noise*

Response of a One-Degree-of-Freedom Linear System 409

There is the same rms response if

$$z_{rms}^2 = \frac{\pi}{2} \frac{f_0 \, G_{\ddot{x}_0} \, Q}{(4\pi^2 f_0^2)^2} = \frac{\pi}{4} \frac{G \, Q}{(2\pi)^4 f_0^3}$$

i.e., if

$$G \approx 2 \, G_{\ddot{x}_0} \quad [8.72]$$

If $\lambda = \dfrac{\pi}{2}$,

$$G = \frac{\pi}{2} \frac{G_{\ddot{x}_0}}{\arctan\left(\dfrac{\pi}{2}\right)}$$

$$z_{rms}^2 \approx \frac{G \, Q \, f_0}{(2\pi f_0)^4} \arctan\left(\frac{\pi}{2}\right) \quad [8.73]$$

Equivalence with a white noise $G_{\ddot{x}_0}$ exists if

$$G \approx G_{\ddot{x}_0} \frac{\pi}{2 \arctan\left(\dfrac{\pi}{2}\right)} \quad [8.74]$$

$$G \approx 1.5647 \, G_{\ddot{x}_0}$$

8.5.4. *Excitation defined by an absolute displacement*

In this case, the relative displacement rms response is given by:

$$z_{rms}^2 = \int_0^\infty G_z(f) \, df = \int_{f_1}^{f_2} G_z(f) \, df$$

where

$$G_z(f) = H_{xz}^2(f) \, G_x(f)$$

410   Random Vibration

In terms of the expressions listed in Table 4.1 of Volume 1:

$$H_{xz} = \Omega^2 \, H_{\ddot{x}z} = \frac{\Omega^2}{\omega_0^2 \sqrt{(1-h^2)^2 + \frac{h^2}{Q^2}}}$$

$$z_{rms}^2 = \int_{f_1}^{f_2} \frac{\Omega^4 \, G_x(f)}{\omega_0^4 \left[(1-h^2)^2 + \frac{h^2}{Q^2}\right]} \, df = \int_{h_1}^{h_2} \frac{f_0 \, h^4 \, G_x(f)}{\left[(1-h^2)^2 + \frac{h^2}{Q^2}\right]} \, dh$$

If the PSD is defined by a straight line segment of arbitrary slope (in linear scales), and with the notations of section 8.5.2,

$$z_{rms}^2 = \int_{h_1}^{h_2} \frac{f_0 \, h^4 \, (A\,f + B)}{\left[(1-h^2)^2 + \frac{h^2}{Q^2}\right]} \, dh$$

yielding

$$z_{rms}^2 = \frac{\pi \, f_0}{4\,\xi} \{A\,f_0\,[I_5(h_2) - I_5(h_1)] + B\,[I_4(h_2) - I_4(h_1)]\}$$

or

$$z_{rms}^2 = \frac{\pi \, f_0}{4\,\xi} \left\{ \frac{G_2 - G_1}{h_2 - h_1} [I_5(h_2) - I_5(h_1)] + \frac{G_1\,h_2 - G_2\,h_1}{h_2 - h_1} [I_4(h_2) - I_4(h_1)] \right\}$$

[8.75]

In the same way, the rms velocity and acceleration are given by:

$$\dot{z}_{rms}^2 = \int_{f_1}^{f_2} (2\pi)^2 \, f^2 \, G_z(f) \, df = (2\pi)^2 \, f_0^3 \int_{h_1}^{h_2} \frac{h^6 \, G_x}{(1-h^2)^2 + \frac{h^2}{Q^2}} \, dh$$

$$\ddot{z}_{rms}^2 = \frac{\pi}{4\xi}(2\pi)^2 f_0^3 \left\{ \frac{G_2 - G_1}{h_2 - h_1}[I_7(h_2) - I_7(h_1)] + \frac{G_1 h_2 - G_2 h_1}{h_2 - h_1}[I_6(h_2) - I_6(h_1)] \right\}$$

[8.76]

and

$$\ddot{z}_{rms}^2 = \int_{f_1}^{f_2} (2\pi f)^4 G_z(f)\, df = (2\pi)^4 f_0^5 \int_{h_1}^{h_2} \frac{h^8 G_x}{(1-h^2)^2 + \frac{h^2}{Q^2}}\, dh$$

$$\ddot{z}_{rms}^2 = \frac{\pi}{4\xi}(2\pi)^4 f_0^5 \left\{ \frac{G_2 - G_1}{h_2 - h_1}[I_9(h_2) - I_9(h_1)] + \frac{G_1 h_2 - G_2 h_1}{h_2 - h_1}[I_8(h_2) - I_8(h_1)] \right\}$$

[8.77]

I_6, I_7, I_8 and I_9 being respectively given by [A6.26], [A6.27], [A6.28] and [A6.29].

If the PSD is constant in the interval (h_1, h_2), the rms values result from the above relationships while setting $G_1 = G_2 = G_{x_0}$.

8.5.5. *Case where the excitation is defined by PSD comprising n straight line segments*

For a PSD made up of n straight line segments, we have

$$z_{rms}^2 = \sum_{i=1}^{n} z_{rms_i}^2 \qquad [8.78]$$

each term z_{rms_i} being calculated using the preceding relations:

– either by using abacuses or curves giving I_b in terms of h, for various values of b [PUL 68];

– or on a computer, by programming the expression for I_b.

When the number of bands is very large, it can be quicker to directly calculate [8.31] using numerical integration than to use these analytical expressions.

In the usual case where the excitation is an acceleration, we determine z_{rms}, $\dot{z}_{rms}$ and $\ddot{z}_{rms}$ starting from relations [8.57], [8.58] and [8.59]. The rms displacement is thus equal to

$$z_{rms}^2 = \frac{\pi}{4\xi(2\pi)^4 f_0^3} \sum_{j=1}^{n} a_j G_j \qquad [8.79]$$

where, for $2 \leq j \leq n-1$,

$$a_j = \frac{{}^{j-1,j}\Delta I_1 - h_{j-1}{}^{j-1,j}\Delta I_0}{h_j - h_{j-1}} - \frac{{}^{j,j+1}\Delta I_1 - h_{j+1}{}^{j,j+1}\Delta I_0}{h_{j+1} - h_j} \qquad [8.80]$$

with

$${}^{i,i+1}\Delta I_p = I_p(h_{i+1}) - I_p(h_i) \qquad [8.81]$$

and, for $j = 1$ and $j = n$ respectively,

$$a_1 = \frac{h_2{}^{12}\Delta I_0 - {}^{12}\Delta I_1}{h_2 - h_1} \quad \text{and} \quad a_n = \frac{{}^{n-1,n}\Delta I_1 - h_{n-1}{}^{n-1,n}\Delta I_0}{h_n - h_{n-1}}$$

Similarly

$$\dot{z}_{rms}^2 = \frac{\pi}{4\xi(2\pi)^2 f_0} \left\{ \frac{h_2{}^{12}\Delta I_2 - {}^{12}\Delta I_3}{h_2 - h_1} G_1 + \sum_{j=2}^{n-1} b_j G_j + \frac{{}^{n-1,n}\Delta I_3 - h_{n-1}{}^{n-1,n}\Delta I_2}{h_n - h_{n-1}} G_n \right\}$$

[8.82]

$$b_j = \frac{{}^{j-1,j}\Delta I_3 - h_{j-1}{}^{j-1,j}\Delta I_2}{h_j - h_{j-1}} - \frac{{}^{j,j+1}\Delta I_3 - h_{j+1}{}^{j,j+1}\Delta I_2}{h_{j+1} - h_j} \qquad [8.83]$$

and

$$\ddot{z}_{rms}^2 = \frac{\pi f_0}{4\xi} \left\{ \frac{h_2{}^{12}\Delta I_4 - {}^{12}\Delta I_5}{h_2 - h_1} G_1 + \sum_{j=2}^{n-1} c_j G_j + \frac{{}^{n-1,n}\Delta I_5 - h_{n-1}{}^{n-1,n}\Delta I_4}{h_n - h_{n-1}} G_n \right\}$$

[8.84]

$$c_j = \frac{{}^{j-1,j}\Delta I_5 - h_{j-1}\,{}^{j-1,j}\Delta I_4}{h_j - h_{j-1}} - \frac{{}^{j,j+1}\Delta I_5 - h_{j+1}\,{}^{j,j+1}\Delta I_4}{h_{j+1} - h_j} \qquad [8.85]$$

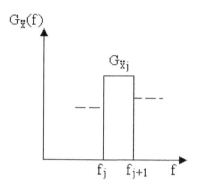

Figure 8.7. *PSD made up of horizontal straight line segments*

When the PSD is made up only of horizontal straight line segments, these relations are simplified in terms of

$$\left. \begin{array}{l} z_{rms}^2 = \dfrac{1}{(2\pi)^4 f_0^3} \dfrac{\pi}{4\xi} \displaystyle\sum_{j=1}^n G_{\ddot{x}j} \left[I_0(h_{j+1}) - I_0(h_j) \right] \\[1em] \dot{z}_{rms}^2 = \dfrac{1}{(2\pi)^2 f_0} \dfrac{\pi}{4\xi} \displaystyle\sum_{j=1}^n G_{\ddot{x}j} \left[I_2(h_{j+1}) - I_2(h_j) \right] \\[1em] \ddot{z}_{rms}^2 = f_0 \dfrac{\pi}{4\xi} \displaystyle\sum_{j=1}^n G_{\ddot{x}j} \left[I_4(h_{j+1}) - I_4(h_j) \right] \end{array} \right\} \qquad [8.86]$$

For example,

$$z_{rms}^2 = \sum_{i=1}^n \frac{G_i}{(2\pi)^4 f_0^3\, 4\xi} \left[\frac{\xi}{\alpha} \ln \frac{h^2 + \alpha\,h + 1}{h^2 - \alpha\,h + 1} + \arctan \frac{2\,h + \alpha}{2\,\xi} + \arctan \frac{2\,h - \alpha}{2\,\xi} \right]_{h_i}^{h_{i+1}}$$

$$[8.87]$$

Particular case where the n levels all have constant width Δf

On this assumption, the excitation being an acceleration, the rms values z_{rms}, $\dot{z}_{rms}$ and $\ddot{z}_{rms}$ can be obtained from:

$$z_{rms}^2 = \frac{1}{(2\pi)^4 f_0^3} \frac{\pi}{4\xi} \left\{ G_{\ddot{x}1} \left[I_0\left(f_1 + \frac{\Delta f}{2}\right) - I_0(f_1) \right] \right.$$

$$\left. + \sum_{j=2}^{n-1} G_{\ddot{x}j} \left[I_0\left(f_j + \frac{\Delta f}{2}\right) - I_0\left(f_j - \frac{\Delta f}{2}\right) \right] + G_{\ddot{x}n} \left[I_0(f_n) - I_0\left(f_n - \frac{\Delta f}{2}\right) \right] \right\} \quad [8.88]$$

$$\dot{z}_{rms} = \frac{1}{(2\pi)^2 f_0} \frac{\pi}{4\xi} \; \{\text{same term as in [8.88] while replacing } I_0 \text{ with } I_2\} \quad [8.89]$$

$$\ddot{z}_{rms} = f_0 \frac{\pi}{4\xi} \; \{\text{same term as in [8.88] while replacing } I_0 \text{ with } I_4\} \quad [8.90]$$

8.6. Rms value of the absolute acceleration of the response

If the excitation is an acceleration $\ddot{x}(t)$,

$$\ddot{y}_{rms}^2 = \int_{f_1}^{f_2} \frac{1 + \left(2\xi \frac{f}{f_0}\right)^2}{\left(1 - \frac{f^2}{f_0^2}\right)^2 + \left(2\xi \frac{f}{f_0}\right)^2} G_{\ddot{x}}(f) \, df \quad [8.91]$$

Let us set $h = \frac{f}{f_0}$. If the PSD is defined by straight line segments of arbitrary slope in linear scales, $G_{\ddot{x}}(f) = A f + B$ and

$$\ddot{y}_{rms}^2 = f_0 \int_{h_1}^{h_2} \frac{1 + (2\xi h)^2}{(1 - h^2)^2 + (2\xi h)^2} (A f_0 h + B) \, dh$$

$$\ddot{y}_{rms}^2 = \frac{\pi f_0}{4\xi} \left\{ \frac{G_{\ddot{x}2} - G_{\ddot{x}1}}{h_2 - h_1} \left[\Delta I_1 + 4\xi^2 \Delta I_3 \right] + \frac{G_{\ddot{x}1} h_2 - G_{\ddot{x}2} h_1}{h_2 - h_1} \left[\Delta I_0 + 4\xi^2 \Delta I_2 \right] \right\}$$

[8.92]

where $\Delta I_p = I_p(h_2) - I_p(h_1)$. For a noise of constant PSD between f_1 and f_2, $G_{\ddot{x}1} = G_{\ddot{x}2} = G_{\ddot{x}0}$ = constant,

$$\ddot{y}_{rms}^2 = \frac{\pi}{4\xi} f_0 G_{\ddot{x}0} \left[I_0(h_2) - I_0(h_1) \right] + \pi \xi f_0 G_{\ddot{x}0} \left[I_2(h_2) - I_2(h_1) \right]$$

$$\ddot{y}_{rms}^2 = \pi f_0 G_{\ddot{x}0} \left\{ \frac{1}{4\xi} \left[I_0(h_2) - I_0(h_1) \right] + \xi \left[I_2(h_2) - I_2(h_1) \right] \right\}$$

[8.93]

Example 8.2.

Under the conditions of the example quoted in section 8.5.3.1, we have $\ddot{y}_{rms}^2 = 39.82$ m/s² (to be compared with $4\pi^2 f_0^2 z_{rms} = 39.62$ m/s²).

8.7. Transitory response of a dynamic system under stationary random excitation

T.K. Caughey and H.J. Stumpf [CAU 61] analyzed, for applications related to the study of structures subjected to earthquakes, the transitory response of a linear one-degree-of-freedom mechanical system subjected to a random excitation having the following characteristics:

– stationary;

– Gaussian;

– zero mean;

– of PSD $G_\ell(\Omega)$.

The solution of the differential equation for the movement

$$\ddot{u}(t) + 2\xi \omega_0 \dot{u}(t) + \omega_0^2 u(t) = \ell(t)$$

[8.94]

can be written in the general form

$$u(t) = u_0 \, e^{-\xi \omega_0 t} \left[\cos \omega_0 \sqrt{1-\xi^2} \, t + \frac{\xi}{\sqrt{1-\xi^2}} \sin \omega_0 \sqrt{1-\xi^2} \, t \right]$$

$$+ \frac{\dot{u}_0}{\omega_0 \sqrt{1-\xi^2}} e^{-\xi \omega_0 t} \sin\left(\omega_0 \sqrt{1-\xi^2} \, t\right) + \omega_0^2 \int_0^t h(t-\alpha) \, \ell(\alpha) \, d\alpha \quad [8.95]$$

The input $\ell(t)$ being Gaussian and the system linear, it was seen that the response $u(t)$ is also Gaussian, with a distribution of the form

$$p(u) = \frac{1}{s(t)\sqrt{2\pi}} e^{-\frac{[u-m(t)]^2}{2s^2(t)}} \quad H$$

which requires knowledge of the mean and the standard deviation. The mean is given by

$$m(t) = \overline{u(t)} = u_0 \, e^{-\xi \omega_0 t} \left[\cos \omega_0 \sqrt{1-\xi^2} \, t + \frac{\xi}{\sqrt{1-\xi^2}} \sin \omega_0 \sqrt{1-\xi^2} \, t \right]$$

$$+ \frac{\dot{u}_0}{\omega_0 \sqrt{1-\xi^2}} e^{-\xi \omega_0 t} \sin\left(\omega_0 \sqrt{1-\xi^2} \, t\right) + \omega_0^2 \int_0^t h(t-\alpha) \, \overline{\ell(\alpha)} \, d\alpha \quad [8.96]$$

where $\overline{\ell(\lambda)} = 0$ since $\ell(t)$ was defined as a function with zero average, yielding

$$m(t) = u_0 \, e^{-\xi \omega_0 t} \left[\cos \omega_0 \sqrt{1-\xi^2} \, t + \frac{\xi}{\sqrt{1-\xi^2}} \sin \omega_0 \sqrt{1-\xi^2} \, t \right]$$

$$+ \frac{\dot{u}_0}{\omega_0 \sqrt{1-\xi^2}} e^{-\xi \omega_0 t} \sin\left(\omega_0 \sqrt{1-\xi^2} \, t\right) \quad [8.97]$$

m(t) depends on u_0 and $\dot{u}_0$. In addition,

$$s^2(t) = E\{[u(t) - m(t)]^2\}$$

$$s^2(t) = \omega_0^4 \int_0^t \int_0^t h(t-\alpha)\, h(t-\mu)\, \overline{\ell(\alpha)\,\ell(\mu)}\, d\alpha\, d\mu$$

T.K. Caughey and H.J. Stumpf [CAU 61] show that

$$s^2(t) = \int_0^\infty H^2(\Omega)\, G_\ell(\Omega) \left\{ 1 + e^{-2\xi\omega_0 t}\, [\ A\ ] \right\} \qquad [8.98]$$

where

$$A = 1 + \frac{2\xi}{\sqrt{1-\xi^2}} \sin\omega_0\sqrt{1-\xi^2}\,t\, \cos\omega_0\sqrt{1-\xi^2}\,t$$

$$-\, e^{-\xi\omega_0 t}\left(2\cos\omega_0\sqrt{1-\xi^2}\,t + \frac{2\xi}{\sqrt{1-\xi^2}} \sin\omega_0\sqrt{1-\xi^2}\,t \right)\cos\Omega t$$

$$-\, e^{-\omega_0\xi t}\, \frac{2\Omega}{\omega_0\sqrt{1-\xi^2}} \sin\omega_0\sqrt{1-\xi^2}\,t\, \sin\Omega t$$

$$+\, \frac{(\omega_0\xi)^2 - \omega_0^2(1-\xi^2) + \Omega^2}{\omega_0^2(1-\xi^2)} \sin\omega_0\sqrt{1-\xi^2}\,t$$

$$H^2(\Omega) = \frac{\omega_0^4}{\left(\omega_0^2 - \Omega^2\right)^2 + \left(2\Omega\omega_0\xi\right)^2} \qquad [8.99]$$

If the function $G(\Omega)$ is smoothed, presents no acute peak and if ξ is small, $s^2(t)$ can approach

$$s^2(\theta) \approx \frac{\pi}{4\xi} \omega_0 \, G_\ell(\omega_0) \left\{ 1 - e^{-2\xi\theta} \left[1 + \frac{2\xi^2}{1-\xi^2} \sin^2\left(\sqrt{1-\xi^2}\,\theta\right) \right. \right.$$

$$\left. \left. + \frac{\xi}{\sqrt{1-\xi^2}} \sin\left(2\sqrt{1-\xi^2}\,\theta\right) \right] \right\} \quad [8.100]$$

where $\theta = \omega_0 t$. It should be noted that, when θ is very large, $s^2(\theta) \to \frac{\pi}{2} \omega_0 \, Q \, G_\ell(\omega_0)$ which is the same as expression [8.17].

For $\xi = 0$, this expression becomes

$$s^2(\theta) \approx \frac{\pi}{4} \omega_0 \, G_\ell(\omega_0) \left[2\theta - \sin(2\theta) \right] \quad [8.101]$$

The variance $s^2(\theta)$ is independent of the initial conditions. Variations of the quantity $\dfrac{2 s^2}{\pi \omega_0 \, G_\ell(\omega_0)}$ in terms of θ are represented in Figure 8.8 for several values of ξ.

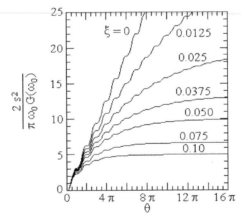

Figure 8.8. *Variance of the transitory response of a one-degree-of-freedom system to a random vibration*

It should be noted that, when ξ increases, the curve tends more and more quickly towards a horizontal asymptote (which is lower because ξ is larger). This asymptote corresponds to the steady state response, which is obtained during some cycles.

NOTE.–

The results are theoretically exact for a white noise vibration. They constitute a good approximation for a system slightly damped subjected to an excitation of which the spectral density is slightly variable close to the natural frequency of the system [LIN 67].

Particular cases

1. The excitation is an acceleration, $\ell(t) = -\dfrac{\ddot{x}(t)}{\omega_0^2}$ and $G_\ell(\Omega) = \dfrac{G_{\ddot{x}}}{\omega_0^4}$; this yields

$$s^2(\theta) \approx \frac{\pi}{2} \frac{G_{\ddot{x}_0}(\omega_0)}{\omega_0^3} Q \left\{ 1 - e^{-2\xi\theta} \left[1 + \frac{2\xi^2}{1-\xi^2} \sin^2\left(\sqrt{1-\xi^2}\,\theta\right) \right. \right.$$

$$\left. \left. + \frac{\xi}{\sqrt{1-\xi^2}} \sin\left(2\sqrt{1-\xi^2}\,\theta\right) \right] \right\} \quad [8.102]$$

and, for $\xi = 0$

$$s^2(\theta) \approx \frac{\pi}{4} \frac{G_{\ddot{x}_0}}{\omega_0^3} (2\theta - \sin 2\theta) \quad [8.103]$$

Being a Gaussian distribution, the probability that the response $u(t)$ exceeds a given value, $k\,s$, at the time t is

$$P(u > k\,s) = \int_{k\,s}^{\infty} \frac{1}{s\sqrt{2\pi}} e^{-z^2} \sqrt{2}\,dz$$

where $z = \dfrac{u - m}{s\sqrt{2}}$ and $dz = \dfrac{du}{s\sqrt{2}}$

$$P = \frac{1}{\sqrt{\pi}} \int_{k\,s}^{\infty} e^{-z^2}\,dz$$

$$P = \frac{1}{2}\left[1 - \operatorname{erf}\left(\frac{k\,s - m}{s\sqrt{2}}\right)\right] \quad [8.104]$$

while the probability that $|u(t)| > k\,s$ is

$$P[|u(t)| > k\,s] = \int_{ks}^{\infty} \frac{1}{s\sqrt{2\pi}} e^{-\frac{(u-m)^2}{2s^2}} du + \int_{-\infty}^{-ks} \frac{1}{s\sqrt{2\pi}} e^{-\frac{(u-m)^2}{2s^2}} du$$

$$P[|u(t)| > k\,s] = 1 - \frac{1}{2}\left[\operatorname{erf}\left(\frac{ks-m}{s\sqrt{2}}\right) + \operatorname{erf}\left(\frac{ks+m}{s\sqrt{2}}\right)\right] \qquad [8.105]$$

2. $u_0 = \dot{u}_0 = 0$ (zero initial conditions).

On this assumption, $m = 0$ and

$$P(|u| > k\,s) = 1 - \operatorname{erf}\left(\frac{k}{\sqrt{2}}\right) \qquad [8.106]$$

Figure 8.9 shows the variations of P with k.

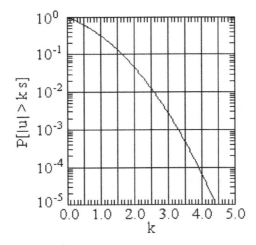

Figure 8.9. *Probability that the response exceeds k times the standard deviation*

Example 8.3.

Let us consider a random acceleration having a PSD of constant level $G_{\ddot{x}_0} = 1$ (m/s^2)^2/Hz applied at time t = 0 to a simple system having as initial conditions

$$u_0 = z_0 = 10^{-3} \text{ m}$$

$$\dot{u}_0 = 0.02 \text{ m/s}$$

The simple system is linear, with one degree of freedom, natural frequency $f_0 = 10$ Hz and factor quality Q = 10.

The mean m is stabilized to zero after approximately 1.5 s (Figure 8.10).

The standard deviation $s(t)$, which starts from zero at t = 0, tends towards a limit equal to the rms value z_{rms} (m is then equal to zero) of the stationary response $z(t)$ of the simple system (Figure 8.11):

$$z_{rms} = \sqrt{\frac{\pi Q G_{\ddot{x}_0}}{2 \omega_0^3}} = 7.9577 \; 10^{-3} \text{ m}$$

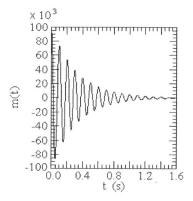
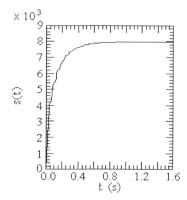

Figure 8.10. *Mean of response versus time*   **Figure 8.11.** *Standard deviation versus time*

The time pattern of probability for $z(t)$ higher than 1.5 z_{rms} tends in stationary mode towards a constant value P_0 approximately equal to 6.54 10^{-1} (Figure 8.12).

Figure 8.13 shows the variations with time of the probability that $z(t)$ is higher than $1.5\,s(t)$, a probability which tends towards the same limit P_0 for sufficiently large t.

Figures 8.14 and 8.15 show the same functions as in Figures 8.12 and 8.13, plotted for k = 1, 2, 3 and 4.

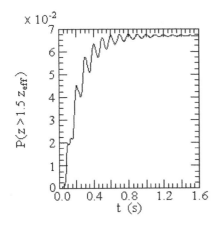

Figure 8.12. *Probability that the response is higher than 1.5 times its stationary rms value*

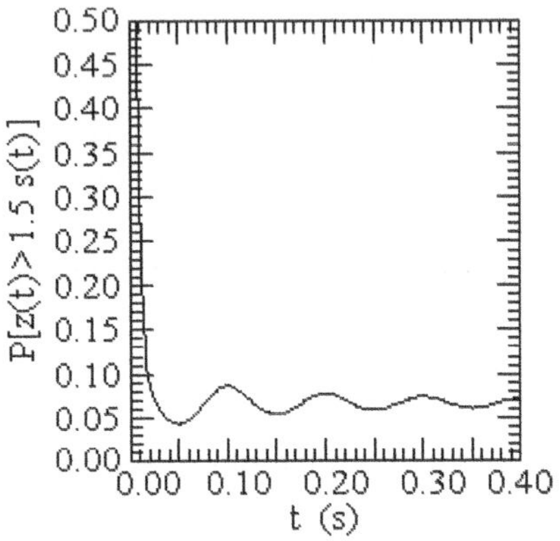

Figure 8.13. *Probability that the response is higher than 1.5 times its standard deviation*

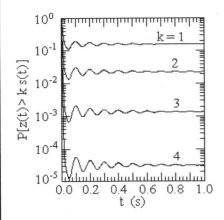
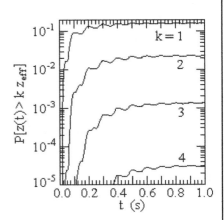

Figure 8.14. *Probability that the response is higher than k times its standard deviation*

Figure 8.15. *Probability that the response is higher than k times its stationary rms value*

8.8. Transitory response of a dynamic system under amplitude modulated white noise excitation

Let us consider a non-stationary random acceleration of the form [BAR 68]

$$\ddot{x}(t) = \varphi(t)\, e(t) \qquad [8.107]$$

where

$e(t)$ is a wideband white noise;

$\varphi(t)$ is a signal of simple shape (rectangular wave, then half-sine wave) intended to modulate $e(t)$ (envelope function); $\varphi(t)$ is of the form $\varphi(t) = \varphi_0\, \Psi(t)$ with

$$\begin{cases} \varphi(t) = \varphi_0 & \text{for } 0 \le t \le \tau \\ \varphi(t) = 0 & \text{elsewhere} \end{cases} \quad \text{or} \quad \begin{cases} \varphi(t) = \varphi_0 \sin\frac{\pi}{\tau}t & \text{for } 0 \le t \le \tau \\ \varphi(t) = 0 & \text{elsewhere} \end{cases}$$

The study of the response of a one-degree-of-freedom linear system to such a signal is mostly concerned with mechanical shock (the durations are sufficiently short that we can neglect the effects of fatigue). These transients are also used to simulate real environments of a similar shape, such as earthquakes, the blast of explosions, launching of missiles, etc.

Modulating the amplitude of the random signal led to a reduction of the energy transmitted to the mechanical system for length of time τ. R.L. Barnoski [BAR 65] [NEA 66] proposed using the dimensionless time parameter $\theta^* = \dfrac{f_0\, \Delta\tau}{Q}$ in which the *effective time interval* $\Delta\tau$ intervenes, in which the energy contained in the modulated pulse is equal to that of the stationary signal over the duration τ. This parameter θ^* is a function of the time necessary for the amplitude response of a simple resonator to decrease to $\dfrac{1}{e}$ times its steady state value, namely $T_0 = \dfrac{Q}{\pi\, f_0}$:

$$\theta^* = \frac{\Delta\tau}{\pi\, T_0} \qquad [8.108]$$

$$\Delta\tau = \frac{1}{\varphi_0^2} \int_0^\infty \varphi^2(t)\, dt = \frac{1}{\varphi_0^2} \int_0^\tau \varphi_0^2\, \Psi^2(t)\, dt$$

$$\Delta\tau = \int_0^\tau \Psi^2(t)\, dt \qquad [8.109]$$

For a square wave, $\Delta\tau = \tau$ and for a half-sine wave, $\Delta\tau = \dfrac{\tau}{2}$.

From an analog simulation, R.L. Barnoski and R.H. MacNeal plotted the ratio β of the maximum response to the rms value of the response versus θ^*, for values of the probability $P_M(\beta)$ (the probability that all the maxima of the response are lower than β times the rms value of the response). The oscillator is assumed to be at rest at the initial time.

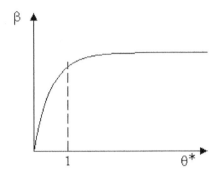

Figure 8.16. *Reduced maximum response versus the dimensionless time parameter*

The curves obtained all have the same characteristic appearance: fast rise of β from the origin to $\theta^* \approx 1$, then slow increase in β in terms of θ^* (Figure 8.16).

The curves plotted in the stationary case constitute an upper limit of the transitory case when the duration of the impulse increases. These results can thus make it possible to estimate the time necessary to effectively achieve stationarity in its response. For $\theta^* > 1$ (a long impulse duration compared to the natural period of the system), the results of the stationary case can constitute a conservative envelope of β. For large θ^* and a great number of response cycles, the response tends to becoming independent of factor Q.

We can note in addition that, when the duration of the impulse increases and becomes long compared with the natural period of the resonator, the response peak tends to become independent of the shape of the modulation.

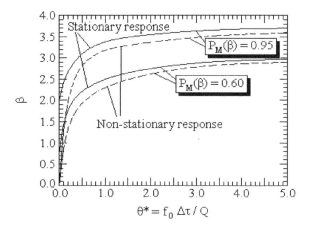

Figure 8.17. *Reduced maximum response to stationary and pulsed random excitation versus the dimensionless time parameter (rectangular envelope, $Q = 20$, $f_0 = 159$ Hz) [BAR 65], [BAR 71]*

Chapter 9

Characteristics of the Response of a One-Degree-of-Freedom Linear System to Random Vibration

9.1. Moments of response of a one-degree-of-freedom linear system: irregularity factor of response

9.1.1. *Moments*

By definition,

$$M_n = \int_0^\infty \Omega^n \, G_u(\Omega) \, d\Omega$$

with

$$G_u(\Omega) = \frac{G_\ell(\Omega)}{\left(1 - \dfrac{\Omega^2}{\omega_0^2}\right)^2 + \left(2\xi\dfrac{\Omega}{\omega_0}\right)^2}$$

Setting $h = \dfrac{\Omega}{\omega_0}$, we can write:

$$M_n = \int_0^\infty \frac{h^n \omega_0^n}{\left(1-h^2\right)^2 + \left(2\xi h\right)^2} \frac{G_\ell(h)}{\omega_0} \omega_0 \, dh$$

i.e.

$$M_n = \omega_0^n \int_0^\infty \frac{h^n G_\ell(h)}{\left(1-h^2\right)^2 + \left(2\xi h\right)^2} \, dh$$

or

$$M_n = (2\pi)^n \int_0^\infty \frac{f^n G_\ell(f)}{\left(1-\frac{f^2}{f_0^2}\right)^2 + \left(2\xi \frac{f}{f_0}\right)^2} \, df \qquad [9.1]$$

For white noise,

$$M_n = \omega_0^n G_{\ell 0} \int_0^\infty \frac{h^n}{\left(1-h^2\right)^2 + \left(2\xi h\right)^2} \, dh$$

$$M_n = \omega_0^n G_{\ell 0} \frac{\pi}{4\xi} \left[I_n(\infty) - I_n(0)\right] \qquad [9.2]$$

The main contribution to the integral comes from the area around the natural frequency ω_0: the results obtained for white noise are a good approximation to actual cases where G_ℓ varies little around ω_0. For noise with a constant PSD between two frequencies f_1 and f_2,

$$M_n = \omega_0^n G_{\ell 0} \frac{\pi}{4\xi} \left[I_n(h_2) - I_n(h_1)\right] \qquad [9.3]$$

I_n being defined in Appendix A6 [LAL 94]. The most useful moments are M_0, M_2 and M_4.

Moment of order 0

$$M_0 = \int_0^\infty G_\ell(h) \frac{dh}{\left(1-h^2\right)^2 + (2\xi h)^2} = u_{rms}^2 \quad [9.4]$$

If the excitation is a white noise, $G_\ell(h)$ = constant, $[I_0(\infty) - I_0(0)] = 1$ and

$$M_0 = u_{rms}^2 = \frac{\pi}{2} Q G_\ell(h) = \frac{\pi}{2} Q \omega_0 G_\ell(\Omega) \quad [9.5]$$

If the excitation is an acceleration, $\ell(t) = -\frac{\ddot{x}(t)}{\omega_0^2}$ and $G_\ell(\Omega) = \frac{G_{\ddot{x}}(\Omega)}{\omega_0^4}$ yielding [KRE 83], [LEY 65]:

$$M_0 = z_{rms}^2 = \frac{\pi Q G_{\ddot{x}0}(\Omega)}{2 \omega_0^3} \quad [9.6]$$

Moment of order 1

$$M_1 = \omega_0 \int_0^\infty \frac{h \, G_\ell(h)}{\left(1-h^2\right)^2 + (2\xi h)^2} dh \quad [9.7]$$

If $G_\ell(h)$ = constant,

$$M_1 = \omega_0 \, G_{\ell 0}(h) \int_0^\infty \frac{h}{\left(1-h^2\right)^2 + (2\xi h)^2} dh \quad [9.8]$$

Knowing [LAL 94] that

$$\int_0^\infty \frac{h}{\left(1-h^2\right)^2 + (2\xi h)^2} dh = \frac{1}{4\xi\sqrt{1-\xi^2}} \left(\pi + \arctan \frac{2\xi\sqrt{1-\xi^2}}{2\xi^2 - 1} \right) \text{ if } \xi \leq \frac{1}{\sqrt{2}}$$

$$\int_0^\infty \frac{h}{\left(1-h^2\right)^2 + (2\xi h)^2} dh = \frac{1}{4\xi\sqrt{1-\xi^2}} \left(\arctan \frac{2\xi\sqrt{1-\xi^2}}{2\xi^2 - 1} \right) \text{ if } \xi \geq \frac{1}{\sqrt{2}}$$

it becomes, in the most usual case ($\xi \leq \frac{1}{\sqrt{2}}$) [VAN 72]

$$M_1 = \frac{\omega_0 \, G_{\ell_0}(h)Q}{2\sqrt{1-\xi^2}} \left(\pi + \arctan \frac{2\xi\sqrt{1-\xi^2}}{2\xi^2 - 1} \right) \quad [9.9]$$

or

$$M_1 = \frac{\omega_0 \, u_{rms}^2}{\sqrt{1-\xi^2}} \left(1 - \frac{1}{\pi} \arctan \frac{2\xi\sqrt{1-\xi^2}}{1-2\xi^2} \right) \quad [9.10]$$

Relationship [9.9] can also be written, since $\arctan x + \arctan \frac{1}{x} = \frac{\pi}{2}$ [CHA 72],

$$M_1 = \frac{\omega_0 \, G_{\ell_0}(h)Q}{2\sqrt{1-\xi^2}} \left(\frac{\pi}{2} + \arctan \frac{1-2\xi^2}{2\xi\sqrt{1-\xi^2}} \right) \quad [9.11]$$

$$M_1 = \frac{\omega_0 \, u_{rms}^2}{\sqrt{1-\xi^2}} \left(\frac{1}{2} - \frac{1}{\pi} \arctan \frac{2\xi^2 - 1}{2\xi\sqrt{1-\xi^2}} \right) \quad [9.12]$$

The ratio $\dfrac{M_1}{\omega_0 \, u_{rms}^2}$ varies little with ξ so long as $\xi \leq 0.1$ and is then close to 1 (Figure 9.1).

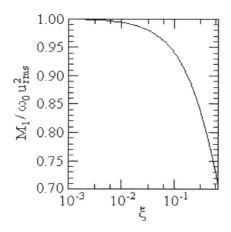

Figure 9.1. *First moment of the response of a one-degree-of-freedom system versus its damping*

Expression [9.12] can be written in the equivalent form [DER 79]:

$$M_1 = \frac{\omega_0 \, u_{rms}^2}{\sqrt{1-\xi^2}} \left(1 - \frac{2}{\pi} \arctan \frac{\xi}{\sqrt{1-\xi^2}} \right) \qquad [9.13]$$

Moment of order 2

$$M_2 = \omega_0^2 \int_0^\infty \frac{h^2 \, G_\ell(h)}{\left(1-h^2\right)^2 + (2\,\xi\,h)^2} \, dh \qquad [9.14]$$

If G_ℓ = constant,

$$M_2 = \omega_0^2 \, G_{\ell 0} \int_0^\infty \frac{h^2}{\left(1-h^2\right)^2 + (2\,\xi\,h)^2} \, dh$$

$$M_2 = \frac{\pi \, \omega_0^2 \, G_{\ell 0}}{4\,\xi} = \frac{\pi \, \omega_0^2 \, G_{\ell 0} \, Q}{2} \qquad [9.15]$$

or

$$M_2 = \omega_0^2 \, u_{rms}^2 = \omega_0^2 \, M_0 \qquad [9.16]$$

We again find the relationship

$$n_0^+ = \frac{1}{2\,\pi} \sqrt{\frac{M_2}{M_0}}$$

($\omega_0 = 2\,\pi\,n_0^+$).

9.1.2. *Irregularity factor of response to noise of a constant PSD*

By definition,

$$r = \frac{M_2}{\sqrt{M_0 \, M_4}} \left(= \frac{\dot{u}_{rms}^2}{u_{rms} \, \ddot{u}_{rms}} \right) \qquad [9.17]$$

According to [9.3], we have, for noise of a constant PSD,

$$r = \frac{\Delta I_2}{\sqrt{\Delta I_0 \, \Delta I_4}} = \frac{I_2(h_2) - I_2(h_1)}{\sqrt{[I_0(h_2) - I_0(h_1)][I_4(h_2) - I_4(h_1)]}} \qquad [9.18]$$

since

$$\Delta I_n = \omega_0^n \, G_{\ell_0}(h) \, \frac{\pi}{4\xi} \left[I_n(h_2) - I_n(h_1) \right] \qquad [9.19]$$

It was seen in addition that [9.5] $M_0 = u_{rms}^2$ and that [9.16] $M_2 = \omega_0^2 \, M_0$, yielding

$$r = \frac{\omega_0^2 \, M_0}{\sqrt{M_0 \, M_4}} = \omega_0^2 \sqrt{\frac{M_0}{M_4}} = \omega_0^2 \, \frac{u_{rms}}{\sqrt{M_4}} \qquad [9.20]$$

For white noise, M_4 is infinite since its calculation assumes integration between zero and infinity of the quantity

$$\frac{h^4}{\left(1 - h^2\right)^2 + \left(2 \, \xi \, h\right)^2},$$

which tends towards a constant (equal to 1) when $h \to \infty$. Figure 9.2 shows this function plotted for $\xi = 0.01$, 0.05 and 0.1.

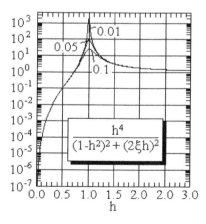

Figure 9.2. *Function to be integrated for calculation of fourth moment versus h*

The integral thus does not converge and the parameter r is zero. The ratio h being also large, possibly when f_0 is small, the parameter r of the response tends in the same way towards zero when the natural frequency of the system decreases.

NOTE.–

The curves $I_0(h)$, $I_2(h)$ and $I_4(h)$ show that [CHA 72]:

– if ξ is small, these functions increase very quickly when h is close to 1;

– I_0 and I_2 tend towards 1 when h becomes large, while I_4 continues to increase.

Thus, ideal white noise can give a good approximation to wideband noise for the calculation of z_{rms} and $\dot{z}_{rms}$, but it cannot be used to approximate $\ddot{z}_{rms}$.

If ξ is small and if h is large compared to 1, the functions I_0 and I_2 are roughly equal to 1.

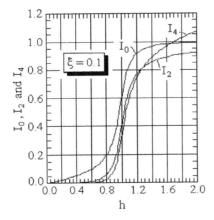
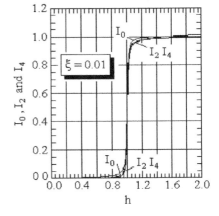

Figure 9.3. *Integrals I_0, I_2 and I_4 versus h, for $\xi = 0.1$*

Figure 9.4. *Integrals I_0, I_2 and I_4 versus h, for $\xi = 0.01$*

9.1.3. Characteristics of irregularity factor of response

If the PSD is constant between two frequencies f_1 and f_2, the parameter r varies with ξ and f_0 (via h).

434   Random Vibration

Example 9.1.

Let us again take the example of section 8.5.3.1. The variations of the rms values z_{rms}, $\dot{z}_{rms}$ and $\ddot{z}_{rms}$ according to f_0 (for $\xi = 0.01$, 0.05 and 0.1) and according to ξ (for $f_0 = 10$ Hz, 100 Hz and 1,000 Hz) are shown in Figures 9.5 to 9.10.

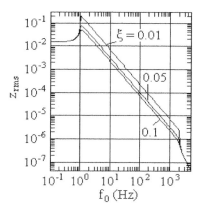

Figure 9.5. *Rms relative displacement of the response of a one-degree-of-freedom system versus its natural frequency*

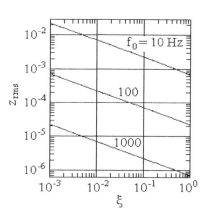

Figure 9.6. *Rms relative displacement of the response of a one-degree-of-freedom system versus its damping factor*

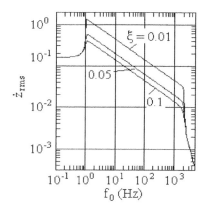

Figure 9.7. *Rms relative velocity of the response of a one-degree-of-freedom system versus its natural frequency*

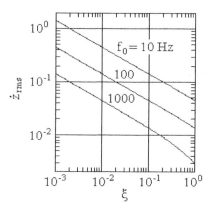

Figure 9.8. *Rms relative velocity of the response of a one-degree-of-freedom system versus its damping*

Characteristics of the Response to Random Vibration   435

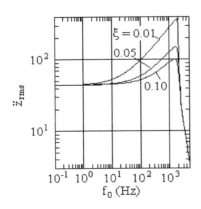

Figure 9.9. *Rms relative acceleration of the response of a one-degree-of-freedom system versus its natural frequency*

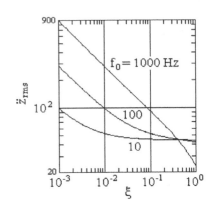

Figure 9.10. *Rms relative acceleration of the response of a one-degree-of-freedom system versus its damping*

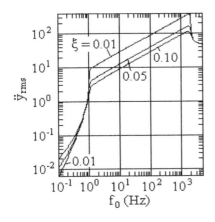

Figure 9.11. *Rms absolute acceleration of the response of a one-degree-of-freedom system versus its natural frequency*

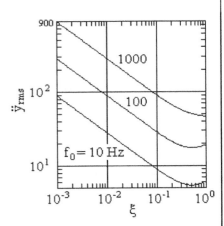

Figure 9.12. *Rms absolute acceleration of the response of a one-degree-of-freedom system versus its damping*

Figures 9.13 and 9.14 show the variations of the parameter r under the same conditions.

It can be seen in Figure 9.13 that, whatever f_0, $r \to 1$ when $\xi \to 0$. The distribution of maxima of the response thus tends, when ξ tends towards zero, towards a Rayleigh distribution. When ξ increases, r decreases with f_0.

Figure 9.14 underlines the existence of a limit independent of ξ for the frequencies $f_0 < f_1$ and $f_0 > f_2$.

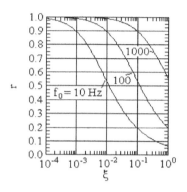

Figure 9.13. *Irregularity factor of the response of a one-degree-of-freedom system versus its damping*

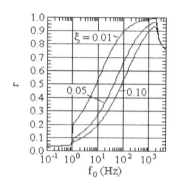

Figure 9.14. *Irregularity factor of the response of a one-degree-of-freedom system versus its natural frequency*

This example shows that:

– r is closer to 1 when ξ is smaller and f_0 is larger;

– when f_0 becomes large compared to f_2, r tends, whatever ξ, towards the same limiting value (0.749). When $f_0 \to \infty$, the PSD input is completely transmitted and the signal response has the same characteristics as the input signal (input $\ddot{x}$, response z). This result can be shown as follows. We saw that, when $h = \dfrac{f}{f_0}$ is small (i.e. when f_0 is large, Figure 9.15), we have

$$I_n \approx \frac{4\xi}{\pi}\left(\frac{h^{n+1}}{n+1} + 2\frac{h^{n+3}}{n+3} + \cdots\right) \approx \frac{4\xi}{\pi}\frac{h^{n+1}}{n+1}$$

Here, n = 0 and $I_0 \approx \dfrac{4\xi}{\pi} h$

$$z_{rms}^2 \approx \dfrac{\pi}{4\xi}\dfrac{G_{\ddot{x}0}}{2\pi\omega_0^3}\left[\dfrac{4\xi}{\pi}(h_2 - h_1)\right] = \dfrac{G_{\ddot{x}0}}{2\pi\omega_0^3}(h_2 - h_1)$$

$$z_{rms}^2 \approx \dfrac{G_{\ddot{x}0}}{2\pi f_0 \omega_0^3}(f_2 - f_1)$$

and

$$\omega_0^4 z_{rms}^2 \approx G_{\ddot{x}0}(f_2 - f_1) = \ddot{x}_{rms}^2$$

In the same way,

$\omega_0^2 \dot{z}_{rms} \approx$ rms value of the first derivative of $\ddot{x}$,

$\omega_0^2 \ddot{z}_{rms} \approx$ rms value of the second derivative of $\ddot{x}$.

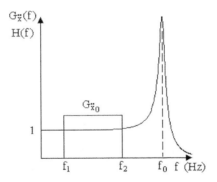

Figure 9.15. *Natural frequency greater than the upper limit of the noise*

It is thus normal that r has as a limit the value

$$r_0 = \dfrac{M_2}{\sqrt{M_0 M_4}} = \dfrac{(\ddot{x})_{rms}^{'2}}{\ddot{x}_{rms}(\ddot{x})_{rms}^{''}}$$

calculated from the PSD of the excitation. From expression [6.53], we obtain

438   Random Vibration

$$r_0 = \frac{\sqrt{5}}{3} \frac{f_2^3 - f_1^3}{\sqrt{(f_2 - f_1)(f_2^5 - f_1^5)}} = 0.749$$

If $f_1 = 10$ Hz and $f_2 = 1,000$ Hz, then $r_0 = 0.749$.

In addition, when f_0 is small compared to f_1 (Figure 9.16), r tends towards a constant value r_0' equal, whatever ξ, to 0.17234 (for the values of f_1 and f_2 selected in this example). Indeed, for large h, i.e. for small f_0, we have

$$\frac{1}{(1-h^2)^2} = \frac{1}{h^4 \left(\frac{1}{h^2} - 1\right)^2} = \frac{1}{h^4} \frac{1}{\left(1 - \frac{1}{h^2}\right)^2}$$

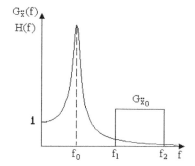

Figure 9.16. *Natural frequency lower than the lower limit of the noise*

Knowing that

$$(1-x)^{-2} \approx 1 - 2x + 2.3\frac{x^2}{2!} - \cdots$$

it results that

$$\frac{h^n}{(1-h^2)^2} \approx \frac{h^n}{h^4}\left(1 - \frac{2}{h^2} + \frac{3}{h^4} + \cdots\right)$$

Characteristics of the Response to Random Vibration   439

$$\int \frac{h^n}{\left(1-h^2\right)^2} dh \approx \frac{h^{n-3}}{n-3} - 2\frac{h^{n-5}}{n-5} + \cdots$$

yielding

$$I_n \approx \frac{4\xi}{\pi}\left(\frac{h^{n-3}}{n-3} - 2\frac{h^{n-5}}{n-5} + \cdots\right)$$

i.e., at first approximation,

$$I_n \approx \frac{4\xi}{\pi}\frac{h^{n-3}}{n-3} \qquad [9.21]$$

$n = 0$ since the excitation is a PSD of constant level G_0 between f_1 and f_2. Thus

$$I_0 \approx \frac{4\xi}{\pi}\frac{h^{-3}}{-3}$$

and

$$z_{rms}^2 \approx \frac{G_{\ddot{x}0}}{6\pi\omega_0^3}\frac{h_2^3 - h_1^3}{h_1^3 h_2^3}$$

$$z_{rms}^2 \approx \frac{G_{\ddot{x}0} f_0^3}{6\pi\omega_0^3}\frac{f_2^3 - f_1^3}{f_1^3 f_2^3}$$

$$z_{rms}^2 \approx x_{rms}^2$$

where x_{rms} is the rms displacement of the excitation $\ddot{x}(t)$, namely

$$x_{rms} = \frac{1}{(2\pi)^2}\sqrt{\frac{G_{\ddot{x}0}\left(f_2^3 - f_1^3\right)}{3 f_1^3 f_2^3}}$$

It could be shown in the same way that $\dot{z}_{rms} = \dot{x}_{rms}$ and $\ddot{z}_{rms} = \ddot{x}_{rms}$ if $\dot{x}_{rms}$ and $\ddot{x}_{rms}$ are the rms velocity and acceleration of the excitation

$$\dot{x}_{rms} = \frac{1}{2\pi}\sqrt{\frac{G_{\ddot{x}0}(f_2 - f_1)}{f_1 f_2}}$$

$$\ddot{x}_{rms} = \sqrt{G_{\ddot{x}0}(f_2 - f_1)}$$

yielding, according to the above relationships:

$$r = \frac{\dot{z}_{rms}^2}{z_{rms}\ddot{z}_{rms}} = \frac{\dot{x}_{rms}^2}{x_{rms}\ddot{x}_{rms}} = \sqrt{\frac{3 f_1 f_2 (f_2 - f_1)}{f_2^3 - f_1^3}} \qquad [9.22]$$

Example 9.2.

If $f_1 = 10$ Hz and $f_2 = 1{,}000$ Hz, we obtain $r = 0.17234$.

Figure 9.17 summarizes these results.

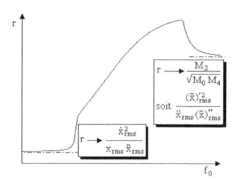

Figure 9.17. *Irregularity factor of response of a one-degree-of-freedom system versus its natural frequency*

Particular case where $\xi = 1$

In this case, and for white noise,

$$M_n = \omega_0^n \, G_{\ddot{x}0}(h) \int_0^\infty \frac{h^n}{\left(1-h^2\right)^2 + 4 h^2} \, dh \qquad [9.23]$$

Let us set

$$J_n = \int \frac{h^n}{(1-h^2)^2 + 4h^2} \, dh = \int \frac{h^n}{(1+h^2)^2} \, dh$$

$$J_n(h) = \frac{h^{n-3}}{n-3} - 2J_{n-2} - J_{n-4} \qquad [9.24]$$

Let us calculate the first terms:

$$J_0 = \int \frac{dh}{(1+h^2)^2}$$

While setting $h = \tan \varphi$, we obtain

$$J_0 = \frac{1}{2}\left(\varphi + \frac{\sin 2\varphi}{2}\right)$$

i.e.

$$J_0 = \frac{1}{2}\left(\arctan h + \frac{h}{1+h^2}\right) \qquad [9.25]$$

$$J_2 = \int \frac{h^2}{(1+h^2)^2} \, dh = \int \frac{dh}{h^2+1} - J_0$$

yielding

$$J_2 = \frac{1}{2}\left(\arctan h - \frac{h}{1+h^2}\right) \qquad [9.26]$$

$$J_4 = h - 2J_2 - J_0$$

$$J_4 = h + \frac{h}{2(1+h^2)} - \frac{3}{2}\arctan h \qquad [9.27]$$

Example 9.3.

$f_1 = 10$ Hz

$f_2 = 1,000$ Hz

Figure 9.18 shows the variations of r with f_0 for $\xi = 1$.

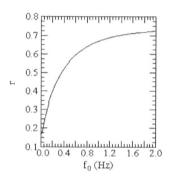

Figure 9.18. *Irregularity factor of the response of a one-degree-of-freedom system versus its natural frequency, for a damping factor equal to 1*

Particular case where $\xi = 0$ (for a white noise)

$$M_n = \omega_0^n \, G_{\ddot{x}_0}(h) \int_0^\infty \frac{h^n}{\left(1-h^2\right)^2} \, dh \qquad [9.28]$$

Let us set [LAL 94]

$$J_n = \int \frac{h^n}{\left(1-h^2\right)^2} \, dh$$

Appendix A6.2 gives the expressions for J_0 and J_2, as well as the formula of recurrence which allows J_4 to be calculated.

NOTE.–

Approximate calculation of r for a PSD with several peaks (for example, the case of a PSD obtained in the response of a several-degrees-of-freedom system).

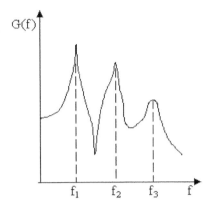
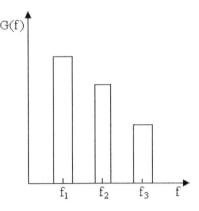

Figure 9.19. *PSD with several peaks*   **Figure 9.20.** *Equivalent "box" spectrum*

J.T. Broch [BRO 70] transforms this PSD into equivalent boxes (the box spectrum containing the same amount of energy centered around the resonance frequencies; see section 9.4.1) and shows that r can be approximated by:

$$r^2 = \frac{\left[1 + \sum_n \left(\frac{f_n}{f_1}\right)^2 \beta_n\right]^2}{\left[1 + \sum_n \beta_n\right]\left[1 + \sum_n \left(\frac{f_n}{f_1}\right)^4 \beta_n\right]}$$ [9.29]

where:

f_1 = frequency of the first resonance

f_n = frequency of nth resonance

$$\beta_n = -\left(\frac{c_n}{c_1}\right)\frac{\Delta f_n}{\Delta f_1}$$

$\frac{c_n}{c_1}$ = energy ratio of the maximum responses to the resonances

Δf_n = width of peak n at –3 dB.

9.1.4. *Case of a band-limited noise*

If $f_1 = 0$, relationship [9.18] is written [CHA 72]:

$$r = \frac{I_2(h_2)}{\sqrt{I_0(h_2)\, I_4(h_2)}} \qquad [9.30]$$

If ξ is small and if h_2 is large compared to 1, we have $I_0 \approx I_2 \approx 1$ yielding

$$r^2 \approx \frac{1}{1 - 4\xi^2 + \dfrac{4\xi h}{\pi}} \qquad [9.31]$$

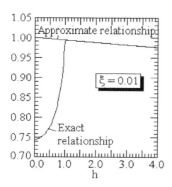

Figure 9.21. *Comparison of the exact and approximate expressions of the irregularity factor of the response to a narrowband noise for a damping ratio equal to 0.01*

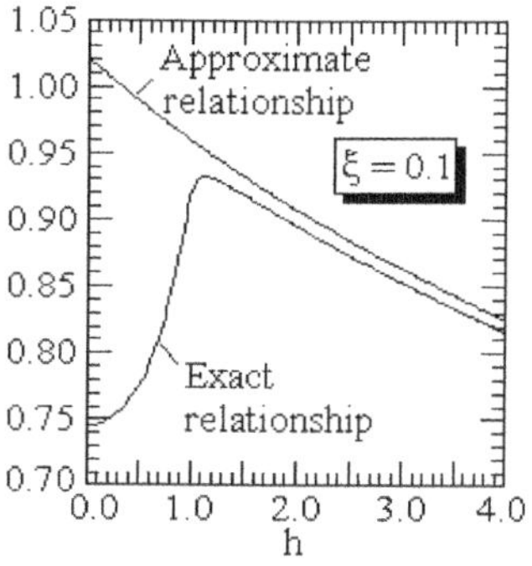

Figure 9.22. *Comparison of the exact and approximate expressions of the irregularity factor of the response to a narrowband noise for a damping ratio equal to 0.1*

NOTE.–

Under these conditions, we have

$$M_n = \frac{\pi}{2} \omega_0^n \, Q \, G_{\ell 0}(h) \, \Delta I_n \qquad [9.32]$$

where

$$\Delta I_n = I_n(h_2) - I_n(h_1)$$

$$M_n = \frac{\pi}{2} \omega_0^{n+1} \, Q \, G_{\ell 0}(\Omega) \, \Delta I_n \qquad [9.33]$$

and, if the noise is defined by an acceleration,

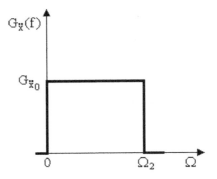

Figure 9.23. *Constant PSD between two frequencies*

$$M_n = \frac{\pi}{2} \omega_0^{n-3} Q G_{\ddot{x}0}(\Omega) \Delta I_n \qquad [9.34]$$

yielding, for [LEY 65]

$$n = 0 \qquad M_0 = \frac{\pi}{2} \frac{Q G_{\ddot{x}0}}{\omega_0^3} \Delta I_0 = z_{rms}^2$$

$$n = 2 \qquad M_2 = \frac{\pi}{2} \frac{Q G_{\ddot{x}0}}{\omega_0} \Delta I_2$$

$$n = 4 \qquad M_4 = \frac{\pi}{2} \omega_0 Q G_{\ddot{x}0} \Delta I_4$$

with $\Delta I_n = I_n(\Omega_2)$ in this particular case.

9.2. Autocorrelation function of response displacement

It is shown that this function is equal to [BAR 68]:

$$R_z(\tau) \approx \frac{\pi}{2} \frac{Q G_{\ddot{x}0}}{\omega_0^3} e^{-\xi \omega_0 \tau} \left(\cos \omega_0 \sqrt{1-\xi^2}\ \tau + \frac{\xi}{\sqrt{1-\xi^2}} \sin \omega_0 \sqrt{1-\xi^2}\ \tau \right)$$

$$[9.35]$$

For $\tau = 0$,

$$R_z(\tau) = z_{rms}^2 = \frac{\pi}{2} \frac{Q G_{\ddot{x}0}}{\omega_0^3} \qquad [9.36]$$

The duration (*time* or *correlation interval*) $\Delta\tau$ necessary for the envelope of $R_z(\tau)$ to decrease 1/e times its initial amplitude is given by

$$\Delta\tau = \frac{Q}{\pi f_0} = \frac{1}{\xi \omega_0} \qquad [9.37]$$

9.3. Average numbers of maxima and minima per second

If $E\left[n^+(a)\right]$ is the average number per second of crossings of a given threshold a with a positive slope, the average number of peaks per second lying between a and a + da can be approximate using the difference [POW 58]:

$$E\left[n^+(a)\right] - E\left[n^+(a+da)\right]$$

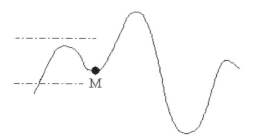

Figure 9.24. *Minimum with positive amplitude*

This method can lead to errors, since it assumes that the average number of minima per second (such as M in Figure 9.24) taking place with a positive amplitude is negligible compared to the average number of peaks per second [LAL 92]. This is the case for narrowband noise.

To evaluate the validity of this approximation in the case of band-limited noise, J.B. Roberts [ROB 66] calculates, using the formulation of S.O. Rice:

– the average number of peaks per second between a and a + da:

$$E\left[n_p^+(a)\,da\right] = -da \int_{-\infty}^{0} c\,f(a,0,c)\,dc \qquad [9.38]$$

– the average number of minima per second between a and a + da:

$$E\left[n_p^-(a)\,da\right] = da \int_{0}^{\infty} c\,f(a,0,c)\,dc \qquad [9.39]$$

where $f(a,b,c)$ is the joint probability density function of the random process. The function $f(a,b,c)\,da\,db\,dc$ is the probability that, at time t, the signal $z(t)$ lies between a and a + da, its first derivative $\dot{z}(t)$ between b and b + db and its second derivative $\ddot{z}(t)$ between c and c + dc, a maximum being defined by a zero derivative (see section 6.1).

$$f(a,0,c) = \frac{1}{\left(8\pi^3\,M\right)^{1/2}} \exp\left[-\frac{\dot{z}_{rms}^2\,\ddot{z}_{rms}^2\,a^2 + 2\,\dot{z}_{rms}^4\,a\,c + \dot{z}_{rms}^2\,z_{rms}^2\,c^2}{2|M|}\right]$$

$$[9.40]$$

$$|M| = z_{rms}^2\,\dot{z}_{rms}^2\,\ddot{z}_{rms}^2 - \dot{z}_{rms}^6 \qquad [9.41]$$

The ratio of the number of minima to the number of peaks is equal to [RIC 39]:

$$R = \frac{1 - \sqrt{\pi}\,\lambda\,e^{\lambda^2}\,(1 - \mathrm{erf}\,\lambda)}{1 + \sqrt{\pi}\,\lambda\,e^{\lambda^2}\,(1 + \mathrm{erf}\,\lambda)} \qquad [9.42]$$

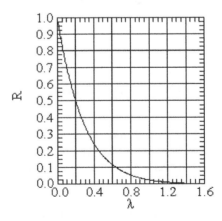

Figure 9.25. *Ratio of the number of minima to the number of peaks*

where

$$\lambda = \frac{a\ r}{z_{rms}\sqrt{2(1-r^2)}}$$

R is thus a function of the only parameter λ, which depends on the reduced level $\frac{a}{z_{rms}}$ and of parameter r. We note in Figure 9.25 that R decreases when λ increases. If R is small, the process is within a narrowband. R is thus a measurement of the regularity of the oscillations.

Figure 9.26 shows R varying with r, for $\frac{a}{z_{rms}}$ varying from 1 to 5.

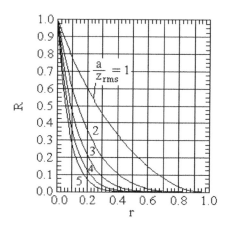
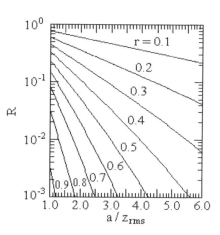

Figure 9.26. *Ratio of the number of minima to the number of peaks versus the irregularity factor*

Figure 9.27. *Ratio of the number of minima to the number of peaks versus the threshold a*

The ratio R decreases when r increases, and this becomes faster when $\frac{a}{z_{rms}}$ is larger. This tendency is underlined in another form by the curves $R\left(\frac{a}{z_{rms}}\right)$, for variable r.

9.4. Equivalence between the transfer functions of a bandpass filter and a one-degree-of-freedom linear system

There are several types of equivalences based on different criteria.

9.4.1. Equivalence suggested by D.M. Aspinwall

The selected assumptions are as follows [ASP 63], [BAR 65], [SMI 64]:

– the bandpass filter and the linear one-degree-of-freedom system are assumed to let the same quantity of power pass in response to white noise excitation: the two responses must thus have same rms value;

– the two filters have the same amplification (respectively at the central frequency and the natural frequency).

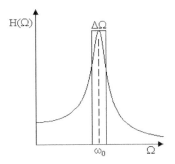

Figure 9.28. *Bandpass filter equivalent to the transfer function of a one-degree-of-freedom system*

The rms value of the response u(t) to the excitation $\ell(t)$ is given by

$$u_{rms}^2 = \int_0^\infty |H_{1u}(\Omega)|^2 \, G_1(\Omega) \, d\Omega$$

For the mechanical system, we have for example

$$|H(\Omega)| \frac{1}{\sqrt{\left[1-\left(\dfrac{\Omega}{\omega_0}\right)^2\right]^2 + 4\xi^2 \left(\dfrac{\Omega}{\omega_0}\right)^2}}$$

450   Random Vibration

and for the bandpass filter:

$$\begin{cases} H(\Omega) = H_0 & \text{for } \Omega_c \in \left(\Omega_c - \dfrac{\Delta\Omega}{2}, \ \Omega_c + \dfrac{\Delta\Omega}{2} \right) \\ H(\Omega) = 0 & \text{elsewhere} \end{cases}$$

The first assumption lays down that $u^2_{rms1dof} = u^2_{rmsWN}$, i.e., if $G_\ell(\Omega) = G_{\ell_0}$ is the PSD of the white noise

$$\frac{\pi}{2} \omega_0 \, Q \, G_{\ell_0} = G_{\ell_0} \, H_0^2 \, \Delta\Omega$$

or

$$\frac{\pi}{2} \omega_0 \, Q = H_0^2 \, \Delta\Omega$$

and the second

$$H_0 = Q = \frac{1}{2\xi}$$

yielding

$$\frac{\pi \, Q \, \omega_0}{2} = Q^2 \, \Delta\Omega$$

$$\Delta\Omega = \frac{\pi}{2} \frac{\omega_0}{Q} \qquad [9.43]$$

or

$$\Delta f = \frac{\pi}{2} \frac{f_0}{Q} = \pi \, f_0 \, \xi = \frac{\pi}{2} \Delta f_R \qquad [9.44]$$

This interval Δf is sometimes called the *mean square bandwidth* of the mode [NEW 75]. If the mechanical system has as a transfer function

$$H(\Omega) = \frac{\sqrt{1+\left(\dfrac{\Omega}{Q\,\omega_0}\right)^2}}{\sqrt{\left[1-\left(\dfrac{\Omega}{\omega_0}\right)^2\right]^2+\left(\dfrac{\Omega}{Q\,\omega_0}\right)^2}}$$

we have, with the same assumptions

$$\frac{\pi}{2}\frac{\omega_0}{Q}\left(1+Q^2\right)G_{\ell_0} = G_{\ell_0}\,H_0^2\,\Delta\Omega$$

and

$$H_0 = Q$$

yielding

$$\Delta\Omega = \frac{\pi}{2}\,\omega_0\,\frac{1+Q^2}{Q^3}$$

$$\Delta f = \frac{\pi}{2}\,f_0\,\frac{1+Q^2}{Q^3} \qquad [9.45]$$

and, if Q is large,

$$\Delta f \approx \frac{\pi}{2}\,\frac{f_0}{Q} \qquad [9.46]$$

The error is lower than 10% for $Q > 4$.

9.4.2. *Equivalence suggested by K.W. Smith*

We look for width B of a rectangular filter centered on f_0, of height Q, such that the area under the curve $H^2(\Omega)$ is equal to that of the rectangle $A_m^2\,B$ ($A_m = Q$) [SMI 64].

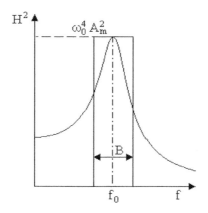

Figure 9.29. *Bandpass filter equivalent to the transfer function of a one-degree-of-freedom system according to K.W. Smith*

Knowing that the PSD G_u of the response is related to that of the excitation by $G_u = H_{\ell u}^2 G_\ell$, we have, for white noise,

$$u_{rms}^2 = \int G_u \, d\Omega = \int H_{\ell u}^2 G_\ell \, d\Omega = G_{\ell 0} \int H_{\ell u}^2 \, d\Omega$$

However

$$\int_0^\infty \frac{d\Omega}{\left(1-h^2\right)^2 + \dfrac{h^2}{Q^2}} = \omega_0 \int_0^\infty \frac{dh}{\left(1-h^2\right)^2 + \dfrac{h^2}{Q^2}} = \frac{\pi \omega_0}{4\xi}$$

yielding

$$A_m^2 \, B = \frac{\pi \omega_0}{4\xi} \qquad [9.47]$$

and, since we must have $A_m = Q$,

$$B = \frac{\pi \omega_0}{2Q} = \pi \xi \omega_0 \qquad [9.48]$$

9.4.3. Rms value of signal filtered by the equivalent bandpass filter

This is equal to

$$u_{rms}^2 = \int_0^\infty H_0^2 \, G_{\ell_0} \, df$$

i.e.

$$u_{rms}^2 = H_0^2 \, G_{\ell_0} \, \Delta f$$

If the vibration is an acceleration, we must replace G_{ℓ_0} with $\dfrac{G_{\ddot{x}_0}}{\omega_0^4}$. It then becomes:

$$u_{rms}^2 = G_{\ddot{x}_0} \frac{\pi}{2} \frac{Q^2}{\omega_0^4} \frac{f_0}{Q}$$

$$\omega_0^2 \, u_{rms} = \sqrt{\frac{\pi}{2} f_0 Q \, G_{\ddot{x}_0}} \qquad [9.49]$$

Example 9.4.

Let us consider a one-degree-of-freedom linear mechanical filter of natural frequency $f_0 = 120$ Hz and quality factor $Q = 22$, subjected to a white random noise of spectral density of acceleration $G_{\ddot{x}_0} = 10$ (m/s²)² Hz between 1 Hz and 1,000 Hz (rms value: $\ddot{x}_{rms} \approx 100$ m/s²). The rms value of the response calculated from [9.49] is equal to

$$\omega_0^2 \, z_{rms} = \sqrt{\frac{\pi}{2} f_0 \, Q \, G_{\ddot{x}_0}} = \sqrt{\frac{\pi}{2} 120 \cdot 22 \cdot 10} = 203.6 \text{ m/s}^2$$

It can be checked that the rms displacement z_{rms} obtained from relations [8.33] and [8.34] is in conformity with the value extracted from this result (i.e. $z_{rms} = 0.358$ mm).

Chapter 10

First Passage at a Given Level of Response of a One-Degree-of-Freedom Linear System to a Random Vibration

10.1. Assumptions

The problem of the definition of the first passage time at a given threshold value is important since it can be associated with the probability of failure in a structure when exceeding a characteristic stress failure limit, an acceptable deformation or a collision between two parts due to excessive displacement response, etc. It is a question of determining, for a given probability, how long a random vibration can be applied before a critical amplitude of the response is observed [GRA 66] or of evaluating the maximum amplitude which can be reached within a given time T.

Let $u(t)$ be the response of a simple mechanical system (a one-degree-of-freedom linear system) to a presumed stationary random excitation $\ell(t)$. The duration T at the end of which $u(t)$ reaches a selected level a for the first time is sought. Level a can be positive only, negative only or either positive or negative.

Consider the whole range of samples of responses $u(t)$ to the excitations (inputs) $\ell(t)$ which constitute the random process.

Each sample reached level a for the first time at the end of time T_i. All of the values T_i obey a statistical law $p(T)$ which would be very interesting to ascertain but which, unfortunately, could not be found accurately in cases of a general nature [GRA 66].

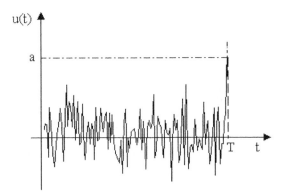

Figure 10.1. *First passage through a given threshold in time T*

Certain simplifying assumptions must therefore be made to continue with the analysis. However, analog and numerical simulations [BAR 65], [CRA 66a] could indicate the form of the solution.

Let $p(T)\,dT$ be the probability that the response $u(t)$ exceeds the threshold a for the first time (since time $t = 0$) in the interval $T \le t \le T + dt$. Simulations show that the probability density of the first crossing, $p(T)$, has a form which at first depends on the initial conditions, according to whether, at $t = 0$, the excitation is already stationary (Figure 10.2, curve A) or whether, before this moment, it is zero (Figure 10.2, curve B).

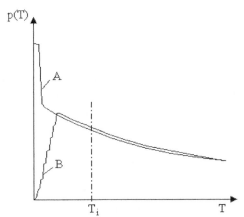

Figure 10.2. *Probability density of first upcrossing as a function of initial conditions*

Whatever the case might be, the remainder of the curve indicates exponential decay which might be approximated analytically by a relation of the form

$$p(T) = \sum_i A_i \, \alpha_i \, e^{-\alpha_i T}$$

$(T \geq 0)$. Estimating $p(T)$ for T small is not a simple problem. On the other hand, as soon as the mean time of the first up-crossing is large enough with respect to the duration during which this study is difficult, the expression for $p(T)$ may be reduced to the approximate form:

$$p(T) = A \, \alpha \, e^{-\alpha T} \qquad [10.1]$$

and the distribution function [VAN 75] to

$$P = A \, e^{-\alpha T} \qquad [10.2]$$

where

- A is the probability of starting, when $t = 0$, below the threshold;
- α is the *limiting decay rate* of the first crossing density [CRA 70];
- $P(T)$ is the probability of no crossing between 0 and T $(T \geq 0)$.

Numerical simulations also show that, for a sufficiently large threshold a, $A \approx 1$ (A depends on the state of the system at $t = 0$, but not on α) [CRA 66a]. On the other hand, for low values of a, A is higher than 1 when noise is applied to $t = 0$ (transitory phase) and A is lower than 1 if, at time $t = 0$, the noise is already stationary.

From work relating to this subject, several classifications may be made:

– assumptions as to the height of maxima, i.e. as to the independence of threshold crossings of the signal or its envelope;

– assumptions as to the nature of the noise: wideband [DIT 71], [TIP 25], [VAN 75] or narrowband [CRA 63] response of a slightly damped one-degree-of-freedom system to a Gaussian stationary white noise [BAR 61], [BAR 65], [BAR 68b], [CRA 66a], [CRA 68], [CRA 70], [CAU 63], [GRA 65], [GRA 66], [LIN 70], [LUT 73], [LYO 60], [MAR 66], [ROB 76], [ROS 62], [VAN 69], [YAN 71].

The study can be carried out:

– by means of (Figure 10.3), at $t = 0$, a *zero start* (response beginning from a rest position) and therefore, in the first few instants, a transitional phase [CHA 72], [CRA 66a], [GRA 65], [LUT 73], [YAN 72];

– for, at $t = 0$, a *stationary start*, an already established mode [CRA 66a], [GRA 66], [LIN 70], [LUT 73], [YAN 71];

– by considering a short *burst* of random signal between two times 0 and t (research of the extreme response);

– by employing analytical methods, leading to theoretical results [GRA 65] [GRA 66], [LIN 70], [YAN 71]; and

– analog [GRA 66] [LUT 73] or numerical simulations [CRA 66a].

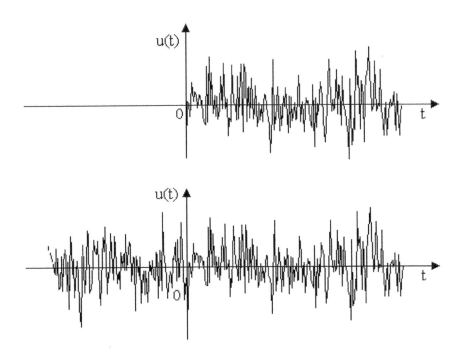

Figure 10.3. *Zero or stationary start*

Several assumptions may be made to determine an approximate value for the parameter α. Recalling the classification proposed by S.H. Crandall [CRA 70], the following cases will be successively examined:

1) Threshold level a is sufficiently high and the threshold excursions are so rare that they can be regarded as statistically independent.

2) The maxima of the response can be assumed to be independent.

3) The threshold upcrossings of the envelope of maxima are independent.

4) The maxima of the envelope of the peaks are independent.

5) The amplitudes of the peaks follow a Markov process.

6) The response peaks are divided into groups for each of which the envelope of the peaks varies slowly.

10.2. Definitions

Consider a response random signal $u(t)$, whose derivative is $\dot{u}(t)$, and let it be placed in a diagram $\dfrac{\dot{u}(t)}{\omega_0}$, $u(t)$ (ω_0 being the natural pulsation of the one-degree-of-freedom system subjected to vibration).

The barriers are classified as follows:

– type B: a barrier with a limit such that $u \leq a$ (Figure 10.4(1));

– type D: a barrier with a symmetric double-passage level such that $|u| \leq a$ (Figure 10.4(2));

– type E: a barrier with a limit concerning the envelope $e(t)$ of the process $u(t)$ (but not the process itself), such that

$$e^2 = u^2 + \frac{\dot{u}^2}{\omega_0^2}$$

Here the range $e < a$ is a circle (Figure 10.4(3)) of radius a. In this phase plan, the trajectory of a response is a clockwise random spiral [CRA 66a].

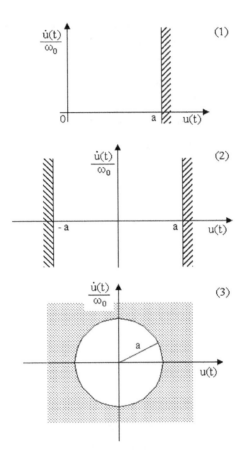

Figure 10.4. *Various types of barriers*

10.3. Statistically independent threshold crossings

When considering high thresholds (higher than 2.5 times the rms value), the times at which the signal response crosses threshold a can be regarded as being distributed according to Poisson's law, i.e. of the form [CRA 70], [GRA 66] [VAN 69]:

$$p(T) = \alpha\, e^{-\alpha T} \qquad [10.3]$$

$(A = 1)$.

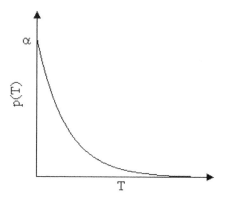

Figure 10.5. *Poisson's law for threshold crossings*

It was seen (Chapter 5) that, for a Gaussian process with zero mean, the number of crossings beyond a threshold a with positive slope can be written

$$n_a^+ = n_0^+ \, e^{-\frac{a^2}{2 u_{rms}^2}}$$

where

- u_{rms} = rms value of the response signal $u(t)$;

- n_0^+ = mean number of passages through zero with a positive slope;

$$n_0^+ = \frac{1}{2\pi} \sqrt{\frac{M_2}{M_0}}$$

when $M_0 = u_{rms}^2$ and $M_2 = \dot{u}_{rms}^2$ [MIL 61]. Depending on the case, the parameter α, independent of its stationary state at $t = 0$, is equal to [RIC 64]:

- $n_a^+ = n_0^+ \, e^{-\frac{a^2}{2 u_{rms}^2}}$ if the barrier is of type B ($u \le a$) [CRA 63], [ROB 76].

$$\alpha_B = n_a^+ = n_0^+ \, e^{-\frac{a^2}{2 u_{rms}^2}} = n_0^+ \, e^{-\frac{v^2}{2}} \qquad [10.4]$$

($v = a/u_{rms}$). This yields the probability of there being no crossing of the threshold during the interval (0, T) [MIL 61]:

$$P(T) = \exp(-\alpha_B T) = \exp\left(-n_0^+ T e^{-\frac{v^2}{2}}\right) \qquad [10.5]$$

$-2 n_a^+$ if the barrier is defined by $|u| < a$ (type D) [GRA 66], [VAN 72]. Then

$$\alpha_D = 2 \alpha_B = 2 n_0^+ e^{-v^2/2} \qquad [10.6]$$

and

$$P(T) = \exp(-\alpha_D T) = \exp\left(-2 n_0^+ T e^{-\frac{v^2}{2}}\right) \qquad [10.7]$$

Returning to the case of the type B barriers, let us estimate the probability $P[u(t) \geq a]$ that the response u is higher than a given threshold a (the probability of exceeding a in the positive direction) [BEN 64], [COL 59]. The probability that $u(t)$ crosses the threshold a with a positive slope in the interval of time t, $t + \Delta t$ is

$$P[u(t) \geq a] = n_a^+ \Delta t \qquad [10.8]$$

if it is assumed that Δt is arbitrarily small, sufficiently so for there to be only one passage in this interval. Let us also assume that the probability of a passage of the threshold during Δt can be regarded as independent of time t which Δt starts.

Let $P_0(t)$ be the probability of threshold a not being exceeded in time 0, t and $P_0(t + \Delta t / t)$ as the conditional probability of a not being exceeded in t, $t + \Delta t$ knowing that a was not exceeded in (0, t):

$$P_0(t + \Delta t) = P_0(t + \Delta t / t) P_0(t) \qquad [10.9]$$

However,

$$P_0(t + \Delta t / t) = P[u(t) < a] = 1 - P[u(t) \geq a] \qquad [10.10]$$

$$P_0(t + \Delta t / t) = 1 - n_a^+ \Delta t \qquad [10.11]$$

yielding

$$P_0(t + \Delta t) = P_0(t)\left(1 - n_a^+ \Delta t\right)$$

or

$$\frac{P_0(t + \Delta t) - P_0(t)}{\Delta t} = - n_a^+ P_0(t) \qquad [10.12]$$

When $\Delta t \to 0$, this becomes

$$\frac{dP_0(t)}{dt} = - n_a^+ P_0(t)$$

yielding

$$P_0(T) = A_0 e^{-n_a T} \qquad [10.13]$$

where A_0 is a constant of integration. Knowing that, at $t = 0$, $P_0(0) = 1$, we have $A_0 = 1$. The probability of exceeding a in (0, T) (distribution function) is thus [LIN 67], [RAC 69], [YAN 71]

$$P(T) = 1 - P_0(T) = 1 - e^{-n_a^+ T} \qquad [10.14]$$

The density of probability $p(T)$ associated with $P(T)$ is

$$p(T) = \frac{dP}{dt} = n_a^+ e^{-n_a^+ T} \qquad [10.15]$$

$p(T)$ depends solely on $M_0 = u_{rms}^2$ and n_0^+, related to M_0 and to M_2. Hence:

– the mean time of the first up-crossing [CRA 63]

$$E(T) = \int_0^\infty T\, p(T)\, dT = \frac{1}{n_a^+} \qquad [10.16]$$

– the variance

$$s_T^2 = \int_0^\infty [T - E(T)]^2 \, p(T) \, dT$$

$$s_T = \frac{1}{n_a^+} \qquad [10.17]$$

If it is assumed that the probability of exceeding threshold a is independent of the initial time, the mean time between upcrossings compatible with $E(T)$ can be assumed to be equal to

$$T_m = \frac{1}{n_a^+} \qquad [10.18]$$

yielding, starting from [10.13]

$$P_0(T) = e^{-\frac{T}{T_m}} \qquad [10.19]$$

NOTES.–

1. *According to Poisson's law, the probability that the failure occurred at $t = 0$ is assumed to be zero even if this probability is finite, but very small for the large thresholds [GRA 66], [YAN 71]. If the probability of passage above the level a at $t = 0$ is of interest, it can however be added to the right member of [10.17]. Always assuming that the response signal is Gaussian with zero mean, this probability is given by*

$$P(u > a) = \frac{1}{u_{rms} \sqrt{2\pi}} \int_a^\infty e^{-\frac{u^2}{2 u_{rms}^2}} \, du$$

Setting $t = \dfrac{u}{u_{rms} \sqrt{2}}$

$$P(u > a) = \frac{1}{\sqrt{\pi}} \int_{\frac{a}{u_{rms}\sqrt{2}}}^\infty e^{-t^2} \, dt$$

which following transformation becomes

$$P(u > a) = \frac{1}{2}\left[1 - erf\left(\frac{a}{u_{rms}\sqrt{2}}\right)\right] \qquad [10.20]$$

It should be noted that all these expressions are independent of the Q factor of the system.

2. Relation [10.17] is important, because it makes it possible to set limits to certain problems when an exact analysis of extreme levels is not possible. The quantity n_a^+ can either be determined analytically or experimentally. For small $n_a^+ T$, i.e. small P, we have:

$$P(T) \leq n_a^+ T$$

(since $e^{-x} \approx 1 - x + \cdots \geq 1 - x$), yielding

$$P \leq T\, n_0^+\, e^{-\frac{a^2}{2u_{rms}^2}}$$

$$T \geq \frac{P}{n_0^+}\, e^{\frac{a^2}{2u_{rms}^2}} \qquad [10.21]$$

G.P. Thrall [THR 64] showed that $n_a^+ T$ constitutes an upper limit of the probability of crossing a positive (or negative) threshold in time T without using the assumption of independence of the upcrossings.

An arbitrary correlation between successive extrema tends to decrease the probability that such peaks exist during a given time period [GRA 66], [THR 64].

Figure 10.6 shows the variations of $\frac{a}{u_{rms}}$ with $n_0^+ T$, for various values of P.

3. *J.N. Yang and M. Shinozuka [YAN 71] express the same results in the form*

$$P(N) = 1 - exp\left[-N\ exp\left(-\frac{a^2}{2\ u_{rms}^2}\right)\right] \qquad [10.22]$$

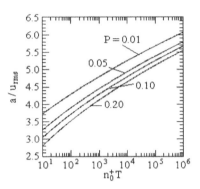

Figure 10.6. *Average threshold reached over timespan T, for various probabilities*

where $P(N)$ is the probability that the first excursion above threshold a takes place in the N first half-cycles, corresponding to a duration T:

$$N = \frac{2T}{T_a}$$

$$T_a = 2\pi\ \frac{u_{rms}}{\dot{u}_{rms}} = \frac{1}{n_0^+} \qquad [10.23]$$

4. *In the case of a Gaussian process with zero mean, M.R. Leadbetter [LEA 69] expresses the probability that, over duration T, the maximum of the process $u(T)$ is larger than a, as shown by:*

$$P[M(T) > a] = 2\ P[u(T) > a]$$

yielding, by considering the variance equal to $u_{rms}^2\ T$,

$$P[M(T) \le a] = \frac{2}{u_{rms} \sqrt{2 \pi T}} \int_0^a \exp\left(-\frac{x^2}{2 u_{rms}^2 T}\right) dx \qquad [10.24]$$

$(a \ge 0)$.

All these calculations assume the independence of the probability of the amplitude of a peak in comparison to those of the preceding peaks. This assumption of peak independence has been criticized. H. Cramer [CRA 66b] rigorously showed that the law of distribution of the upcrossings tends asymptotically towards Poisson's law when threshold a tends towards infinity [LEA 69]. For thresholds observed in practice, this formulation leads to an error the importance of which largely depends on the bandwidth of the process. Numerical simulations show that these expressions are correct for large a:

– there is broad agreement when $a \ge 2\, u_{rms}$ for the wideband processes, whose PSD is relatively uniform on a broad frequency range,

– there is agreement with simulation when $a \ge 3\, u_{rms}$ for the processes having a bandwidth equal to an octave.

O. Ditlevsen [DIT 71] observes that, for the wideband processes, the error factor falls outside of safety and that assuming Poisson's law, there is no tolerance over the time that has actually elapsed above the threshold level.

For the much narrower band processes, the calculations carried out with these relations lead to first upcrossing times that are much shorter than the experimental times. S.H. Crandall and W.D. Mark [CRA 63] estimate that, in a narrowband noise, the peaks tend to form groups and cannot be regarded as being independent. This grouping tends to decrease n_a^+ and thus to increase $p(T)$. The error is therefore safety-orientated [CRA 66a] [GRA 66]. The variations increase when the bandwidth decreases and decrease with a. The mean rate n_a^+ is asymptotically exact, when $\frac{a}{u_{rms}} \to \infty$, and as such can be used as a reference to compare other estimates [MIL 61].

It is important in this case to take account of the statistical dependence of the occurrences of the upcrossings. Y.K. Lin [LIN 70] establishes approximate first passage probabilities by considering that the threshold passages constitute a continuous random process and by assuming several models for their distribution. J.N. Yang and M. Shinozuka [YAN 71] obtain other approximate relations (upper and lower limits) by considering a punctual representation of the process of successive maxima and minima (narrowband process).

Example 10.1.

Let us consider a Gaussian narrowband process with zero mean, central frequency 200 Hz and rms value u_{rms}. We want to determine the time T_0 for which $u(t)$ is lower than the level $a = 4.5\ u_{rms}$ with a probability of 90%.

$$n_a^+ = n_0^+\ e^{-\dfrac{a^2}{2\,u_{rms}^2}}$$

$$n_a^+ = 200\ e^{-\dfrac{(4.5)^2}{2}}$$

$$T_0 = -\dfrac{\ln(0.9)}{200}\ e^{\dfrac{(4.5)^2}{2}}$$

$$T_0 \approx 13\ s$$

The probability that $u > 4.5\ u_{rms}$ at $t = 0$ is

$$P(u > a) = \dfrac{1}{2}\left[1 - \text{erf}\left(\dfrac{4.5}{\sqrt{2}}\right)\right]$$

$$P(u > a) \approx 3.4\ 10^{-6}$$

(a low value compared to 0.1).

10.4. Statistically independent response maxima

It is assumed here that the response process $u(t)$ is a zero mean, narrowband process and that, firstly, we wish to determine the peak distribution of $u(t)$ [CRA 63]. The mean number of crossings of the threshold a with positive slope, per unit time, for a stationary process, is equal to n_a^+. For two close levels, the quantity

$$n_a^+ - n_{a+da}^+ = -\dfrac{dn_a^+}{da}\ da$$ is the mean number of peaks per unit time between a and

a + da (false in a strict sense but, in a narrowband process, the probability of positive minima is negligible). For the same reason, the mean number of peaks on a given level per unit time is equal to the mean number of cycles per unit time (r = 1), i.e. to n_0^+. The fraction of the peaks located between a and a + da is given by

$$p(a)\,da = -\frac{1}{n_0^+}\frac{dn_a^+}{da}\,da \qquad [10.25]$$

where p(a) is the probability density of peaks on the level a among all the peaks between 0 and the infinite. Yielding

$$p(a) = \frac{1}{u_{rms}}\,e^{-\frac{a^2}{2u_{rms}^2}} \qquad [10.26]$$

(Rayleigh's distribution) and [CRA 63], [YAN 71]

$$P\big(|u| < a\,u_{rms}\big) = 1 - e^{-\frac{a^2}{2u_{rms}^2}} \qquad [10.27]$$

Over a timespan T, there is a mean number N of positive and negative peaks ($N = 2\,n_0^+\,T$). Disregard dispersion relating to N and assume that the amplitudes of N peaks are statistically independent. The probability that all N peaks of $|u(t)|$ are smaller than a u_{rms} is then, when $v = \dfrac{a}{u_{rms}}$

$$P = \{P[|u(t)| < a\,u_{rms}]\}^N$$

$$P = \left(1 - e^{-v^2/2}\right)^{2\,n_0^+\,T} \qquad [10.28]$$

and the probability that these peaks are higher than a u_{rms} is $Q = 1 - P$. The probability density associated with this Q factor is:

$$p(T) = \frac{dQ}{dT} = -\frac{dP}{dT}$$

however,

$$\ln P = 2 n_0^+ T \ln\left(1 - e^{-v^2/2}\right)$$

$$\frac{dP}{P} = 2 n_0^+ \ln\left(1 - e^{-\frac{v^2}{2}}\right) dT$$

$$p = -\frac{dP}{dT} = -2 n_0^+ \ln\left(1 - e^{-v^2/2}\right) P \qquad [10.29]$$

where P can be written

$$P = \exp\left[2 n_0^+ T \ln\left(1 - e^{-v^2/2}\right)\right] \qquad [10.30]$$

$p(T)$ is thus of the form $p = \alpha\, e^{-\alpha T}$ with

$$\alpha = -2 n_0^+ \ln(1 - e^{-\frac{v^2}{2}}) \qquad [10.31]$$

It should be noted that α and hence $p(T)$ are independent of the Q factor of the system.

Figure 10.7 shows the variations of the ratio ρ_1 of the α values calculated on the basis of the first two assumptions [10.4] and [10.31], which has as its expression

$$\rho_1 = \frac{-2 n_0^+ \ln(1 - e^{-\frac{v^2}{2}})}{2 n_0^+ e^{-\frac{v^2}{2}}}$$

$$\rho_1 = \frac{-\ln(1 - e^{-\frac{v^2}{2}})}{e^{-\frac{v^2}{2}}} \qquad [10.32]$$

This ratio very quickly tends towards 1. It is almost equal to 1 for $v > 3$. The value of α calculated on the basis of assumption 2 thus converges very quickly towards the value of α calculated by assuming that the upcrossings follow a Poisson law, which as has been shown, is a rigorous law for v when very large.

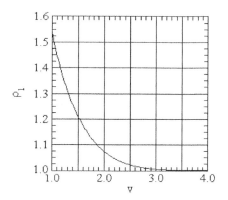

Figure 10.7. *Ratio of the α constants calculated on the basis of [10.4] and [10.31]*

Assumption 2 can however be criticized. A narrowband noise arises as a quasi-sinusoidal oscillation whose mean frequency is the natural frequency of the system and whose amplitude and phase vary randomly (Figure 10.8).

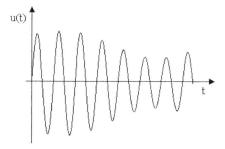

Figure 10.8. *Narrowband noise*

The maxima have an amplitude which does not vary a great deal from one peak to another, so much so that the peaks cannot be considered to have independent amplitudes, especially when the Q factor is large.

10.5. Independent threshold crossings by the envelope of maxima

To take account of the above criticism, response crossings of the threshold $u(t)$ are estimated as starting from crossings (with positive slope) of the same threshold by the envelope $R(t)$ of maxima of $u(t)$.

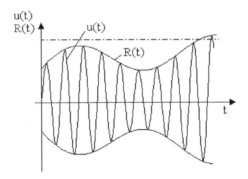

Figure 10.9. *Narrowband noise and its envelope*

The upcrossings of threshold a are regarded here as independent $[|R(t)| > a]$, tantamount to a Poisson distribution.

Let $\dot{R}(t)$ be the process derivative of $R(t)$. The joint probability density of the envelope $R(t)$ and its derivative $\dot{R}(t)$ is [CRA 67]

$$p(R, \dot{R}) = \frac{R}{R_{rms}^2} e^{-R^2/2 R_{rms}^2} \frac{1}{\sqrt{2\pi} \dot{R}_{rms}} e^{-\dot{R}^2/2 \dot{R}_{rms}^2}$$

where R_{rms} and $\dot{R}_{rms}$ are respectively the rms values of $R(t)$ and of $\dot{R}(t)$. It is shown that $\dot{R}_{rms} = q_E \dot{u}_{rms}$ where $\dot{u}_{rms} = 2\pi n_0^+ u_{rms}$ (n_0^+ being the mean frequency of the response) and $q_E = \sqrt{1 - \frac{M_1^2}{M_0 M_2}}$.

Hence

$$\dot{R}_{rms} = \sqrt{\frac{M_0 M_2 - M_1^2}{M_0}}$$

(assuming a response with zero mean, so that $R_{rms} = u_{rms}$). From this yielding, by replacing M_0, M_1 and M_2 with their expressions established in the case of a response to a white noise (relations [10.6], [10.10] and [10.16]), we obtain:

$$\dot{R}_{rms} = \sqrt{2\pi}\, u_{rms}\, n_0^+\, g(\xi)$$

and

$$g(\xi) = \sqrt{2\pi}\left[1 - \frac{1}{1-\xi^2}\left(1 - \frac{1}{\pi}\arctan\frac{2\xi\sqrt{1-\xi^2}}{1-2\xi^2}\right)^2\right]^{1/2} = \sqrt{2\pi}\, q_E \quad [10.33]$$

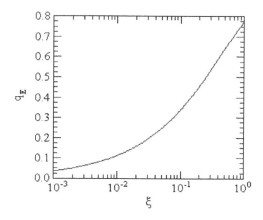

Figure 10.10. *Parameter q_E as a function of the damping ratio*

Example 10.2.

For $\xi = 0.05$, we obtain $q_E = 0.2456$ and $g(\xi) = 0.6156$.

It was shown that the mean number by unit time of crossings above a given level a by a signal u(t) is equal to $n_a^+ = n_0^+\, e^{-a^2/2\, u_{rms}^2}$. The mean frequency with which the envelope crosses level a is therefore

$$m_a^+ = \int_0^\infty p(R, \dot{R}) \, \dot{R} \, d\dot{R} = \frac{1}{\sqrt{2\pi}} \frac{\dot{R}_{rms}}{u_{rms}^2} a \, e^{-a^2/2 u_{rms}^2} \qquad [10.34]$$

or

$$\alpha = m_a^+ = n_a^+ \, g(\xi) \, \frac{a}{u_{rms}} \qquad [10.35]$$

It should be noted here that α depends on damping factor ξ. As above, variations in the ratio ρ_2 for the values of α obtained on the basis of assumptions 3 [10.35] and 1 [10.4] varying with v are plotted, as shown (Figure 10.11), for several values of ξ.

$$\rho_2 = \frac{n_a^+ \, g(\xi) \, \dfrac{a}{u_{rms}}}{2 \, n_0^+ \, e^{-\frac{v^2}{2}}} = \frac{n_0^+ \, e^{-\frac{v^2}{2}} \, g(\xi) \, v}{2 \, n_0^+ \, e^{-\frac{v^2}{2}}}$$

$$\rho_2 = \frac{v}{2} g(\xi) = \frac{v}{2} \sqrt{2\pi} \left[1 - \frac{1}{1-\xi^2} \left(1 - \frac{1}{\pi} \arctan \frac{2\xi\sqrt{1-\xi^2}}{1-2\xi^2} \right)^2 \right]^{1/2}$$

$$\rho_2 = \frac{v}{2} \sqrt{2\pi} \left[1 - \frac{1}{4(1-\xi^2)} \left(1 - \frac{2}{\pi} \arctan \frac{2\xi^2 - 1}{2\xi\sqrt{1-\xi^2}} \right)^2 \right]^{1/2} \qquad [10.36]$$

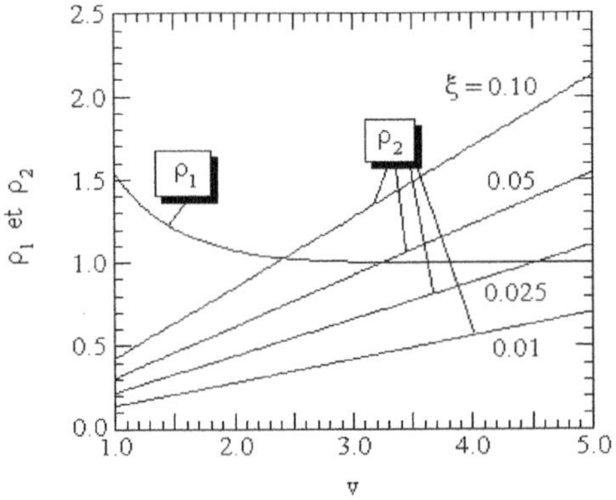

Figure 10.11. *Coefficients α in assumptions 2 and 3 in relation to α as in assumption 1*

It should be noted that the ratio $\rho_2(v)$ exceeds value 1 when v is sufficiently large.

For high values of v, assumption 3 is less suited than 1 and 2. For weaker v, it is on the contrary better suited.

Relation [10.34] can be also written

$$m_a^+ = \frac{\dot{R}_{rms}}{\sqrt{2\pi}} \frac{a}{u_{rms}^2} e^{-\frac{a^2}{2 u_{rms}^2}} \qquad [10.37]$$

or,

$$m_a^+ = \frac{1}{\sqrt{2\pi}} \frac{\dot{R}_{rms}}{u_{rms}} v e^{-\frac{v^2}{2}} \qquad [10.38]$$

Knowing that $q_E = \frac{\dot{R}_{rms}}{\dot{u}_{rms}}$ and that $\frac{\dot{u}_{rms}}{u_{rms}} = \sqrt{\frac{M_2}{M_0}} = 2\pi n_0^+$, this expression becomes

$$m_a^+ = \sqrt{2\pi} \, q_E \, n_0^+ \, v \, e^{-\frac{v^2}{2}} = \sqrt{2\pi} \, q_E \, v \, n_a^+ \qquad [10.39]$$

or [DEE 71], [VAN 69], [VAN 75]

$$m_a^+ = \frac{v}{M_0} \sqrt{\frac{M_0 M_2 - M_1^2}{2\pi}} \, e^{-\frac{v^2}{2}} \qquad [10.40]$$

yielding another form of the first passage probability of the type E threshold [THR 64], [VAN 75]:

$$P(T) = \exp\left[-\sqrt{2\pi} \, q_E \, v \, n_a^+ \, T\right] \qquad [10.41]$$

and

$$\alpha = \sqrt{2\pi} \, q_E \, v \, n_a^+ \qquad [10.42]$$

10.6. Independent envelope peaks

10.6.1. S.H. Crandall method

If the threshold level a of the response is sufficiently large, we can consider that, in a given time, there are as many crossings of this threshold by the envelope R(t) with positive slope than the maxima of the envelope. This means that the envelope does not have any peak below a [CRA 70]. For an arbitrary threshold $b \geq a$, the peak distribution of the envelope, as in assumption 2, is therefore dictated by the form

$$P = \text{prob}(\text{peak of envelope} < b) = 1 - \frac{m_b^+}{m_0^+} \qquad [10.43]$$

where

– m_b^+ is the mean number by unit time of upcrossings of the threshold b by the envelope R(t);

– m_0^+ is the expected frequency of R(t).

It was shown that [10.35]

$$m_b^+ = n_b^+ \, g(\xi) \, \frac{b}{u_{rms}}$$

To simplify, S.H. Crandall defines m_0^+ as the value of m_b^+ for $b = u_{rms}$ (the envelope never intersects the axis $u = 0$; another reference should therefore be taken) yielding

$$v_0^+ = n_b^+ \, g(\xi)$$

$$m_0^+ = n_0^+ \, e^{-\frac{u_{rms}^2}{2 u_{rms}^2}} \, g(\xi) = n_0^+ \, g(\xi) \, e^{-\frac{1}{2}} \qquad [10.44]$$

Let $v = b/u_{rms}$.

$$\frac{m_b^+}{m_0^+} = \frac{n_b^+ \, g(\xi) \, v}{n_0^+ \, e^{-1/2} \, g(\xi)} = \frac{n_0^+ \, e^{-\frac{v^2}{2}} \, g(\xi) \, v}{n_0^+ \, e^{-1/2} \, g(\xi)} = v \, e^{-\frac{1}{2}(v^2 - 1)} \qquad [10.45]$$

$$P = 1 - v \, e^{-\frac{1}{2}(v^2 - 1)} \qquad [10.46]$$

The probability density is such that

$$p(v) = \frac{dP}{dv}$$

$$\ln(1 - P) = \ln(v) - \frac{1}{2}(v^2 - 1)$$

$$\frac{d(1 - P)}{1 - P} = -\frac{dP}{1 - P} = \frac{dv}{v} - v \, dv$$

$$p = \frac{dP}{dv} = \left(\frac{1}{v} - v\right) v \, e^{-\frac{1}{2}(v^2 - 1)}$$

$$\left.\begin{array}{l} p(v) = \left(v^2 - 1\right) e^{-\frac{1}{2}(v^2 - 1)} \quad \text{if } v \geq 1 \\ p(v) = 0 \quad \text{if } v < 1 \end{array}\right\} \qquad [10.47]$$

Over one duration T, the average number of peaks of the envelope is $N = m_0^+ \, T$. Amplitudes of maxima supposedly being statistically independent yield the probability that the N peaks are lower than threshold b:

$$P_N = [P(1 \text{ peak} < b)]^N$$

$$P_N = \left(1 - \frac{m_b^+}{m_0^+}\right)^{v_0^+ \, T} \qquad [10.48]$$

$$P_N = \left[1 - v\, e^{-\frac{1}{2}(v^2-1)}\right]^{\frac{\omega_0}{2\pi} g(\xi) \frac{T}{\sqrt{e}}} \qquad [10.49]$$

P_N has the form $e^{-\alpha T}$ with

$$\alpha = -m_0^+ \ln\left(1 - \frac{m_b^+}{m_0^+}\right)$$

$$\alpha = -\frac{\omega_0}{2\pi\sqrt{e}} g(\xi) \ln\left[1 - v\, e^{-\frac{1}{2}(v^2-1)}\right] \qquad [10.50]$$

This is a type E barrier. For large v, we have:

$$\alpha \approx -m_b^+ \qquad [10.51]$$

α is then:

– independent of the level of reference retained to define m_0^+;

– identical to the value obtained on the basis of assumption 3;

– infinite for v = 1, a consequence of the assumption that there is no peak below the reference level v = 1.

The curves in Figures 10.12 and 10.13 show variations of the ratio ρ_3 of the values of α given by relationships [10.50] and [10.4]:

$$\rho_3 = \frac{-g(\xi) \ln\left[1 - v\, e^{-\frac{1}{2}(v^2-1)}\right]}{2\sqrt{e}\, e^{-\frac{v^2}{2}}} \qquad [10.52]$$

10.6.2. *D.M. Aspinwall method*

The distribution over time of the response peaks of a mechanical resonator excited by white noise is not known. D.M. Aspinwall [ASP 63] [BAR 65] defined an approximate method, making it possible to estimate the probability that the response maximum of the envelope of the peaks exceeds a given level in a given time. It is based on the average number N of maxima of the envelope per second in noise of constant PSD between two frequencies f_a and f_b, equal to [7.86] [RIC 44]:

$$N = 0.64110 \, (f_b - f_a) \qquad [10.53]$$

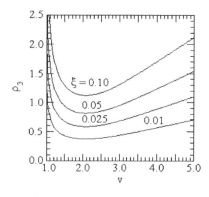
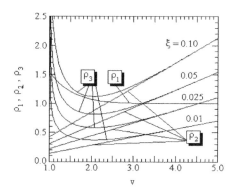

Figure 10.12. *Ratio of the constants α calculated from [10.50] and [10.4]*

Figure 10.13. *Coefficients α in assumptions 2, 3 and 4 in relation to α as in assumption 1*

This vibration can be regarded as the response of an ideal bandpass filter (f_a, f_b) to white noise input. In order to be able to use this value of N in the case of a resonator, D.M. Aspinwall defined an equivalence between a bandpass filter and mechanical resonator being based on two criteria:

– the two filters let the same quantity of power pass through them in response to white noise excitation (the rms response is therefore the same);

– they have the same amplification (the central frequency of the bandpass filter being equal to the natural frequency).

It was shown in section 9.4 that, on the basis of these assumptions, the mechanical filter equivalent to the bandpass filter has as a natural frequency f_0 the central frequency f_c of the bandpass filter and for Q factor: $Q = \dfrac{\pi}{2} \dfrac{f_0}{\Delta f}$.

D.M. Aspinwall then assumed that:

– the mean number of maxima of the envelope is a good representation of the number of maxima actually observed;

– the heights of maxima of the envelope are independent random variables in time T.

Following these assumptions, the distribution law of the highest peak amplitude of the envelope during a time T follows a law of the form:

$$F_M(v) = [F_E(v)]^{NT} \quad [10.54]$$

where:

– $F_E(v)$ is the distribution of the probability law giving the height of a maximum of the envelope chosen randomly (the probability that a randomly chosen maximum is lower than the reduced threshold v);

– $F_M(v)$ is the probability that all N T peaks occurring over the duration T are lower than threshold v.

N is given by [10.53] for a bandpass filter (f_a, f_b). The equivalent mechanical filter is such that:

$$\Delta f = f_b - f_a = \frac{\pi}{2} \frac{f_0}{Q}$$

yielding

$$N = 0.641 \frac{\pi}{2} \frac{f_0}{Q} \approx \frac{f_0}{Q}$$

and

$$NT \approx \frac{f_0 T}{Q} \quad [10.55]$$

The ratio $\frac{f_0}{Q}$ is the mean number of maxima of the envelope in time T [NEA 66].

S.O. Rice proposes a curve showing the variations of $F_E(v)$ according to α as well as a way of approaching the relationship for v > 2.5 ([7.88]):

$$F_E(v) \approx 1 - \frac{\sqrt{\frac{\pi}{6}}}{0.64110} \, v \, e^{-\frac{v^2}{2}} \qquad [10.56]$$

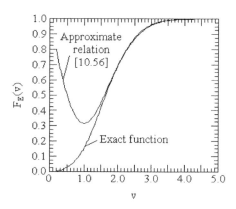

Figure 10.14. *Probability that a randomly chosen maximum is lower than threshold v*

Therefore, for $v > 2.5$,

$$F_M(v) \approx \left(1 - \frac{\sqrt{\frac{\pi}{6}}}{0.64110} \, v \, e^{-\frac{v^2}{2}}\right)^{\frac{f_0 T}{Q}} \qquad [10.57]$$

An analog simulation has shown that this relation gives only approximate results. The variations arise when taking an average value for N into account instead of the maxima's real number, from the approximation related to the equivalence of mechanical and electric filters and from the assumption of independence of maxima of the envelope which is not necessarily justified (N and F_E are undoubtedly different). The variations observed are greater for high values of Q.

R.L. Barnoski [BAR 61] noted, after a complementary study (analog simulation), that:

– for a given value $\frac{f_0 T}{Q}$, the experimental values of v are higher than those calculated by D.M. Aspinwall;

– for very large $f_0 T$, the values of v are lower than those predicted by D.M. Aspinwall (for $2 \le Q \le 50$). For $f_0 T$ large, v is almost independent of Q.

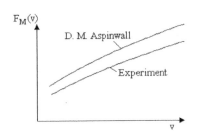
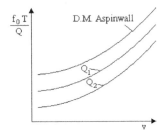

Figure 10.15. *Probability that all peaks over a duration T are lower than threshold v*

Figure 10.16. *Mean number of maxima of envelope lower than a threshold v*

D.M. Aspinwall proposes a relationship which better harmonizes theoretical and experimental results by replacing in [10.57] the parameter Q of the mechanical system with

$$Q^* = 0.2 Q + 3 \qquad [10.58]$$

$$F_M(v) \approx \left(1 - \frac{\sqrt{\frac{\pi}{6}}}{0.64110} v e^{-\frac{v^2}{2}}\right)^{\frac{f_0 T}{0.2 Q + 3}} \qquad [10.59]$$

For a system having a given factor Q, in Figures 10.17 to 10.20 the value of $F_M(v)$ may be seen after calculation of the exponent $\frac{f_0 T}{0.2 Q + 3}$ for v, given f_0 and T. These curves respectively show:

– $F_M(v)$ for several values of $\frac{f_0 T}{Q^*}$ (1, 2, 5, 10, 20, 50, 100, 250, 500, 1,000 and 5,000);

– F_M as a function of $\frac{f_0 T}{Q^*}$, for $v = 2.5, 3, 3.5, 4, 4.5$ and 5;

– $\frac{f_0 T}{Q^*}$ as a function of v, for $F_M = 0.05, 0.2, 0.5, 0.8, 0.9$ and 0.99;

$-\dfrac{f_0\,T}{Q^*}$ as a function of F_M, for $v = 2.5\,,\,3,\,3.5,\,4,\,4.5$ and 5.

G.P. Thrall [THR 64] proposes a theoretical limit for the product $f_0\,T$ as a function of the probability of exceeding a threshold v in a time T of the form

$$f_0\,T \geq \dfrac{1 - F_M(v)}{2}\,e^{\dfrac{v^2}{2}} \qquad [10.60]$$

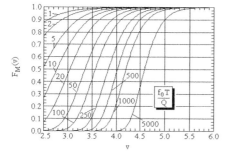

Figure 10.17. *Probability that all the peaks over duration T are lower than threshold v*

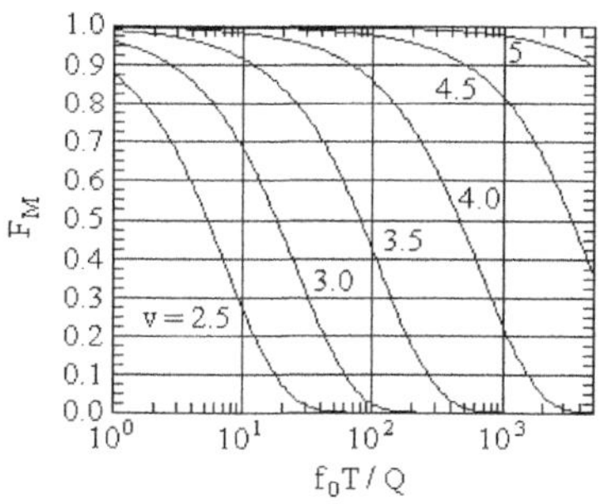

Figure 10.18. *Probability that all the peaks over duration T are lower than threshold v versus the mean number of maxima of the envelope*

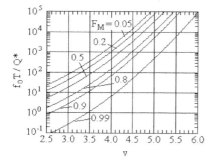
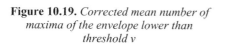
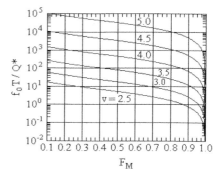

Figure 10.19. *Corrected mean number of maxima of the envelope lower than threshold v*

Figure 10.20. *Corrected mean number of maxima of the envelope versus the probability that all the peaks over duration T are lower than threshold v*

This limit appears to be correct as for when T is large. Figures 10.21 and 10.22 show the curve defined by this relationship and those resulting from [10.57] for $Q = 5$ and $Q = 50$, for $F_M(v) = 0.95$ and $F_M(v) = 0.99$, respectively.

The expression obtained by D.M. Aspinwall can be written in the form

$$P = e^{-\alpha T} \qquad [10.61]$$

where

$$\alpha = -\frac{f_0}{Q} \ln\left(1 - \frac{\sqrt{\frac{\pi}{6}}}{0.64110} e^{-\frac{v^2}{2}}\right) \qquad [10.62]$$

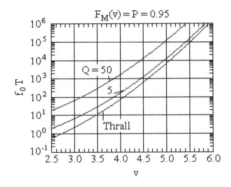

Figure 10.21. *Mean number of cycles versus threshold, for a probability of 0.95*

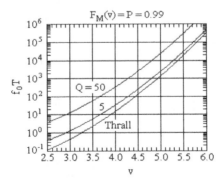

Figure 10.22. *Mean number of cycles versus threshold, for a probability of 0.99*

(type E barrier). The ratio ρ_4 of this value of α to that obtained on the basis of assumption 1 [10.4] is equal to:

$$\rho_4 = \frac{-2\xi f_0 \ln\left(1 - \frac{\sqrt{\frac{\pi}{6}}}{0.64110} e^{-\frac{v^2}{2}}\right)}{2\frac{\omega_0}{2\pi} e^{-\frac{v^2}{2}}}$$

$$\rho_4 = \frac{-\xi \ln\left(1 - \frac{\sqrt{\frac{\pi}{6}}}{0.64110} e^{-\frac{v^2}{2}}\right)}{e^{-\frac{v^2}{2}}} \qquad [10.63]$$

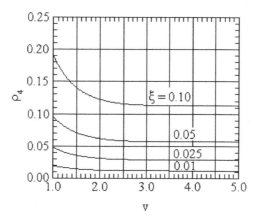

Figure 10.23. *Ratio of the constants α calculated from [10.62] and [10.4]*

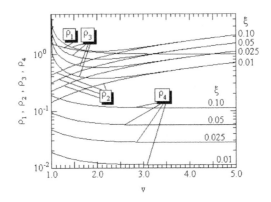

Figure 10.24. α *coefficients on the basis of assumptions 2, 3, 4 and 5, related to α in assumption 1*

10.7. Markov process assumption

10.7.1. W.D. Mark assumption

Much of the work in connection with first passage time relates to the response of a slightly damped resonator to white noise, which can be characterized by a second order Markov process [MAR 66], [WAN 45]. This process was sometimes approached by a first order Markov process, in order for solutions to be deduced more easily [GRA 65], [MAR 66], [SIE 51], [SLE 61].

Definition

A Markov process is a name given to a process in which the distribution of probability at any one time t depends solely on the distribution at any previous time [CRA 83], [SVE 80]. The structure of a Markov process is wholly determined for each and every future instant by the distribution at a particular initial instant and by a probability density function of transition.

The importance of the Markov process lies in the fact that there is a formal technique to obtain a partial differential equation satisfied by the probability density function of transition of the process.

The response of a simple system excited by a random vibration is presented in the form of a quasi-sinusoidal frequency signal equal to the natural frequency of the system, the amplitude and phase of which vary randomly. The peaks have an amplitude modulated by a continuous curve. Assuming that the amplitudes of the peaks constitute a one-dimensional continuous Markov process with a discrete parameter (time), then the probability that $v = \dfrac{a_{max}}{u_{rms}}$ [a_{max} = peak value of $u(t)$] is lower than a value v_0 is given by [BAR 68a], [MAR 66]:

$$P(v \leq v_0) \approx A_0 \, e^{-\alpha_0 \, \omega_0 \, T Q} \qquad [10.64]$$

provided that $T > \tau_{cor}$, where

– A_0 is a constant which depends on the initial conditions. A_0 can in general be taken as equal to 1;

– Q is the quality factor of the system;

– $\omega_0 = 2 \pi f_0$ is its natural pulsation;

– T is the duration of the considered signal;

– τ_{cor} is the smallest value of T for which the correlation function $R_x(T)$ can be regarded as negligible for all $T > \tau_{cor}$. In the majority of practical applications, for which $v_0 \geq 2.5$, the condition $T > \tau_{cor}$ is always respected;

– α_0 is a parameter function of the Q factor and threshold v_0.

This yields

$$P \approx e^{-\alpha T} \qquad [10.65]$$

with $\alpha = \alpha_0 \, \omega_0 \, Q$. For sufficiently large v_0 and for $Q \gg \dfrac{1}{2}$, W.D. Mark [MAR 66] gives:

$$\alpha_0 \approx \dfrac{\mu}{2 \pi Q} \, e^{-\dfrac{v_0^2}{2}} \, \mathrm{erf} \left\{ \left[\dfrac{v_0^2}{2} \tanh\left(\dfrac{\pi}{2 \mu Q} \right) \right]^{1/2} \right\} \qquad [10.66]$$

where

– $\mu = 1$ when the threshold is defined by $v < v_0$;

488   Random Vibration

- $\mu = 2$ when it is defined by $|v| < v_0$;
- $\text{erf}(\ )$ is the error function[1].

The same law applies to the envelope of $u(t)$ with

$$\alpha_0 \approx \frac{v_0^2}{2\,Q^2}\,e^{-\frac{v_0^2}{2}}\left[1+\frac{2.1!}{v_0^2}+\frac{2^2.2!}{v_0^4}+\cdots+\frac{2^n.n!}{v_0^{2n}}+\cdots\right]^{-1} \qquad [10.67]$$

$(n \geq 1)$. Figures 10.25 to 10.34 show respectively:

- α_0 as a function of Q, for $v_0 = 2, 3, 4$ and 5 and $\mu = 1$ and 2;
- α_0 as a function of v_0, for $Q = 5, 10, 20$ and 50, $\mu = 1$;
- $f_0\,T$ as a function of v_0, for $Q = 5, 10, 20$ and 50 and $P_0 = 0.95$ ($\mu = 1$);

$$f_0\,T = \frac{-\ln P_0}{2\,\pi\,Q\,\alpha_0} \qquad [10.68]$$

- $\dfrac{f_0\,T}{Q}$ as a function of v_0, for $Q = 5, 10, 20$ and 50, $P_0 = 0.95$ and $\mu = 1$;

$$\frac{f_0\,T}{Q} = \frac{-\ln P_0}{2\,\pi\,Q^2\,\alpha_0} \qquad [10.69]$$

- α_0 as a function of $f_0\,T$, for $Q = 5, 10, 20$ and 50 and $P_0 = 0.95$ [BAR 68a];

$$\alpha_0 = \frac{-\ln P_0}{2\,\pi\,Q\,f_0\,T} \qquad [10.70]$$

---

[1] Error function E_1 defined in Appendix A4.1.

$-\alpha_0$ as a function of Q, for $f_0 T = 1$, 10, 10^2, 10^3 and 10^4, $\mu = 1$ and $P_0 = 0.95$.

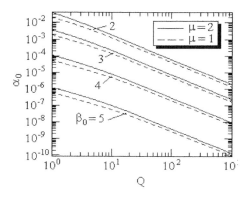

Figure 10.25. *Constant α_0 versus Q factor*

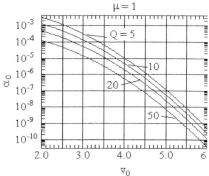

Figure 10.26. *Constant α_0 versus threshold*

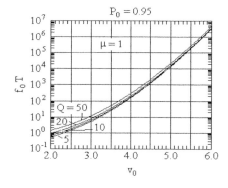

Figure 10.27. *Mean number of cycles versus the threshold for a probability of 0.95*

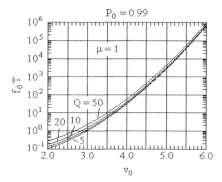

Figure 10.28. *Mean number of cycles versus the threshold for a probability of 0.99*

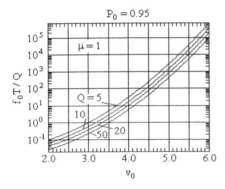

Figure 10.29. *Mean number of maxima of the envelope lower than threshold v_0 for probability of 0.95*

Figure 10.30. *Mean number of maxima of the envelope lower than threshold v_0 for probability of 0.99*

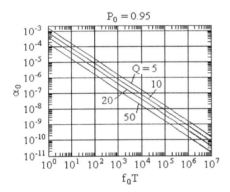
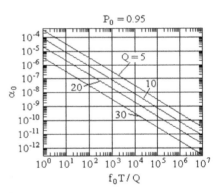

Figure 10.31. *Constant α_0 versus the number of cycles*

Figure 10.32. *Constant α_0 versus $f_0 T/Q$*

In addition, the ratio ρ_5 of the values of α calculated in this section [10.66] on the basis of assumption 1 [10.4] is given by:

$$\rho_5 = \mu \operatorname{erf}\left\{\left[\frac{v_0^2}{2}\tanh\left(\frac{\pi}{2\mu Q}\right)\right]^{1/2}\right\} \qquad [10.71]$$

(knowing that $\alpha = \alpha_0 \omega_0 Q$). It should be noted that $\rho_5 \to 1$ when $v_0 \to \infty$ (Figure 10.35).

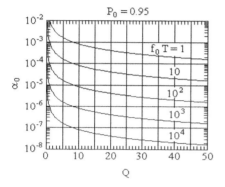

Figure 10.33. *Constant α_0 versus the Q factor, for some values of $f_0 T$*

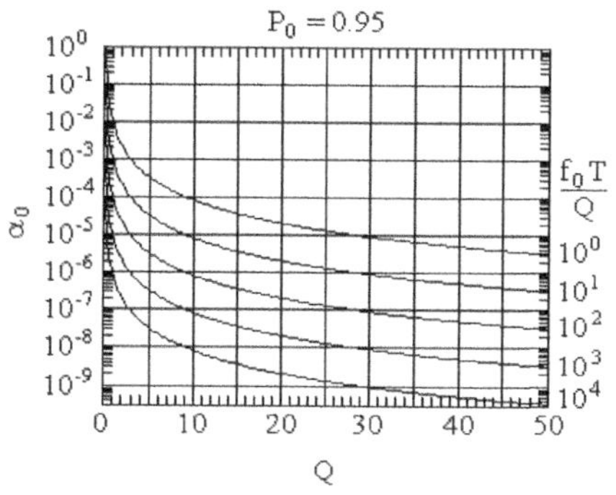

Figure 10.34. *Constant α_0 versus the Q factor, for some values of $f_0 T / Q$*

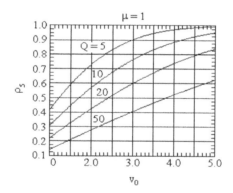

Figure 10.35. *Ratio of constants α calculated from [10.66] and [10.4]*

Application

We propose evaluating the probability of collision between two one-degree-of-freedom linear systems fixed on the same support (Figure 10.36) and subjected to stationary broadband white noise.

Let z_1 and z_2 be the relative response displacement of the masses m_1 and m_2. The parameter of interest here is the sum $z_1 + z_2$. Let

$$|v_0| = \frac{\left|(z_1 + z_2)_{max}\right|}{s_{z_1 z_2}} \qquad [10.72]$$

where

$$s_{z_1 z_2}^2 = E\left[(z_1 + z_2)^2\right] = z_{ms1}^2 + 2\rho\, z_{rms1}\, z_{rms2} + z_{rms2}^2 \qquad [10.73]$$

and ρ is the coefficient of correlation. The probability that the maximum value of v in the selected time interval T is lower than or equal to v_0 is sought.

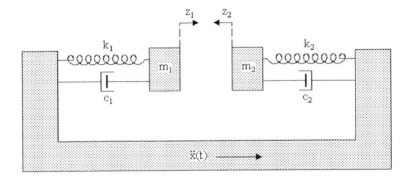

Figure 10.36. *Two one-degree-od-freedom systems on the same support*

The results, obtained by a digital simulation, are presented in the form of curves giving the mean value of $|v_0|$ as a function of the ratio of the natural frequencies f_2/f_1, for various values of the product $f_1\, T$ and various Q factors of the resonators (Figure 10.37) [BAR 68b].

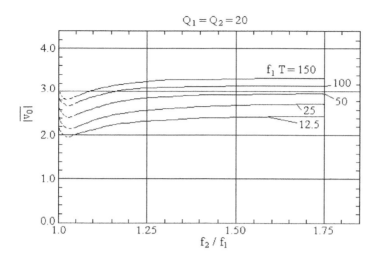

Figure 10.37. *Mean distance between masses [BAR 68b]*

10.7.2. J.N. Yang and M. Shinozuka approximation

J.N. Yang and M. Shinozuka [YAN 71] assume that the process is Markovian, i.e. that peak n depends solely on the preceding peak n − 1 and is independent of all the peaks which occurred before (n − 1)th peak. For a narrowband noise centered around Ω_0, they thus obtain the probability of the first excursion

$$P(N) = 1 - \left(1 - \frac{\bar{q}}{1-q_0}\right)^N \approx 1 - \exp\left[-\frac{N\bar{q}}{1-q_0}\right] \qquad [10.74]$$

where N was defined previously $\left(2\,n_0^+\,T\right)$

$$\bar{q} = \int_a^\infty \int_0^a f\left(x, y, \frac{T_0}{2}\right) dx\, dy \qquad [10.75]$$

$$q_0 = e^{-\frac{a^2}{2\,u_{rms}^2}}$$

$$T_0 = \frac{2\pi}{\Omega_0}$$

$$f\left(x, y, \frac{T_0}{2}\right) = \frac{x\,y}{u_{rms}^4\left[1 - k^2(T_0/2)\right]} I_0\left\{\frac{x\,y\,k_0(T_0/2)}{u_{rms}^2\left[1 - k_0^2(T_0/2)\right]}\right\}$$

$$\exp\left\{-\frac{\left(x^2 + y^2\right)}{2\,u_{rms}^2\left[1 - k_0^2(T_0/2)\right]}\right\} \qquad [10.76]$$

where $I_0(\ )$ is the zero-order modified Bessel function of the first kind

$$k_0(T_0/2) = \sqrt{\rho_0^2(T_0/2) + \lambda_0^2(T_0/2)} \qquad [10.77]$$

$$\rho_0\left(\frac{T_0}{2}\right) = \frac{2}{u_{rms}^2}\int_0^\infty G(\Omega)\cos(\Omega - \omega_0)\frac{T_0}{2}\,d\Omega \qquad [10.78]$$

$$\lambda_0\left(\frac{T_0}{2}\right) = \frac{2}{u_{rms}^2}\int_0^\infty G(\Omega)\sin(\Omega - \omega_0)\frac{T_0}{2}\,d\Omega \qquad [10.79]$$

When the PSD $G(\Omega)$ of the narrowband noise is symmetric with respect to Ω_0, $\lambda_0\left(\frac{T_0}{2}\right) = 0$. This is roughly the same case for the response of a slightly damped one-degree-of-freedom system to white noise.

10.8. E.H. Vanmarcke model

10.8.1. *Assumption of a two state Markov process*

E.H. Vanmarck [VAN 75] considers a two state description of the fluctuations of the random variable $u(t)$ of a wideband process with respect to the specified level a (a being a type B barrier).

Let T_0 and T_1 be the successive time intervals of last below a (safe area) and above a, respectively.

The sum of $T_0 + T_1$ is a time between two successive upcrossings (with positive slopes). Let us assume that the precise times of the upcrossings are variables that are independent and identically distributed. Then, mean times $E[T_0]$ and $E[T_1]$ are such that [VAN 75]

First Passage of a Given Level of Response   495

$$E(T_0 + T_1) = \frac{1}{n_a^+} = \frac{1}{n_0^+} \exp\left(\frac{a^2}{2 u_{rms}^2}\right) = \frac{1}{n_0^+} \exp\left(\frac{v^2}{2}\right) \quad [10.80]$$

$$\frac{E(T_1)}{E(T_0 + T_1)} = \frac{1}{\sqrt{2\pi}} \int_v^\infty e^{-\frac{v^2}{2}} d(v) \equiv \Phi(v) \quad [10.81]$$

$\Phi(v)$ is the complementary Gauss distribution function. It is connected to the error function[2] by

$$\Phi(v) = \frac{1}{2}\left[1 - \operatorname{erf}\left(\frac{v}{\sqrt{2}}\right)\right] \quad [10.82]$$

or, for large v, by:

$$\Phi(v) = \frac{1}{\sqrt{2\pi} \, v} e^{-\frac{v^2}{2}} \left[1 - v^{-2} + 3 v^{-4} - \cdots\right] \quad [10.83]$$

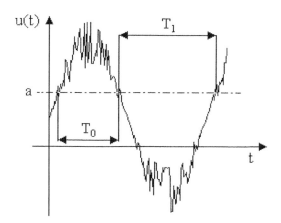

Figure 10.38. *Ranges of relative response displacement*

To simplify, it is assumed that the time intervals T_0 have a common exponential distribution of mean $E(T_0) = \frac{1}{\alpha_{1B}}$, the probability of not crossing the level a can be approached in terms of

---

2 Error function E_1 defined in Appendix A4.1.

$$P(T) = A_B \exp\left[-\alpha_{1B} T\right] = [1 - \Phi(v)] \exp\left\{-\frac{n_0^+ T e^{-v^2/2}}{1 - \Phi(v)}\right\} \quad [10.84]$$

where A_B, the probability that the signal is lower than the threshold at the start time ($T = 0$) under stationary starting conditions, is equal to

$$A_B = \frac{E(T_0)}{E(T_0) + E(T_1)} = 1 - \Phi(v) \quad [10.85]$$

The constant α_{1B} is related to the constant α_B used in the assumption in section 10.3 (equation [10.4]) and more specifically in the relationship

$$\frac{\alpha_{1B}}{\alpha_B} = \frac{1}{1 - \Phi(v)}$$

yielding

$$\alpha_{1B} = \frac{n_0^+ e^{-\frac{v^2}{2}}}{1 - \Phi(v)} \quad [10.86]$$

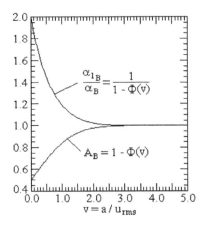

Figure 10.39. *Probability that signal is lower than threshold v at the start time for a type B barrier*

First Passage of a Given Level of Response   497

Figure 10.39 shows how A_B and α_{1B}/α_B tend towards 1 when $v = a/u_{rms}$ is sufficiently large.

Type D barrier

Similarly, in this case,

$$P(T) = A_D \exp(-\alpha_{1D} T) \qquad [10.87]$$

$$P(T) = [1 - 2\Phi(v)] \exp\left\{-\frac{2 n_0^+ T e^{-v^2/2}}{1 - 2\Phi(v)}\right\} \qquad [10.88]$$

$(\alpha_{1D} = 2\alpha_{1B})$. Figure 10.40 shows the variations of A_D and of $\dfrac{\alpha_{1D}}{\alpha_D}$ in relation to $v = \dfrac{a}{u_{rms}}$.

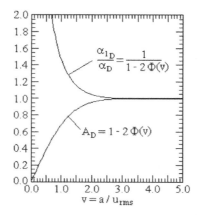

Figure 10.40. *Probability that the signal is lower than threshold v at the start time for a type D barrier*

Type E barrier

The reasoning is similar to the above examples. Set $R(t)$ as the envelope of the signal, T_0' as the time during which the envelope is below the threshold and T_1' as the time during which it is above it; the mean time between two crossings is then [VAN 75]:

498   Random Vibration

$$E(T'_0 + T'_1) = \frac{1}{v_a^+} = \left(\sqrt{2\pi}\, q_E\, v\, n_0^+\right)^{-1} e^{v^2/2} \qquad [10.89]$$

$$\frac{E[T'_1]}{E[T'_0 + T'_1]} = \int_v^\infty v\, e^{-\frac{v^2}{2}}\, dv = e^{-\frac{v^2}{2}} \qquad [10.90]$$

From these two relations,

$$E[T'_1] = \frac{1}{\sqrt{2\pi}\, q_E\, v\, n_0^+} \qquad [10.91]$$

$$E[T'_0] = \frac{e^{v^2/2} - 1}{\sqrt{2\pi}\, q_E\, v\, n_0^+} \qquad [10.92]$$

We can obtain an approximation of the first passage time distribution by assuming that the intervals T'_0 are distributed exponentially with a mean equal to

$$E(T'_0) = \frac{1}{\alpha_{E1}} \qquad [10.93]$$

Under random starting conditions ($t = 0$), we obtain:

$$P_E(T) = A_E \exp\left[-\alpha_{E1}\, T\right] \qquad [10.94]$$

$$P_E(T) = \left[1 - e^{-\frac{v^2}{2}}\right] \exp\left[-\frac{\sqrt{2\pi}\, q_E\, v\, n_0^+\, T}{e^{v^2/2} - 1}\right] \qquad [10.95]$$

the probability of a start ($t = 0$) in the safe range (for $u < a$) being

$$A_E = \frac{E(T'_0)}{E(T'_0 + T'_1)} = 1 - \exp\left(-\frac{v^2}{2}\right) \qquad [10.96]$$

It can be noted [VAN 75] that:

$$\frac{E(T_1)}{E(T'_1)} = q_E\left(1 - v^{-2} + 3v^{-4} - \cdots\right)^{-1} \to q_E \qquad [10.97]$$

q_E is approximately equal to the ratio of mean times of the upcrossings of the process and of its envelope above the threshold when v is large (for the high thresholds).

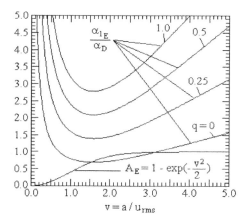

Figure 10.41. *Probability that the signal is lower than threshold v at the start time for a type E barrier*

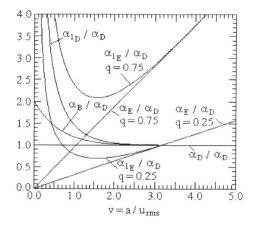

Figure 10.42. *Comparison of mean occurrence rates*

In Figure 10.41, the functions A_E and $\dfrac{\alpha_{1E}}{\alpha_D}$ are plotted as a function of v, knowing that

$$\frac{\alpha_{1E}}{\alpha_D} = \sqrt{\frac{\pi}{2}}\, q_E\, v\, \frac{1}{1 - e^{-v^2/2}} \qquad [10.98]$$

$$\alpha_{1E} = \frac{\sqrt{2\pi} \, q_E \, v \, n_0^+ \, e^{-v^2/2}}{1 - e^{-v^2/2}} \qquad [10.99]$$

Figure 10.42 recapitulates the principal results of the above sections.

10.8.2. *Approximation based on the mean clump size*

This approximation attempts to correct the effects of the assumption of independence for threshold upcrossings. It may be noted from Figure 10.42 that α_D and α_E are very different for v small and v large. This result can be explained by starting with the following:

– the process $|u(t)|$ can exceed the selected threshold only if the envelope is already higher than the threshold;

– the number of D-crossings probably occurring during a single excursion of the envelope depends on the bandwidth of the process and on the level of the considered threshold;

– for the narrowband processes and the weak thresholds, or when the product q_E v is small, the D crossings tend to occur in clumps which immediately follow the individual E-crossings. The average time between D-crossings in a group is equal to $1 / 2 n_0^+$.

Mean clump size

The envelope $R(t)$ of the peaks of a narrowband process is a curve which varies slowly. Each crossing above a given threshold a is followed by the crossing of this threshold by a group of peaks (*clump*) of $|u(t)|$.

Clump size refers to the number of crossings of the threshold a by the process $|u(t)|$ which immediately follows a crossing of a by $R(t)$ (Figure 10.43) [LYO 60].

R.H. Lyon defines the *mean clump size* as the ratio of the number n_a^+ of threshold crossings u = a (by unit time) and the number m_a^+ of crossings of a by the envelope, i.e. $\dfrac{n_a^+}{m_a^+}$ (if the threshold is defined by $|u| = a$, the ratio to be considered is $2 \dfrac{n_a^+}{m_a^+}$).

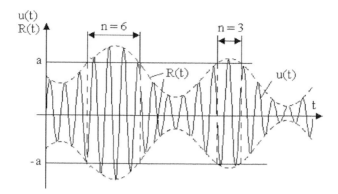

Figure 10.43. *Clumps of response peaks*

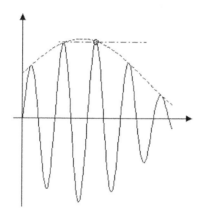

Figure 10.44. *Case of a high threshold for which the definition of the clump size no longer makes sense*

For the high thresholds, this definition leads to a result having little or no sense, as m_a^+ may be higher than n_a^+. R.H. Lyon shows that, under these conditions, the mean number of times that the envelope exceeds a threshold a is[3] given by

$$M_a = \sqrt{\frac{b_2}{2\pi}} \frac{a}{u_{rms}} \exp\left(-\frac{a^2}{2 u_{rms}^2}\right) \qquad [10.100]$$

---

[3] This approximate relation is equivalent to [10.35] if $g(\xi) = \sqrt{2\pi\xi}$.

(where $R_E(0)$ is the autocorrelation function of the envelope $R_E(\tau)$ for $\tau = 0$) and that the mean clump size is

$$\langle cs \rangle = \frac{\omega_m \, u_{rms}}{a \, \sqrt{2 \pi b_2}} = \frac{n_0^+ \, u_{rms} \, \sqrt{2\pi}}{a \, \sqrt{b_2}} \qquad [10.101]$$

$\omega_m = 2 \pi n_0^+ =$ mean pulsation of the narrowband noise.

For a one-degree-of-freedom system (f_0, ξ), we have

$$R_E(0) = \xi \, \omega_m^2 \qquad [10.102]$$

yielding

$$\langle cs \rangle = \sqrt{\frac{Q}{\pi}} = \sqrt{\frac{Q}{\pi}} \, \frac{u_{rms}}{a} \qquad [10.103]$$

The mean clump size varies then in the same way as $\sqrt{Q}$. For $Q = 50$, the groups having an amplitude equal to $2 \, u_{rms}$ or larger will contain just two cycles on average.

The ratio $2 \, n_a^+ / m_a^+$ can be interpreted as the mean size of a D-crossing group. This concept is particularly useful when the mean $E(T_1')$ of the durations T_1' of the threshold crossing a by the envelope is several times larger than $1/2 n_0^+$. However, it can be seen that the number of E-crossings must always be at least as large as the number of D-crossing groups (all the E-crossings are not followed by a D-crossing in the following half-cycle).

For the wideband processes and the high thresholds, i.e. when the product q_E v is large ($q_E = \frac{\dot{R}_{rms}}{\dot{u}_{rms}} = \sqrt{1 - \frac{M_1^2}{M_0 M_2}}$, $v = \frac{a}{u_{rms}}$), the number of E-crossings can be much larger than the number of D-crossings. Estimation of the fraction ρ_D of E-crossings which are not immediately followed by a passage in the following half-cycle is possible [VAN 69]. A D-crossing is assumed to be certain when $T_1' \geq \dfrac{1}{2 n_0^+}$.

If $T_1' < \dfrac{1}{2 n_0^+}$, the probability of a D-crossing occurring during the time interval T_1' is estimated as equal to $\dfrac{T_1'}{\left(2 n_0^+\right)^{-1}} = 2 n_0^+ T_1'$. Hence,

$$p_D = \int_0^{1/n_0^+} \left(1 - 2 n_0^+ T_1'\right) f(T_1') dT_1' \qquad [10.104]$$

an expression in which $f(T_1')$ is the probability density function of T_1'. The exponential distribution (already used) of mean $E[T_1'] = \dfrac{1}{\sqrt{2\pi}\, q_E\, v\, n_0^+}$ is appropriate for calculations yielding

$$p_D = 1 - \dfrac{1}{\sqrt{\pi/2}\, q_E\, v}\left[1 - \exp\left(-\sqrt{\dfrac{\pi}{2}}\, q_E\, v\right)\right] \qquad [10.105]$$

The difference $1 - p_D$ is the fraction of "validated" E-crossings, i.e. those that are immediately followed by at least one D-crossing. Setting α_{2D} as the mean frequency of the validated E-crossings or the mean rate of occurrence of the D-crossing clumps, this rate can be expressed as:

$$\alpha_{2D} = (1 - p_D)\, m_a^+ = 2 n_a^+ \left[1 - \exp\left(-\sqrt{\dfrac{\pi}{2}}\, q_E\, v\right)\right] \qquad [10.106]$$

where m_a^+ is given by [10.39]. An estimate of the mean number of D-crossings by clump (mean clump size) is then

$$E(N_D) = \dfrac{2 n_a^+}{\alpha_{2D}} = \left[1 - \exp\left(-\sqrt{\dfrac{\pi}{2}}\, q_E\, v\right)\right]^{-1} \qquad [10.107]$$

This mean number tends towards 1 when v increases.

NOTE.–

A similar result for the type B barrier is obtained, when replacing $2 n_0^+$, $2 n_a^+$, $\sqrt{\pi/2} \, q_E \, v$ respectively with n_0^+, n_a^+, and $\sqrt{2\pi} \, q_E \, v$ in [10.106].

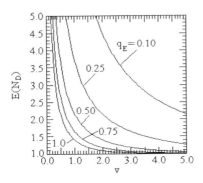

Figure 10.45. *Mean number of D-crossings per clump*

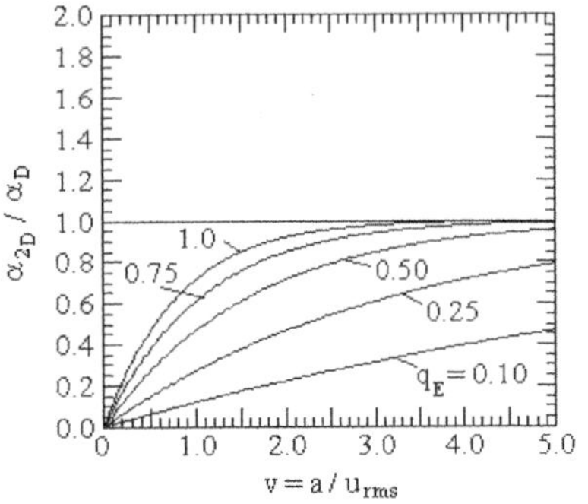

Figure 10.46. *Mean rate of occurrence of groups of D-crossing clumps*

Figure 10.45 shows the variations of $E(N_D)$ with the reduced parameter v, for various values of q_E. If the mean size of the clumps is large, i.e. when the product $q_E \, v$ is small, we have

$$E(N_D) \approx \frac{1}{\sqrt{\pi/2} \, q_E \, v} \qquad [10.108]$$

The distribution of first passage times for a type D barrier can be approximated by assuming that the points corresponding to the D-crossing clumps constitute a Poisson process with a mean rate α_{2D} given by [10.106]. This gives:

$$P_D(T) = \exp\left[-\alpha_{2D} \, T\right] = \exp\left\{-2 \, n_a^+ \, T \left[1 - \exp\left(-\sqrt{\frac{\pi}{2}} \, q_E \, v\right)\right]\right\} \qquad [10.109]$$

The decrease rate α_{2D} approaches the expected asymptotic value $(2 \, n_a^+)$ when v tends towards infinity (Figure 10.46). When v tends towards zero,

$$\alpha_{2D} \to m_a^+$$

(i.e. towards α_E given by [10.42]). The above estimate can be improved (with the low thresholds) while being less strictly observed over the real time duration of the groups. It is enough to consider a two state process by taking T_0^* and T_1^* as mean values such that:

$$E(T_1^*) = \frac{1}{2 n_0^+} E(N_D)$$

$$E(T_0^*) = \frac{1}{\alpha_{2D}} - E(T_1^*)$$

$$E(T_0^*) = \frac{1}{2 n_0^+} E(N_D)\left(e^{v^2/2} - 1\right)$$

The new estimate of the distribution of first passage times for the type D barriers depends on the parameters α_{3D} and

$$A_{3D} = \frac{E(T_0^*)}{E(T_0^* + T_1^*)} = A_E \qquad [10.110]$$

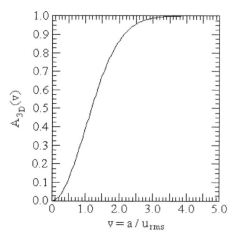

Figure 10.47. *Probability that the response remains below the type D barrier in the first half-cycle of the movement*

The parameter $A_{3D} = A_E$, plotted in Figure 10.47, must be interpreted as the probability that $u(t)$ remains below the type D barrier in the first half-cycle of the movement.

Hence

$$P_D(T) = A_{3D} \exp\left[-\alpha_{3D} T\right] \qquad [10.111]$$

$$P_D(T) = \left(1 - e^{-\frac{v^2}{2}}\right) \exp\left\{-2 n_a^+ T \left(\frac{1 - e^{-\sqrt{\pi/2}\, q_E\, v}}{1 - e^{-v^2/2}}\right)\right\} \qquad [10.112]$$

$$\alpha_{3D} = 2 n_0^+ e^{-v^2/2} \left(\frac{1 - e^{-\sqrt{\pi/2}\, q_E\, v}}{1 - e^{-v^2/2}}\right) \qquad [10.113]$$

$P_D(T)$ is the probability of obtaining a level a in the response of a one-degree-of-freedom system for a length of time T [DEE 71]. As shown in Figure 10.48, α_{3D} tends towards $\alpha_D = 2 n_a^+$ when v tends towards infinity. If v tends towards zero, α_{3D} tends towards α_{1E} and the ratio $\dfrac{\alpha_{3D}}{\alpha_{1D}}$ tends towards $2 q_E$.

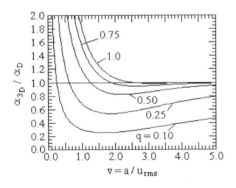

Figure 10.48. *Mean rate of occurrence of the type D groups of passage on the basis of a two state process assumption*

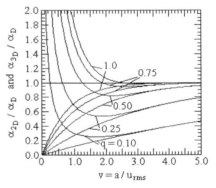

Figure 10.49. *Comparison of the mean rates of occurrence*

The value of α obtained with this method is definitely better than that deduced on the basis of assumption 1 ($\alpha = n_a$ or n_a^+) [10.4]. Nevertheless, it still arrives at values for T that fall short of reality (conservative results) [VAN 72].

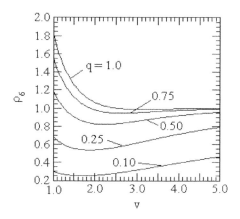

Figure 10.50. *Ratio of constants α calculated from [10.113] and [10.4]*

Figure 10.50 shows the variations as a function of v of the ratio

$$\rho_6 = \frac{2 n_0^+ e^{-\frac{v^2}{2}} \left[1 - \exp\left(-\sqrt{\frac{\pi}{2}} q_E v\right)\right]}{2 n_0^+ e^{-\frac{v^2}{2}} \left[1 - e^{-\frac{v^2}{2}}\right]} \qquad [10.114]$$

for the value of α_{3D} [10.113] and that which results from assumption 1 [10.4], for q_E equal to 0.1, 0.25, 0.5 and 0.75, respectively.

NOTES.–

1. *The E passages and the associated D-passage groups have a tendency to gather together in groups. When this effect, not taken into account up to now, occurs, the true rate of decrease is smaller than α_{3D} and [10.113] constitutes a conservative estimate of $P_D(t)$.*

2. These studies can be extended to the case of the non-stationary random phenomena by considering PSD functions of time $G(\Omega, t)$, from which the spectral moments $M_n(t)$ can be deduced and a mean rate of upcrossings as a function of time $\alpha(t)$:

$$P(T) = A \exp\left[-\int_0^T \alpha(t)\, dt\right] \qquad [10.115]$$

The above method gives results that agree with results from simulations when it is extended to the case of transient signals with [10.115].

Summary chart

Assumption	Constant α	Relation
Statistically independent threshold crossings	$\alpha_B = n_a^+ = n_0^+\, e^{-\frac{v^2}{2}}$	[10.4]
	$\alpha_D = 2\, \alpha_B = 2\, n_0^+\, e^{-v^2/2}$	[10.6]
Statistically independent response maxima	$\alpha = -2\, n_0^+\, \ln(1 - e^{-\frac{v^2}{2}})$	[10.31]
Independent threshold crossings by the envelope of maxima	$\alpha = m_a^+ = n_a^+\, g(\xi)\ \ v = \sqrt{2\pi}\, q_E\, v\, n_a^+$ where $$g(\xi) = \sqrt{2\pi}\left[1 - \frac{1}{1-\xi^2}\left(1 - \frac{1}{\pi}\arctan\frac{2\xi\sqrt{1-\xi^2}}{1-2\xi^2}\right)^2\right]^{1/2}$$	[10.35] [10.42] [10.33]
Independent envelope maxima (S.H. Crandall)	$\alpha = -\dfrac{\omega_0}{2\pi\sqrt{e}}\, g(\xi)\, \ln\left[1 - v\, e^{-\frac{1}{2}(v^2-1)}\right]$	[10.50]
Independent envelope maxima (D.M. Aspinwall)	$\alpha = -\dfrac{f_0}{Q}\, \ln\left(1 - \dfrac{\sqrt{\dfrac{\pi}{6}}}{0.64110}\, e^{-\frac{v^2}{2}}\right)$	[10.62]

Assumption	Constant α	Relation
Markov process	$\alpha = \alpha_0 \; \omega_0 \; Q \approx \dfrac{\mu \, \omega_0}{2 \, \pi} \; e^{-\dfrac{v_0^2}{2}} \; \mathrm{erf}\left\{\left[\dfrac{v_0^2}{2} \, \tanh\left(\dfrac{\pi}{2 \, \mu \, Q}\right)\right]^{1/2}\right\}$	[10.66]
Two state Markov process	$\alpha_{1B} = \dfrac{n_0^+ \, e^{-\dfrac{v^2}{2}}}{1 - \Phi(v)}$	[10.86]
	$\alpha_{1D} = 2 \, \alpha_{1B}$ and $\alpha_{1E} = \dfrac{\sqrt{2 \, \pi} \; q_E \, v \, n_0^+ \, e^{-v^2/2}}{1 - e^{-v^2/2}}$	[10.99]
Approximation based on the mean size of the groups of peaks	$\alpha_{2D} = (1 - \rho_D) \, m_a^+ = 2 \, n_a^+\left[1 - \exp\left(-\sqrt{\dfrac{\pi}{2}} \; q_E \, v\right)\right]$	[10.106]
	$\alpha_{3D} = 2 \, n_0^+ \, e^{-v^2/2}\left(\dfrac{1 - e^{-\sqrt{\pi/2} \; q_E \, v}}{1 - e^{-v^2/2}}\right)$	[10.113]

Table 10.1. *Main results*

Appendix

A1. Laws of probability

A1.1. *Gauss's law*

This law is also called the *Laplace-Gauss* or *normal law*.

Probability density	$p(x) = \dfrac{1}{s\sqrt{2\pi}} e^{-\dfrac{1}{2}\left(\dfrac{x-m}{s}\right)^2}$ m = mean s = standard deviation The law is referred to as reduced centered normal if $m = 0$ and $s = 1$.
Distribution function	$F(X) = P(x < X) = \dfrac{1}{s\sqrt{2\pi}} \int_{-\infty}^{X} e^{-\dfrac{1}{2}\left(\dfrac{x-m}{s}\right)^2} dx$ Reduced variable: $t = \dfrac{x - m}{s}$ If E_1 is the error function $F(T) = \dfrac{1}{2}\left[1 + E_1\left(\dfrac{T}{\sqrt{2}}\right)\right]$ where $T = \dfrac{X - m}{s}$

Mean	m
Variance (central moment of order 2)	s^2
Central moments	$\mu_k = \dfrac{\sigma^k}{\sqrt{2\pi}} \int_{-\infty}^{+\infty} t^k e^{-\frac{t^2}{2}} dt$ k even ($k = 2r$): $\mu_{2r} = \dfrac{2r!}{2^r r!} \sigma^{2r}$ k odd ($k = 2r+1$) $\mu_{2r+1} = 0$
Moment of order 3 and skewness	0
Kurtosis	$\mu'_4 = \dfrac{\mu_4}{s^4} = 3$
Median and mode	0

$P(m - s \leq x \leq m + s) = 0.683$	$P(x \leq m + 1.28155\,s) = 0.90$
$P(m - 1.64\,s \leq x \leq m + 1.64\,s) = 0.90$	$P(x \leq m + 1.64485\,s) = 0.95$
$P(m - 1.96\,s \leq x \leq m + 1.96\,s) = 0.95$	$P(x \leq m + 2\,s) = 0.9772$
$P(m - 2.0\,s \leq x \leq m + 2.0\,s) = 0.954$	$P(x \leq m + 2.32635\,s) = 0.99$
$P(m - 3.0\,s \leq x \leq m + 3.0\,s) = 0.997$	$P(x \leq m + 2.5\,s) = 0.9938$
$P(m - 3.29\,s \leq x \leq m + 3.29\,s) = 0.999$	$P(x \leq m + 3.0\,s) = 0.99865$

A1.2. *Log-normal law*

Probability density	$$p(x) = \frac{1}{x\, s_y \sqrt{2\pi}}\, e^{-\frac{1}{2}\left(\frac{\ln x - m_y}{s_y}\right)^2}$$ m_y = mean s_y = standard deviation of the normal random variable $y = \ln x$ $$p(y)\,dy = \frac{1}{s_y \sqrt{2\pi}}\, e^{-\frac{1}{2}\left(\frac{y - m_y}{s_y}\right)^2} dy$$ The log-normal law is thus obtained from the normal law by the change of variable $x = e^y$.
Distribution function	$$F(X) = P(x < X) = \int_0^X \frac{1}{x\, s_y \sqrt{2\pi}}\, e^{-\frac{1}{2}\left(\frac{\ln x - m_y}{s_y}\right)^2} dx$$
Mean	$$m = E(x) = e^{m_y + \frac{s_y^2}{2}}$$
Variance (central moment of order 2)	$s^2 = e^{2m_y + s_y^2}\left(e^{s_y^2} - 1\right) = m^2\left(e^{s_y^2} - 1\right)$ $s^2 = [E(x)]^2\, v^2$ $s^2 = \left(e^{s_y^2} - 1\right)\left[e^{m_y + \frac{s_y^2}{2}}\right]^2$ $v = \sqrt{e^{s_y^2} - 1}$ v = variation coefficient
Expressions for m_y and s_y^2 with respect to $E(x)$ and s^2	$m_y = \ln[E(x)] - \frac{1}{2}\ln\left[1 + \frac{s^2}{E^2(x)}\right]$ $s_y^2 = \ln\left[1 + \frac{s^2}{E^2(x)}\right]$ $m_y = \ln[E^2(x)] - \frac{1}{2}\ln[s^2 + E^2(x)]$ $s_y^2 = -\ln[E^2(x)] + \ln[s^2 + E^2(x)]$ It should be noted that the transformation $x = e^y$ applies neither to $E(x)$ nor to s^2.

Median	$\tilde{x} = e^{m_y}$
Variation coefficient	$m_y = \ln m - \frac{1}{2}\ln(1+v^2)$ $s_y^2 = \ln(1+v^2)$ $v = \sqrt{e^{s_y^2} - 1}$ If v is the variation coefficient of x, we also have $$m = \tilde{x}\sqrt{1+v^2}.$$ If two log-normal distributions have the same variation coefficient, they have equal values of s_y (and vice versa).
Moment of order j	$\lambda'_j = e^{jm_j + \frac{1}{2}j^2 s_y^2}$
Central moment of order 3	$\lambda_3 = \left(e^{s_y^2} - 1\right)^{3/2} e^{\frac{3}{2}(2m + s_y^2)} \left(e^{s_y^2} + 2\right)^2 = m^3\left(v^6 + 3v^4\right)$
Skewness	$\lambda'_3 = \frac{\lambda_3}{s^3} = \left(e^{s_y^2} - 1\right)\left(e^{s_y^2} + 2\right)^2 = v^3 + 3v$
Central moment of order 4	$\lambda_4 = m^4\left(v^{12} + 6v^{10} + 15v^8 + 16v^6 + 3v^4\right)$
Kurtosis	$\frac{\lambda_4}{s^4} = v^8 + 6v^6 + 15v^4 + 16v^2 + 3$
Mode	$M = e^{m_y - s_y^2}$ For this value, the probability density has a maximum equal to $$\frac{1}{s_y\sqrt{2\pi}} e^{-m_y + \frac{s_y^2}{2}}$$

Log-normal law references: [AIT 81], [CAL 69], [KOZ 64], [PAR 59], [WIR 81], [WIR 83].

NOTES.–

1. *Another definition can be: a random variable x follows a log-normal law if and only if $y = \ln x$ is normally distributed, with average m_y and variance s_y^2.*

2. *This law has several names: the Galton, Mc Alister, Kapteyn, Gibrat law or the logarithmic-normal or logarithmo-normal law.*

3. *The definition of the log-normal law can be given starting from base 10 logarithms ($y = \log_{10} x$):*

$$p(x) = \frac{1}{x \, s_y \sqrt{2 \pi \ln 10}} e^{-\frac{1}{2}\left(\frac{\log_{10} x - m_y}{s_y}\right)^2} \qquad [A1.1]$$

With this definition for base 10 logarithms, we have:

$$m_y = \log_{10} \tilde{x} \qquad [A1.2]$$

$$m = 10^{\left[m_y + \frac{1}{2}\log_{10}\left(s_y^2/0.434\right)\right]} \qquad [A1.3]$$

$$v = \sqrt{10^{s_y^2/0.434} - 1} \qquad [A1.4]$$

$$m_y = \log_{10} x - \frac{1}{2} \log_{10}\left(1 + v^2\right) \qquad [A1.5]$$

$$s_y^2 = 0.434 \, \log_{10}\left(1 + v^2\right) \qquad [A1.6]$$

Hereafter, we will consider only the definition based on Napierian logarithms.

4. *Some authors make the variable change defined by $y = 20 \log x$, y being expressed in decibels. We then have:*

$$v = \sqrt{e^{s_y^2/75.44} - 1} \qquad [A1.7]$$

since $\left(\frac{20}{\ln 10}\right)^2 \approx 75.44$.

5. *Depending on the values of the parameters m_y and s_y, it can sometimes be difficult to imagine a priori which law is best adjusted to a range of experimental*

values. A method making it possible to choose between the normal law and the lognormal law consists of calculating:

– the variation coefficient $v = \dfrac{s}{m}$;

– the skewness $\dfrac{\lambda_3}{s^3}$;

– the kurtosis $\dfrac{\lambda_4}{s^4}$;

knowing that

$$E(x) = m = \frac{\sum_i x_i}{n} \qquad [\text{A1.8}]$$

$$s^2 = \frac{\sum_i (x_i - m)^2}{n} \qquad [\text{A1.9}]$$

$$\lambda_3 = \frac{\sum_i (x_i - m)^3}{n} \qquad [\text{A1.10}]$$

and

$$\lambda_4 = \frac{\sum_i (x_i - m)^4}{n} \qquad [\text{A1.11}]$$

If the skewness is close to zero and the kurtosis is close to 3, the normal law is that which is best adjusted. If $v < 0.2$ and $\dfrac{\lambda'_3}{v} \approx 3$, the log-normal law is preferable.

A1.3. *Exponential law*

This law is often used with reliability where it expresses the time expired up to failure (or the time interval between two consecutive failures).

Probability density	$p(x) = \lambda e^{-\lambda x}$
Distribution function	$F(X) = P(x < X) = 1 - e^{-\lambda X}$
Mean	$m_1 = E(x) = \dfrac{1}{\lambda}$
Moments	$m_n = \dfrac{n!}{\lambda^n} = \dfrac{n}{\lambda} m_{n-1}$
Variance (central moment of order 2)	$s^2 = \dfrac{1}{\lambda^2}$
Central moments	$\mu_n = 1 + \dfrac{n}{\lambda} \mu_{n-1}$
Variation coefficient	$v = 1$
Moment of order 3 (skewness)	$\mu'_3 = \dfrac{\mu_3}{s^3} = \lambda^3 + 3$
Kurtosis	$\mu'_4 = \lambda^4 + 4\lambda^3 + 12$

A1.4. *Poisson's law*

It is said that a random variable X is a Poisson variable if its possible values are countable to infinity $x_0, x_1, x_2 ..., x_k ...$, the probability that $X = x_k$ being given by:

$$p_k = P(X = x_k) = e^{-\lambda} \frac{\lambda^k}{k!} \qquad [A1.12]$$

where λ is an arbitrary positive number.

We can also define this in a similar way as a variable able to take all on the integer values, a countable infinity, 0, 1, 2, 3..., k..., the value k having the probability:

$$p_k \equiv P(X = k) = e^{-\lambda} \frac{\lambda^k}{k!} \qquad [A1.13]$$

The random variable is here a number of events (we saw that, with an exponential law, the variable is the time interval between two events).

Distribution function	$F(X) = P(0 \leq x < X) = \sum_{k=0}^{n} e^{-\lambda} \dfrac{\lambda^k}{k!}$ $(n < X \leq n+1)$
Mean	$m_1 = E(x) = \sum_{k=0}^{\infty} k\, e^{-\lambda} \dfrac{\lambda^k}{k!} = \lambda$
Moment of order 2	$m_2 = \lambda(\lambda+1)$
Variance (central moment of order 2)	$s^2 = \lambda$
Central moments	$\mu_3 = \lambda$ $\mu_4 = \lambda + 3\lambda^2$
Variation coefficient	$v = \dfrac{1}{\sqrt{\lambda}}$
Skewness	$\mu'_3 = \dfrac{1}{\sqrt{\lambda}}$
Kurtosis	$\mu'_4 = \dfrac{1+3\lambda}{\lambda}$

The set of possible values n and their probability p_k constitutes the Poisson's law for parameter λ. This law is a discrete law. It is shown that Poisson's law is the limit of the binomial distribution when the probability p of this last law is equal to $\dfrac{\lambda}{k}$ and when k tends towards infinity.

NOTES.–

1. *Skewness μ'_3 always being positive, the Poisson distribution is dissymmetric, more spread out on the right.*

2. *If λ tends towards infinity, μ'_3 tends towards zero and μ'_4 tends towards 3. There is a convergence from Poisson's law towards the Gaussian law. When λ is large, the Poisson distribution is very close to a normal distribution.*

A1.5. *Chi-square law*

Given ν random variables $u_1, u_2, ..., u_\nu$, assumed to be independent reduced normal, i.e. such that:

$$f(u_i) = \frac{1}{\sqrt{2\pi}} e^{-\frac{u_i^2}{2}} du_i \qquad [A1.14]$$

we call a chi-square law with ν degrees of freedom (ν independent variables) the probability law of the variable χ_ν^2 defined by:

$$\chi_\nu^2 = u_1^2 + u_2^2 + \cdots + u_\nu^2 = \sum_{i=1}^{\nu} u_i^2 \qquad [A1.15]$$

The variables u_i being continuous, the variable χ^2 is continuous in $(0, \infty)$.

NOTES.–

1. *The variable χ^2 can also be defined starting from ν independent non-reduced normal random variables x_i whose averages are respectively equal to $m_i = E(x_i)$ and the standard deviations s_i, while referring back to the preceding definition with the reduced variables $u_i = \frac{x_i - m_i}{s_i}$ and the sum*

$$\chi_\nu^2 = \sum_{i=1}^{\nu} u_i^2.$$

2. *The sum of the squares of independent non-reduced normal random variables does not follow a chi-square law.*

Probability density	$p(\chi_\nu^2) = \frac{1}{2^{\nu/2} \Gamma(\nu/2)} (\chi_\nu^2)^{\frac{\nu}{2}-1} e^{-\frac{\chi_\nu^2}{2}}$ ν = number of degrees of freedom Γ = Euler function of the second kind (gamma function)
Mean	$E(\chi_\nu^2) = \sum_{i=1}^{\nu} E(u_i^2) = \nu$

Moment of order 2	$m_2 = \nu(2+\nu)$
Standard deviation	$s = \sqrt{2\nu}$
Central moments	$\mu_2 = 2\nu \quad \mu_3 = 8\nu \quad \mu_4 = 12\nu(\nu+4)$
Variation coefficient	$V = \sqrt{\dfrac{2}{\nu}}$
Skewness	$\mu'_3 = 2\sqrt{\dfrac{2}{\nu}}$
Kurtosis	$\mu'_4 = 3\,\dfrac{\nu+4}{\nu}$
Mode	$M = \nu - 2$

This law is comparable to a normal law when ν is greater than approximately 30.

A1.6. Rayleigh's law

Probability density	$p(x) = \dfrac{x}{\nu^2}\, e^{-\dfrac{x^2}{2\nu^2}}$ ν is a constant
Distribution function	$F(X) = P(x < X) = 1 - e^{-\dfrac{x^2}{2\nu^2}}$
Mean	$m = \sqrt{\dfrac{\pi}{2}}\,\nu$
Median	$X = \nu\sqrt{2\ln 2} \approx 1.1774\,\nu$
Rms value	$\sqrt{E(x^2)} = \nu\sqrt{2}$
Variance	$s^2 = \left(2 - \dfrac{\pi}{2}\right)\nu^2 = \left(\dfrac{4}{\pi} - 1\right)m^2$

Moment of order k	If k is odd (k = 2r − 1)	$m_{2r-1} = \dfrac{(2r)!}{2^r \, r!} \sqrt{\dfrac{\pi}{2}} \, v^{2r-1}$
	If k is even (k = 2 r)	$m_{2r} = 2^r \, r! \, v^{2r}$
Central moments		$\mu_0 = 1 \qquad\qquad \mu_1 = 0$ $\mu_2 = \left(2 - \dfrac{\pi}{2}\right) v^2 \qquad \mu_3 = \sqrt{\dfrac{\pi}{2}} \, (\pi - 3) \, v^3$
Variation coefficient		$v = \sqrt{\dfrac{4}{\pi} - 1}$
Skewness		$a = \sqrt{\dfrac{\pi}{2}} \, \dfrac{\pi - 3}{\left(2 - \dfrac{\pi}{2}\right)^{3/2}} \approx 0.6311$ $\mu_4 = \left(8 - \dfrac{3\pi^2}{4}\right) v^4$
Kurtosis		$b = \dfrac{32 - 3\pi^2}{(4 - \pi)^2} \approx 3.2451$
Mode		$M = v$

Reduced law

If we set $u = \dfrac{x}{s}$, it results that, knowing that $p(x) = \dfrac{x}{s^2} e^{-\dfrac{x^2}{2\sigma^2}}$,

$$p(x) = \frac{1}{s} \frac{x}{s} e^{-\dfrac{x^2}{2s^2}} = \frac{1}{s} u \, e^{-\dfrac{u^2}{2}} = \frac{1}{s} p(u) \qquad\qquad [A1.16]$$

$$p(u) \, du = p(u) \, d\left(\frac{x}{s}\right) = s \, p(x) \, \frac{dx}{s} = p(x) \, dx \qquad\qquad [A1.17]$$

$\dfrac{x}{v}$	$\text{prob}\left(\dfrac{x}{v} > \dfrac{X}{v}\right)$
1	0.60653
1.5	0.32465
2	0.13534
2.5	4.3937 10^{-2}
3	1.1109 10^{-2}
3.5	2.1875 10^{-3}
4	3.355 10^{-4}
4.5	4.01 10^{-5}
5	3.7 10^{-6}

Table A1.1. *Particular values of the Rayleigh distribution*

A1.7. *Weibull distribution*

Probability density	$p(x) = \begin{cases} \dfrac{\alpha}{v-\varepsilon}\left(\dfrac{x-\varepsilon}{v-\varepsilon}\right)^{\alpha-1} \exp\left[-\left(\dfrac{x-\varepsilon}{v-\varepsilon}\right)^{\alpha}\right] & x > \varepsilon \\ 0 & x < \varepsilon \end{cases}$ α and v = positive constants $\lambda = v - \varepsilon$ = *scale* parameter α = *shape* parameter ε = *location* parameter
Distribution function	$F(X) = \begin{cases} 1 - \exp\left[-\left(\dfrac{X-\varepsilon}{v-\varepsilon}\right)^{\alpha}\right] & X > \varepsilon \\ 0 & X < \varepsilon \end{cases}$
Mean	$m = \varepsilon + (v - \varepsilon)\, \Gamma\!\left(1 + \dfrac{1}{\alpha}\right)$
Median	$X = \varepsilon + (v - \varepsilon)(\ln 2)^{1/\alpha}$
Variance	$s^2 = (v - \varepsilon)^2 \left[\Gamma\!\left(1 + \dfrac{2}{\alpha}\right) - \Gamma^2\!\left(1 + \dfrac{1}{\alpha}\right)\right]$
Mode	$M = \varepsilon + (v - \varepsilon)\left(1 - \dfrac{1}{\alpha}\right)^{1/\alpha}$

Weilbull distribution references: [KOZ 64], [PAR 59].

NOTE.–

We sometimes use the constant $\eta = v - \varepsilon$ in the above expressions.

A1.8. *Normal Laplace-Gauss law with n variables*

Let us set $x_1, x_2 ..., x_n$ n random variables with zero average. The normal law with n variable x_i is defined by its probability density:

$$p(x_1, x_2, \cdots, x_n) = (2\pi)^{-n/2} |M|^{-1/2} \exp\left[-\frac{1}{2|M|} \sum_{i,j}^{n} M_{ij} x_i x_j\right] \quad [A1.18]$$

where $|M|$ is the determinant of the square matrix:

$$\|M\| = \begin{Vmatrix} \mu_{11} & \mu_{12} & \cdots & \mu_{1n} \\ \mu_{21} & \mu_{22} & \cdots & \mu_{2n} \\ \vdots & \vdots & \ddots & \vdots \\ \mu_{n1} & \mu_{n2} & \cdots & \mu_{nn} \end{Vmatrix} \quad [A1.19]$$

$\mu_{ij} = E(x_i, x_j)$ = moments of order 2 of the random variables

M_{ij} = cofactor of μ_{ij} in $|M|$.

Examples

1. n = 1

$$p(x_1) = (2\pi)^{-1/2} |M|^{-1/2} \exp\left(-\frac{1}{2} \frac{M_{11}}{|M|} x_1^2\right)$$

with

$\|M\| = \|\mu_{11}\|$

$|M| = \mu_{11}$

$$M_{11} = 1$$

$$\mu_{11} = E(x_1^2) = s^2$$

yielding

$$p(x_1) = \frac{1}{s\sqrt{2\pi}} e^{-\frac{x_1^2}{2s^2}} \qquad [A1.20]$$

which is the probability density of a one-dimensional normal law as defined previously.

2. n = 2

$$p(x_1, x_2) = (2\pi)^{-1} |M|^{-1/2} \exp\left[-\frac{1}{2|M|}\left(M_{11} x_1^2\right.\right.$$
$$\left.\left. + M_{12} x_1 x_2 + M_{21} x_2 x_1 + M_{22} x_2^2\right)\right] \qquad [A1.21]$$

with

$$\|M\| = \begin{Vmatrix} \mu_{11} & \mu_{12} \\ \mu_{21} & \mu_{22} \end{Vmatrix} \qquad |M| = \mu_{11}\mu_{22} - \mu_{12}\mu_{21}$$

$$\mu_{11} = E(x_1^2) = s_1^2 \qquad M_{11} = \mu_{22}$$

$$\mu_{12} = E(x_1 x_2) = E(x_2 x_1) = \mu_{21} = \rho\, s_1 s_2 \qquad M_{12} = -\mu_{21} = -\mu_{12} = M_{21}$$

$$\mu_{22} = E(x_2^2) = s_2^2 \qquad M_{22} = \mu_{11}$$

ρ is the coefficient of linear correlation between the variables x_1 and x_2, defined by:

$$\rho = \frac{\text{cov}(x_1, x_2)}{s(x_1)\, s(x_2)} \qquad [A1.22]$$

where $\text{cov}(x_1, x_2)$ is the covariance between the two variables x_1 and x_2:

$$\mathrm{cov}(X_1, X_2) = \int\int_{-\infty}^{+\infty} [x_1 - E(X_1)][x_2 - E(X_2)] \, p(x_1, x_2) \, dx_1 \, dx_2 \quad [A1.23]$$

The covariance can be negative, zero or positive. It is zero when x_1 and x_2 are completely independent variables. Conversely, a zero covariance is not a sufficient condition for x_1 and x_2 to be independent.

It is shown that ρ is included in the interval $[-1, 1]$. $\rho = 1$ is a necessary and sufficient condition of linear dependence between x_1 and x_2.

This yields

$$p(x_1, x_2) = \frac{1}{2\pi s_1 s_2 \sqrt{1-\rho^2}} \exp\left\{-\frac{1}{2(1-\rho^2)}\left[\left(\frac{x_1}{s_1}\right)^2 - 2\rho\frac{x_1 x_2}{s_1 s_2} + \left(\frac{x_2}{s_2}\right)^2\right]\right\}$$
[A1.24]

NOTE.–

If the averages were not zero, we would have

$$p(x_1, x_2) = \frac{1}{2\pi s_1 s_2 \sqrt{1-\rho^2}} \exp\left\{-\frac{1}{2(1-\rho^2)}\left[\left(\frac{x_1 - E(x_1)}{s_1}\right)^2 \right.\right.$$

$$\left.\left. -2\rho\frac{x_1 - E(x_1)}{s_1}\frac{x_2 - E(x_2)}{s_2} + \left(\frac{x_2 - E(x_2)}{s_2}\right)^2\right]\right\} \quad [A1.25]$$

If $\rho = 0$, we can write $p(x_1, x_2) = p(x_1) p(x_2)$ where $p(x_1) = \frac{1}{s_1\sqrt{2\pi}} e^{-\frac{x_1^2}{2s_1^2}}$ and $p(x_2) = \frac{1}{s_2\sqrt{2\pi}} e^{-\frac{x_2^2}{2s_2^2}}$. x_1 and x_2 are independent random variables.

It is easily shown that, by using the reduced centered variables $t_1 = \dfrac{x_1 - E(x_1)}{s_1}$ and $t_2 = \dfrac{x_2 - E(x_2)}{s_2}$,

$$\int_{-\infty}^{+\infty}\int_{-\infty}^{+\infty} p(x_1\ x_2)\, dx_1\, dx_2 = 1 \qquad [A1.26]$$

Indeed, with these variables,

$$p(t_1, t_2) = \frac{1}{2\pi s_1 s_2 \sqrt{1-\rho^2}} \exp\left[-\frac{1}{2(1-\rho^2)}\left(t_1^2 - 2\rho t_1 t_2 + t_2^2\right)\right] \qquad [A1.27]$$

and

$$t_1^2 - 2\rho t_1 t_2 + t_2^2 = (t_1 - \rho t_2)^2 + (1-\rho^2) t_2^2 \qquad [A1.28]$$

Let us set $u = \dfrac{t_1 - \rho t_2}{\sqrt{1-\rho^2}}$ and calculate

$$\frac{1}{2\pi}\int_{-\infty}^{+\infty}\int_{-\infty}^{+\infty} e^{-u^2/2}\, e^{-t_2^2/2}\, dt_2\, du \qquad [A1.29]$$

i.e.

$$\frac{1}{\sqrt{2\pi}}\int_{-\infty}^{+\infty} e^{-u^2/2}\, du \cdot \frac{1}{\sqrt{2\pi}}\int_{-\infty}^{+\infty} e^{-t_2^2/2}\, dt_2 \qquad [A1.30]$$

We thus have

$$\int_{-\infty}^{+\infty} p(u)\, du \int_{-\infty}^{+\infty} p(t_2)\, dt_2 = 1 \qquad [A1.31]$$

NOTE.–

It is shown that, if the terms μ_{ij} are zero when $i \neq j$, i.e. if all the correlation coefficients of the variables x_i and x_j are zero ($i \neq j$), we have:

$$\|M\| = \begin{Vmatrix} \mu_{11} & 0 & \cdots & 0 \\ 0 & \ddots & \cdots & 0 \\ 0 & \cdots & \ddots & 0 \\ 0 & 0 & 0 & \mu_{nn} \end{Vmatrix} \quad \text{[A1.32]}$$

$$|M| = \prod_{i=1}^{n} \mu_{ii}$$

$$M_{ij} = 0 \quad \text{if } i \neq j$$

$$M_{ij} = \frac{|M|}{\mu_{ii}} \quad \text{if } i = j$$

and

$$p(x_1, x_2, \cdots, x_n) = (2\pi)^{-n/2} |M|^{-1/2} \exp\left(-\frac{1}{2} \sum_{i=1}^{n} \frac{x_i^2}{\mu_{ii}}\right) \quad \text{[A1.33]}$$

$$p(x_1, x_2, \cdots x_n) = \prod_{i=1}^{n} p(x_i) \quad \text{[A1.34]}$$

For normally distributed random variables, it is sufficient that the cross-correlation functions are zero for these variables to be independent.

A1.9. *Student law*

The Student law with n degrees of freedom of the random variable x whose probable value would be zero for probability density:

$$p(x) = \frac{1}{\sqrt{2\pi}} \frac{\Gamma\left(\frac{n+1}{2}\right)}{\Gamma\left(\frac{n}{2}\right)} \left(1 + \frac{x^2}{n}\right)^{-\frac{n+1}{2}} \quad \text{[A1.35]}$$

A1.10. Gumbel law

Let x be a random variable belonging to $]-\infty, +\infty[$ and following a Gumbel law.

Probability density	$p(x) = \dfrac{1}{s} e^{-\frac{x-x_0}{s}} e^{-e^{-\frac{x-x_0}{s}}}$ x_0 = location parameter s = scale parameter
Reduced variable	$u = \dfrac{x - x_0}{s}$
Probability density	$p(u) = e^{-u} e^{-e^{-u}}$
Distribution function	$F(u) = e^{-e^{-u}}$
Mean	$x_0 + s\,\varepsilon$ ε = Euler's constant (= 0.57721566490...)
Median	$x_0 - s \ln[\ln(2)]$
Mode	x_0
Variance	$\dfrac{\pi^2}{6} s^2$
Skewness	≈ 1.139
Kurtosis	5.4

A2. $1/n^{th}$ octave analysis

Some signal processing tools make it possible to express the PSDs calculated in dB from an analysis into the third octave. We propose here to give the relations which make it possible to go from such a representation to the traditional representation. We will place ourselves in the more general case of a distribution of the points in the $1/n^{th}$ octave.

Appendix   529

A2.1. *Center frequencies*

A2.1.1. *Calculation of the limits in 1/n^{th} octave intervals*

By definition, an octave is the interval between two frequencies f_1 and f_2 such that $\dfrac{f_2}{f_1} = 2$. In the 1/n^{th} octave, we have

$$\frac{f_2}{f_1} = 2^{1/n} \qquad [A2.1]$$

i.e.

$$\log f_2 = \log f_1 + \frac{\log 2}{n} \qquad [A2.2]$$

Example A2.1.

Analysis in the 1/3 octave between $f_1 = 5$ Hz and $f_2 = 10$ Hz.

$$\log f_a = \log 5 + \frac{\log 2}{3} = 0.7993$$

$$f_a = 6.3 \text{ Hz}$$

$$\log f_b = \log f_a + \frac{\log 2}{3} = 0.8997$$

$$f_b = 7.937 \text{ Hz}$$

$$\log f_c = \log f_b + \frac{\log 2}{3} = 1$$

$$f_c = f_2 = 10 \text{ Hz}$$

A2.1.2. Width of the interval Δf centered around f

The width of this interval is equal to

Δf = upper limit − lower limit

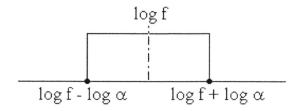

Figure A2.1. *Frequency interval*

Let α be a constant characteristic of width Δf (Figure A2.1) such that:

$$(\log f + \log \alpha) - (\log f - \log \alpha) = \frac{\log 2}{n} \qquad [A2.3]$$

yielding

$$\alpha = 2^{1/2n} \qquad [A2.4]$$

We deduce

$$\Delta f = \alpha\, f - \frac{f}{\alpha} \qquad [A2.5]$$

$$\Delta f = \left(\alpha - \frac{1}{\alpha}\right) f \qquad [A2.6]$$

This value of Δf is particularly useful for the calculation of the rms value of a vibration defined by a PSD expressed in dB.

Example A2.2.

For $n = 3$, it results that $\alpha \approx 1.122462\ldots$ and $\Delta f \approx 0.231563\ldots f$. At 5 Hz, we have $\Delta f = 1.15$ Hz.

A2.2. Ordinates

We propose here to convert the decibels into unit of amplitude $[(m/s^2)^2/Hz]$. We have, if $\ddot{x}_{rms}$ is the rms value of the signal filtered by the filter $(f, \Delta f)$ defined above:

$$N(dB) = 20 \log \frac{\ddot{x}_{rms}}{\ddot{x}_{ref}} \qquad [A2.7]$$

$\ddot{x}_{ref}$ is a reference value. If the parameter studied is an acceleration, the reference value is by convention equal to $1 \; \mu m/s^2 = 10^{-6} \; m/s^2$ (we sometimes find $10^{-5} \; m/s^2$ in certain publications).

Table A2.1 lists the reference values quoted by ISO Standard 1683.2.

Parameter	Formulate (dB)	Reference level
Sound pressure level	$20 \log(p/p_0)$	20 µPa in air 1 µPa in other media
Acceleration level	$20 \log(\ddot{x}/\ddot{x}_0)$	$1 \; \mu m/s^2$
Velocity	$20 \log(v/v_0)$	1 nm/s
Displacement	$20 \log(x/x_0)$	1 µm
Force level	$20 \log(F/F_0)$	1 µN
Power level	$10 \log(P/P_0)$	1 pW
Intensity level	$10 \log(I/I_0)$	$1 \; pW/m^2$
Energy density level	$10 \log(W/W_0)$	$1 \; pJ/m^3$
Energy	$10 \log(E/E_0)$	1 pJ

Table A2.1. *Reference values (ISO Standard 1683 [ISO 94])*

This yields

$$\ddot{x}_{rms} = \ddot{x}_{ref} \; 10^{\frac{N}{20}} \qquad [A2.8]$$

The amplitude of the corresponding PSD is equal to

532   Random Vibration

$$G = \frac{\ddot{x}_{rms}^2}{\Delta f} = \frac{\ddot{x}_{ref}^2 \; 10^{\frac{N}{10}}}{\Delta f} \qquad [A2.9]$$

$$G = \frac{2^{1/2n}}{2^{1/n} - 1} \; \frac{\ddot{x}_{ref}^2 \; 10^{\frac{N}{10}}}{f} \qquad [A2.10]$$

or

$$N = 10 \left[\log \frac{2^{1/n} - 1}{2^{1/2n}} \; \frac{f \; G}{\ddot{x}_{ref}} \right] \qquad [A2.11]$$

Example A2.3.

If $\ddot{x}_{ref} = 10^{-5}$ m/s^2

$$G = \frac{10^{\frac{N}{10} - 10}}{\Delta f}$$

and if n = 3

$$G = \frac{2^{1/6}}{2^{1/3} - 1} \; \frac{10^{\frac{N}{10} - 10}}{f}$$

$$G \approx \frac{10^{\frac{N}{10} - 10}}{0.23 \; f}$$

If, at 5 Hz, the spectrum gives N = 50 dB,

$$G = \frac{2^{1/6}}{2^{1/3} - 1} \; \frac{10^{\frac{50}{10} - 10}}{5}$$

$$G \approx 8.6369 \; 10^{-6} \; (m/s^2)^2/Hz.$$

A3. Conversion of an acoustic spectrum into a PSD

A3.1. *Need*

When the real environment is an acoustic noise, it is possible to evaluate the vibratory levels induced by this noise in a structure and the stresses which result from it using finite element calculation software.

At the stage of the writing of specifications, we do not normally have such a model of the structure. It is nevertheless very important to obtain an evaluation of the vibratory levels for the dimensioning of the material.

To carry out this estimate, F Spann and P. Patt [SPA 84] proposed an approximate method based once again on calculation of the response of a one-degree-of-freedom system (Figure A3.1).

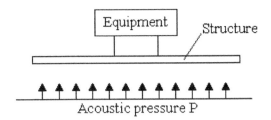

Figure A3.1. *Model for the evaluation of the effects of acoustic pressure*

Let us set:

P = acoustic pressure;

G_P = power spectral density of the pressure;

A = area exposed to the pressure;

β = effectiveness vibroacoustic factor;

M = mass of the specimen and support unit.

The method consists of:

– transforming the spectrum of the pressure expressed into dB into a PSD G_P expressed in $(N/m^2)^2/Hz$;

– calculating, in each frequency interval (in general in the third octave), the response of an equivalent one-degree-of-freedom system from the value of the PSD pressure, the area A exposed to the pressure P and the effective mass M;

– smoothing the spectrum obtained.

A3.2. *Calculation of the pressure spectral density*

By definition, the number N of dB is given by

$$N = 20 \log_{10} \frac{P}{P_0} \qquad [A3.1]$$

where P_0 = reference pressure = $2 \cdot 10^{-5}$ N/m^2 and P = rms pressure = $\sqrt{G_P \, \Delta f}$.

For a 1/nth octave filter centered on the frequency f_c, we have

$$\Delta f = \left(2^{1/2n} - \frac{1}{2^{1/2n}} \right) f_c \qquad [A3.2]$$

yielding

$$G_P = \frac{\left(P_0 \, 10^{N/20} \right)^2}{\Delta f} \qquad [A3.3]$$

In the particular case of an analysis in the third octave, we would have

$$\Delta f = \left(2^{1/6} - \frac{1}{2^{1/6}} \right) f_c \approx 0.23 \, f_c \approx \frac{f_c}{4.32} \qquad [A3.4]$$

and

$$G_P = \frac{\left(P_0 \, 10^{N/20} \right)^2}{\left(2^{1/6} - \frac{1}{2^{1/6}} \right) f_c} \qquad [A3.5]$$

A3.3. *Response of an equivalent one-degree-of-freedom system*

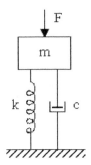

Figure A3.2. *One-degree-of-freedom system subjected to a force*

Let us consider the one-degree-of-freedom linear system in Figure A3.2, excited by a force F applied to mass m. The transfer function of this system is equal to:

$$H(f) = \frac{\ddot{z}}{F} = \frac{\ddot{y}}{F} = \frac{h^2}{m\left[\left(1-h^2\right)^2 + h^2/Q^2\right]^{1/2}} \quad [A3.6]$$

y and z being respectively the absolute response and the relative response of the mass m, and

$$h = \frac{f}{f_0}$$

At resonance, h = 1 and

$$H = \frac{Q}{m} \quad [A3.7]$$

The PSD G_F of the transmitted force is given by:

$$G_F = (\beta A)^2 G_P \quad [A3.8]$$

(F = β A P) and the PSD of the response ÿ to the force F applied to the one-degree-of-freedom system is equal, at resonance, to:

$$G_{\ddot{y}} = H^2 G_F \qquad [A3.9]$$

$$G_{\ddot{y}} = \frac{Q^2}{m^2} (\beta A)^2 G_P \qquad [A3.10]$$

$$G_{\ddot{y}} = \beta^2 \left(\frac{A}{m}\right)^2 Q^2 \frac{\left(P_0 \, 10^{N/20}\right)^2}{\left(2^{1/2n} - \frac{1}{2^{1/2n}}\right) f_c} \qquad [A3.11]$$

In the case of the third octave analysis,

$$G_{\ddot{y}} = \beta^2 \left(\frac{A}{m}\right)^2 Q^2 \frac{\left(P_0 \, 10^{N/20}\right)^2}{\left(2^{1/6} - \frac{1}{2^{1/6}}\right) f_c} \qquad [A3.12]$$

F. Spann and P. Patt set $Q = 4.5$ and $\beta = 2.5$, yielding

$$G_{\ddot{y}} = 126.6 \left(\frac{A}{m}\right)^2 G_P \qquad [A3.13]$$

A4. Mathematical functions

The object of this appendix is to provide tools facilitating the evaluation of some mathematical expressions, primarily integrals, intervening very frequently in calculations related to the analysis of random vibrations and their effect on a one-degree-of-system mechanical system.

A4.1. *Error function*

This function, also called the *probability integral*, is the subject of two definitions.

A4.1.1. *First definition*

The error function is expressed:

$$E_1(x) = \frac{2}{\sqrt{\pi}} \int_0^x e^{-t^2} dt \qquad [A4.1]$$

If $x \to \infty$, $E_1(x)$ tends towards $E_{1\infty}$ which is equal to

$$E_{1\infty} = \frac{2}{\sqrt{\pi}} \int_0^\infty e^{-t^2} dt = 1 \qquad [A4.2]$$

and if $x = 0$, $E_1(0) = 0$. If we set $t = \dfrac{u}{\sqrt{2}}$, it becomes

$$E_1\left(\frac{x}{\sqrt{2}}\right) = \frac{2}{\sqrt{\pi}} \int_0^x e^{-\frac{u^2}{2}} \frac{du}{\sqrt{2}}$$

$$E_1\left(\frac{x}{\sqrt{2}}\right) = \sqrt{\frac{2}{\pi}} \int_0^x e^{-\frac{u^2}{2}} du \qquad [A4.3]$$

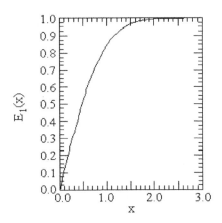

Figure A4.1. *Error function $E_1(x)$*

We can express a series development of $E_1(x)$ by integrating the series development of e^{-t^2} between 0 and x:

$$E_1(x) = \frac{2}{\sqrt{\pi}} \left[x - \frac{x^3}{1!\,3} + \frac{x^5}{2!\,5} - \cdots + (-1)^n \frac{x^{2n+1}}{n!\,(2n+1)} + \cdots \right] \qquad [A4.4]$$

This series converges for any x. For large x, we can obtain the asymptotic development according to [ANG 61], [CRA 63]:

$$E_1(x) \approx 1 - \frac{e^{-x^2}}{x\sqrt{\pi}}\left[1 - \frac{1}{2x^2} + \frac{1.3}{2^2 x^4} - \frac{1.3.5}{2^3 x^6} + \cdots + (-1)^{n-1}\frac{1.3.5\cdots(2n-3)}{2^{n-1} x^{2n-2}} + \cdots \right]$$

[A4.5]

For sufficiently large x, we have

$$E_1(x) \approx 1 - \frac{e^{-x^2}}{x\sqrt{\pi}}$$

[A4.6]

If $x = 1.6$, $E_1(x) = 0.976$, whilst the value approximated by the expression above is

$$E_1(x) \approx 0.973.$$

For $x = 1.8$, $E_1(x) \approx 0.9891$ instead of 0.890.

The ratio of two successive terms, equal to $\frac{2n-1}{x^2}$, is close to 1 when n is close to x^2. This remark makes it possible to limit the calculation by minimizing the error on $E_1(x)$.

NOTE.–

$E(x)$ is the error function and $[1 - E(x)]$, noted $erfc(x)$, is the "complementary error function".

$$erfc(x) = \frac{2}{\sqrt{\pi}} \int_x^\infty e^{-t^2}\, dt$$

[A4.7]

Function $E_1(x) = \dfrac{2}{\sqrt{\pi}} \int_0^x e^{-t^2}\, dt$

Approximate calculation of E_1

The error function can be estimated using the following approximate relationships [ABR 70], [HAS 55]:

$$E_1(x) = 1 - (a_1 t + a_2 t^2 + a_3 t^3 + a_4 t^4 + a_5 t^5)\, e^{-x^2} + \varepsilon(x)$$

[A4.8]

where

$$t = \frac{1}{1+px} \quad (0 \le x < \infty)$$

$$|\varepsilon(x)| \le 1.5 \ 10^{-7}$$

x	$E_1(x)$	ΔE_1	x	$E_1(x)$	ΔE_1	x	$E_1(x)$	ΔE_1
0.025	0.02820	0.02820	0.850	0.77067	0.01399	1.675	0.98215	0.00177
0.050	0.05637	0.02817	0.875	0.78408	0.01341	1.700	0.98379	0.00164
0.075	0.08447	0.02810	0.900	0.79691	0.01283	1.725	0.98529	0.00150
0.100	0.11246	0.02799	0.925	0.80918	0.01227	1.750	0.98667	0.00138
0.125	0.14032	0.02786	0.950	0.82089	0.01171	1.775	0.98793	0.00126
0.150	0.16800	0.02768	0.975	0.83206	0.01117	1.800	0.989090	0.00116
0.175	0.19547	0.02747	1.000	0.84270	0.01064	1.825	0.99015	0.00106
0.200	0.22270	0.02723	1.025	0.85282	0.01012	1.850	0.99111	0.00096
0.225	0.24967	0.02697	1.050	0.86244	0.00962	1.875	0.99199	0.00088
0.250	0.27633	0.02666	1.075	0.87156	0.00912	1.900	0.99279	0.00080
0.275	0.30266	0.02633	1.100	0.88021	0.00865	1.925	0.99352	0.00073
0.300	0.32863	0.02597	1.125	0.88839	0.00818	1.950	0.99418	0.00066
0.325	0.35421	0.02558	1.150	0.89612	0.00773	1.975	0.99478	0.00060
0.350	0.37938	0.02517	1.175	0.90343	0.00731	2.000	0.99532	0.00054
0.375	0.40412	0.02474	1.200	0.91031	0.00688	2.025	0.99781	0.00049
0.400	0.42839	0.02427	1.225	0.91680	0.00649	2.050	0.99626	0.00045
0.425	0.45219	0.2380	1.250	0.92290	0.00610	2.075	0.99666	0.00040
0.450	0.47548	0.02329	1.275	0.929863	0.00573	2.100	0.99702	0.00036
0.475	0.49826	0.02278	1.300	0.93401	0.00538	2.125	0.99735	0.00033
0.500	0.520500	0.02224	1.325	0.93905	0.00504	2.150	0.99764	0.00029
0.525	0.54219	0.02169	1.350	0.094376	0.00472	2.175	0.99790	0.00026
0.550	0.56332	0.02113	1.375	0.94817	0.00441	2.200	0.99814	0.00024
0.575	0.58388	0.02056	1.400	0.95229	0.00412	2.225	0.99835	0.00021
0.600	0.60386	0.01998	1.425	0.95612	0.00383	2.250	0.99854	0.00019
0.625	0.62324	0.01938	1.450	0.95970	0.00356	2.275	0.99871	0.00017
0.650	0.64203	0.01879	1.475	0.96302	0.00332	2.300	0.99886	0.00015
0.675	0.66022	0.01819	1.500	0.96611	0.00309	2.325	0.99899	0.00013
0.700	0.67780	0.01758	1.525	0.96897	0.00286	2.350	0.99911	0.00012
0.725	0.69478	0.01698	1.550	0.97162	0.00265	2.375	0.99922	0.00011
0.750	0.711156	0.01638	1.575	0.97408	0.00246	2.400	0.99931	0.00009
0.775	0.72693	0.01577	1.600	0.97635	0.00227	2.425	0.99940	0.00009
0.800	0.74210	0.01517	1.625	0.97844	0.00209	2.450	0.99947	0.00007
0.825	0.75668	0.01458	1.650	0.98038	0.00194	2.475	0.99954	0.00007
						2.500	0.99959	0.00005

Table A4.1. *Error function $E_1(x)$*

$p = 0.3275911$ $\qquad a_3 = 1.421413741$
$a_1 = 0.254829592$ $\qquad a_4 = -1.453152027$
$a_2 = -0.284496736$ $\qquad a_5 = 1.061495429$

$$E_1(x) = 1 - (a_1 t + a_2 t^2 + a_3 t^3) e^{-x^2} + \varepsilon(x) \qquad [A4.9]$$

$$t = \frac{1}{1 + p x} \qquad |\varepsilon(x)| \leq 2.5\ 10^{-5}$$

$p = 0.47047$ $\qquad a_2 = -0.095879$
$a_1 = 0.3480242$ $\qquad a_3 = 0.7478556$

Other approximate relationships of this type have been proposed [HAS 55] [SPA 87], with developments of the 3rd, 4th and 5th order. C. Hastings also suggests the expression

$$E_1(x) = 1 - \frac{1}{(1 + a_1 x + a_2 x^2 + a_3 x^3 + a_4 x^4 + a_5 x^5 + a_6 x^6)^{16}} \qquad [A4.10]$$

$a_1 = 0.0705230784$ $\qquad a_4 = 0.0001520143$
$a_2 = 0.0422820123$ $\qquad a_5 = 0.0002765672$
$a_3 = 0.0092705272$ $\qquad a_6 = 0.0000430638$

$(0 \leq x \leq \infty)$

Derivatives

$$\frac{d E_1(x)}{dx} = \frac{2}{\sqrt{\pi}} e^{-x^2} \qquad [A4.11]$$

$$\frac{d^2 E_1(x)}{dx^2} = -\frac{4}{\sqrt{\pi}} x\ e^{-x^2} \qquad [A4.12]$$

Approximate formula

The approximate relationship [DEV 62]

$$E_1(x) = \sqrt{1 - \exp\left(-\frac{4x^2}{\pi}\right)} \qquad [A4.13]$$

gives results of a sufficient precision for many applications (error lower than some thousandths, regardless of the value of x).

NOTE.–

The probability $P = \frac{1}{\sqrt{2\pi}} \int_{-\infty}^{x} e^{-\frac{t^2}{2}} dt$ *(normal distribution) can be calculated numerically using the approximate relations of this error function from*

$$P = \frac{1}{2}\left[1 + E_1\left(\frac{x}{\sqrt{2}}\right)\right] \qquad [A4.14]$$

A4.1.2. *Second definition*

The error function is often defined by [PAP 65], [PIE 70]:

$$E_2(x) = \frac{1}{\sqrt{2\pi}} \int_{0}^{x} e^{-\frac{t^2}{2}} dt \qquad [A4.15]$$

With this definition

$$E_2(x) = \frac{E_1\left(\frac{x}{\sqrt{2}}\right)}{2} \qquad [A4.16]$$

yielding

$$E_1(x) = 2 E_2(x\sqrt{2}) \qquad [A4.17]$$

Applications

$$\frac{1}{\alpha\sqrt{2\pi}}\int_{x_1}^{x_2} e^{-\frac{(x-\beta)^2}{2\alpha^2}}\,dx = E_2\left(\frac{x_2-\beta}{\alpha}\right) - E_2\left(\frac{x_1-\beta}{\alpha}\right) \qquad [A4.18]$$

where α and β are two arbitrary constants [PIE 70] and

$$\int_{-\infty}^{\infty} e^{-\frac{t^2}{2}}\,dt = \sqrt{2\pi} \qquad [A4.19]$$

Properties of $E_2(x)$

$E_2(x)$ tends towards 0.5 when $x \to \infty$:

$$E_2 = \frac{1}{\sqrt{2\pi}} \int_0^{\infty} e^{-\frac{t^2}{2}}\,dt = 0.5 \qquad [A4.20]$$

$E_2(0) = 0$

$E_2(-x) = -E_2(x)$

Function $E_2(x) = \dfrac{1}{\sqrt{2\pi}} \int_0^{x} e^{-\frac{t^2}{2}}\,dt$

Approximate calculation of $E_2(x)$

The function $E_2(x)$ can be approximated, for $x > 0$, by the expression defined as follows [LAM 76], [PAP 65]:

$$E_2(x) \approx \frac{1}{2}\left[1 - (at + bt^2 + ct^3 + dt^4 + et^5)e^{-\frac{x^2}{2}}\right] \qquad [A4.21]$$

where

$$t = \frac{1}{1+0.2316418\ x} \qquad a = 0.254829592 \qquad b = -0.284496736$$

$$c = 1.421413741 \qquad d = -1.453152027 \qquad e = 1.061405429$$

The approximation is very good (at least 5 decimal points).

NOTE.–

With these notations, the function $E_2(x)$ is none other than the integral between 0 and x of the Gauss function:

$$G(x) = \frac{1}{\sqrt{2\pi}} e^{-\frac{x^2}{2}}$$

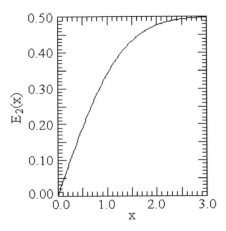

Figure A4.2. *Error function $E_2(x)$*

Figure A4.3 shows the variations of $G(x)$ and of $E_2(x)$ for $0 \le x \le 3$. We thus have:

$$\int_0^x \exp\left(-\frac{u^2}{2\sigma^2}\right) du = \sqrt{\frac{\pi}{2}}\ \sigma\ E_1\left(\frac{x}{\sigma\sqrt{2}}\right) \qquad \text{[A4.22]}$$

x	$E_2(x)$	ΔE_2	x	$E_2(x)$	ΔE_2	x	$E_2(x)$	ΔE_2
0.05	0.01994	0.01994	1.40	0.41924	0.00775	2.75	0.49702	0.00049
0.10	0.03983	0.01989	1.45	0.42647	0.00723	2.80	0.49744	0.00042
0.15	0.05962	0.01979	1.50	0.43319	0.00672	2.85	0.49781	0.00037
0.20	0.07926	0.01964	1.55	0.43943	0.00624	2.90	0.49813	0.00032
0.25	0.09871	0.01945	1.60	0.44520	0.00577	2.95	0.49981	0.00028
0.30	0.11791	0.01920	1.65	0.45053	0.00533	3.00	0.49865	0.00024
0.35	0.13683	0.01892	1.70	0.45543	0.00490	3.05	0.49886	0.00021
0.40	0.15542	0.01859	1.75	0.45994	0.00451	3.10	0.49903	0.00017
0.45	0.17364	0.01822	1.80	0.46407	0.00413	3.15	0.49918	0.00015
0.50	0.19146	0.01782	1.85	0.46784	0.00377	3.20	0.49931	0.00013
0.55	0.20884	0.01738	1.90	0.47128	0.00344	3.25	0.49942	0.00011
0.60	0.22575	0.01691	1.95	0.47441	0.00313	3.30	0.49952	0.00010
0.65	0.24215	0.01640	2.00	0.47725	0.00284	3.35	0.49960	0.00008
0.70	0.25804	0.01589	2.05	0.47982	0.00257	3.40	0.49966	0.00006
0.75	0.27337	0.01533	2.10	0.48214	0.00232	3.45	0.49972	0.00006
0.80	0.28814	0.01477	2.15	0.48422	0.00208	3.50	0.49977	0.00005
0.85	0.30234	0.01420	2.20	0.48610	0.00188	3.55	0.49841	0.00004
0.90	0.31594	0.01360	2.25	0.48778	0.00168	3.60	0.49984	0.00003
0.95	0.32894	0.01300	2.30	0.48928	0.00150	3.65	0.49987	0.00003
1.00	0.34134	0.01240	2.35	0.49061	0.00133	3.70	0.49989	0.00002
1.05	0.35314	0.01180	2.40	0.49180	0.00119	3.75	0.49991	0.00002
1.10	0.36433	0.01119	2.45	0.49286	0.00106	3.80	0.49993	0.00002
1.15	0.37493	0.01060	2.50	0.49379	0.00093	3.85	0.49994	0.00001
1.20	0.38493	0.01000	2.55	0.49461	0.00082	3.90	0.49995	0.00001
1.25	0.39435	0.00942	2.60	0.49534	0.00072	3.95	0.049996	0.00001
1.30	0.40320	0.00885	2.65	0.49598	0.00064	4.00	0.49997	0.00001
1.35	0.41149	0.00829	2.70	0.49653	0.00055			

Table A4.2. *Error function $E_2(x)$*

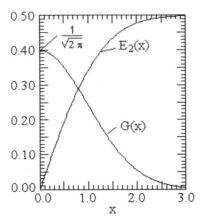

Figure A4.3. *Comparison of the error function $E_2(x)$ and of $G(x)$*

Calculation of x for $E_2(x) = E_0$

The method below applies if x is positive and where $0 < E_0 < 0.5$ [LAM 80]. We calculate successively:

$$z = \sqrt{-2\ \ln(1 - 2\ E_0)}$$

and

$$x = g_0 + g_1\ z + g_2\ z^2 + \cdots + g_{10}\ z^{10} \qquad [A4.23]$$

where

$g_0 = 6.55864 \ \ 10^{-4}$ $\qquad g_6 = -1.17213 \ \ 10^{-2}$
$g_1 = -0.02069$ $\qquad g_7 = 2.10941 \ \ 10^{-3}$
$g_2 = 0.737563$ $\qquad g_8 = -2.18541 \ \ 10^{-4}$
$g_3 = -0.207071$ $\qquad g_9 = 1.23163 \ \ 10^{-5}$
$g_4 = -2.06851 \ \ 10^{-2}$ $\qquad g_{10} = -2.93138 \ \ 10^{-7}$
$g_5 = 0.03444$

For negative values, we will use the property

$$E_2(-x) = -E_2(x)$$

NOTE.–

To calculate x from the given E_1 set $E_2 = \dfrac{E_1}{2}$, calculate x, and then $\dfrac{x}{\sqrt{2}}$.

A4.2. Calculation of the integral $\int e^{ax}/x^n\, dx$

We have [DWI 66]:

$$\int \frac{e^{ax}}{x}\, dx = \ln|x| + \frac{ax}{1!} + \frac{a^2 x^2}{2\cdot 2!} + \frac{a^3 x^3}{3\cdot 3!} + \cdots + \frac{a^n x^n}{n\cdot n!} + \cdots \qquad [A4.24]$$

yielding, since

$$\int \frac{e^{ax}}{x^n}\, dx = -\frac{e^{ax}}{(n-1)x^{n-1}} + \frac{a}{n-1}\int \frac{e^{ax}}{x^{n-1}}\, dx \qquad [A4.25]$$

$$\int \frac{e^{ax}}{x^n}\, dx = -\frac{e^{ax}}{(n-1)x^{n-1}} - \frac{a\, e^{ax}}{(n-1)(n-2)x^{n-2}} - \cdots - \frac{a^{n-2} e^{ax}}{(n-1)!\, x} + \frac{a^{n-1}}{(n-1)!}\int \frac{e^{ax}}{x}\, dx$$

$$[A4.26]$$

A4.3. Euler's constant

Definition

$$\varepsilon = \lim_{n\to\infty}\left[1 + \frac{1}{2} + \cdots + \frac{1}{n} - \ln n\right] \qquad [A4.27]$$

$$\varepsilon \approx 0.57721566490\ldots$$

An approximate value is given by [ANG 61]:

$$\varepsilon \approx \frac{1}{2}\left(\sqrt[3]{10} - 1\right)$$

i.e.

$$\varepsilon \approx 0.5772173\ldots$$

Applications

It is shown that [DAV 64]:

$$\int_0^\infty \ln \lambda \, e^{-\lambda} \, d\lambda = -\varepsilon \qquad [A4.28]$$

and that

$$\int_0^\infty (\ln \lambda)^2 \, e^{-\lambda} \, d\lambda = \frac{\pi}{6} + \varepsilon^2 \qquad [A4.29]$$

A5. Complements to the transfer functions

A5.1. *Error related to digitization of transfer function*

The transfer function is defined by a certain number of points. According to this number, the peak of this function can be more or less truncated and the measurement of the resonance frequency and Q factor distorted [NEU 70].

Any complex system with separate modes is comparable in the vicinity of a resonance frequency to a one-degree-of-freedom system of quality factor Q.

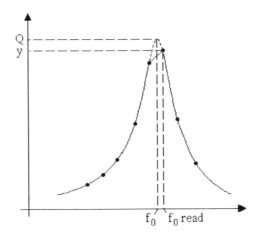

Figure A5.1. *Transfer function of a one-degree-of-freedom system close to resonance*

Let us set y as the value of the quality factor read on the curve, Q being the true value.

Let us set $\beta = \dfrac{y}{Q}$ and $\alpha = \dfrac{f}{f_0} = \dfrac{\text{read resonance frequency}}{\text{true resonance frequency}}$. When α is different from 1, we can set $\alpha = 1 - \delta$, if δ is the relative deviation on the value of the resonance frequency. For $\delta = 0$, we have $\alpha = 1$ and $\beta = 1$. For $\delta \neq 0$, β is less than 1. The resolution error is equal to $\varepsilon_R = 1 - \beta$. The amplitude of the transfer function away from resonance is given by y such that:

$$y^2 = \frac{1 + \dfrac{\alpha^2}{Q^2}}{\left(1-\alpha^2\right)^2 + \dfrac{\alpha^2}{Q^2}} = \frac{Q^2 + \alpha^2}{Q^2\left(1-\alpha^2\right)^2 + \alpha^2} \quad \text{[A5.1]}$$

$$\beta^2 = \frac{y^2}{Q^2} = \frac{1 + \dfrac{\alpha^2}{Q^2}}{Q^2\left(1-\alpha^2\right)^2 + \alpha^2} = \frac{1 + \dfrac{(1-\alpha)^2}{Q^2}}{Q^2 \delta^2 + (1-\delta)^2} \quad \text{[A5.2]}$$

For large Q, we have

$$\beta^2 \approx \frac{1}{Q^2\left(1-\alpha^2\right)^2 + \alpha^2} \quad \text{[A5.3]}$$

i.e., replacing α with $1 - \delta$ and assuming Q^2 to be large compared to 1,

$$\beta^2 \approx \frac{1}{1 - 2\delta + 4 Q^2 \delta^2} \quad \text{[A5.4]}$$

and

$$\varepsilon_R = 1 - \beta \approx 1 - \frac{1}{\sqrt{1 - 2\delta + 4 Q^2 \delta^2}} \quad \text{[A5.5]}$$

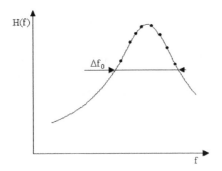

Figure A5.2. *Digitization of n points of the transfer function between the half-power points*

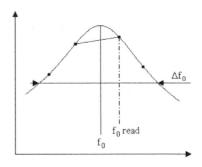

Figure A5.3. *Effect of too low a sampling rate*

If $4 Q^2 \delta^2 \gg 2\delta$, i.e. if $Q^2 \gg \dfrac{1}{2\delta}$,

$$\varepsilon_R \approx 1 - \left(1 + 4 Q^2 \delta^2\right)^{-1/2} \quad [A5.6]$$

$$\varepsilon_R \approx 2 Q^2 \delta^2 \quad [A5.7]$$

Let us assume that there are n points in the interval Δf_0 between the half-power points, i.e. $n-1$ intervals. We have:

$$\delta = 1 - \alpha = 1 - \dfrac{f}{f_0} \quad [A5.8]$$

$$f_0 - f = \dfrac{\Delta f_0}{2(n-1)} \quad [A5.9]$$

yielding

$$\delta = \dfrac{\Delta f_0}{2(n-1) f_0} = \dfrac{1}{2(n-1) Q} \quad [A5.10]$$

i.e., since $\varepsilon_R \approx 2\alpha^2 \delta^2$,

$$\varepsilon_R \approx \frac{1}{2(n-1)^2} \qquad [A5.11]$$

Figure A5.4 shows variations of the error ε_R versus the number of points n in Δf_0. To measure the Q factor with an error less than 2%, it is necessary for n to be greater than 6 points.

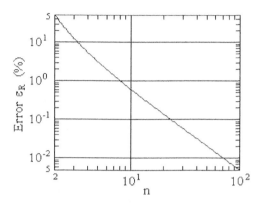

Figure A5.4. *Error of resolution versus number of points in Δf*

NOTE.–

In the case of random vibrations, the frequency increment Δf is related to the sampling frequency f_s by the relationship

$$\Delta f = \frac{f_s}{2M} \qquad [A5.12]$$

where M is the total number of points representing the spectrum. Ideally, the increment Δf should be a very small fraction of the bandwidth Δf_0 around the resonance. The number of points M is limited by the memory size of the calculator and the frequency f_s should be at least twice as large as the highest frequency of the analyzed signal, to avoid aliasing errors (Shannon's theorem). Too large a Δf leads to a small value of n and therefore to an error to the Q factor measurement. Decreasing f_s to reduce Δf (with M constant) can lead to poor representation of the temporal signal and thus to an inaccuracy in the amplitude of the spectrum at *high frequencies. It is recommended to choose a sampling frequency greater than 6 times the largest frequency to be analyzed [TAY 75].*

A5.2. *Use of a fast swept sine for the measurement of transfer functions*

The measurement of a transfer function starting from a traditional swept sine test leads to a test of relatively long duration and, in addition, requires material with a great measurement dynamics.

Transfer functions can also be measured from random vibration tests or by using shocks, the test duration obviously being in this latter case very short. On this assumption, the choice of the form of shock to use is important because, the transfer function being calculated from the ratio of the Fourier transforms of the response (in a point of the structure) and excitation, it is necessary that this latter transform does not present a zero or too small an amplitude in a certain range of frequency. In the presence of noise, the low levels in the denominator lead to uncertainties in the transfer function [WHI 69].

The interest of the fast linear swept sine lies in two points:

– the Fourier transform of a linear swept sine has a roughly constant amplitude in the swept frequency range. W.H. Reed, A.W. Hall, L.E. Barker [REE 60], then R.G. White [WHI 72] and R.J. White and R.J. Pinnington [WHI 82] showed that the average module of the Fourier transform of a linear swept sine is equal to:

$$\overline{|\ddot{X}(\omega)|} = \frac{|\ddot{x}_m|}{2\sqrt{b}} \qquad [A5.13]$$

where $\ddot{x}_m$ = amplitude of acceleration defining the swept sine

$$b = \frac{f_2 - f_1}{t_b} = \text{sweep rate}$$

and that, more generally,

$$\overline{|\ddot{X}|} = \frac{|\ddot{x}_m|}{2\sqrt{\dot{f}}} \qquad [A5.14]$$

where $\dot{f}$ is the sweep rate for an arbitrary law,

Example A5.1.

Linear sweep: 10 Hz to 200 Hz

Durations: 1 s – 0.5 s – 0.1 s and 10 ms

$\ddot{x}_m = 10$ ms^{-2}

Depending on the case, relationship [A5.14] gives 0.3627, 0.256, 0.1147 or 0.03627 (m/s).

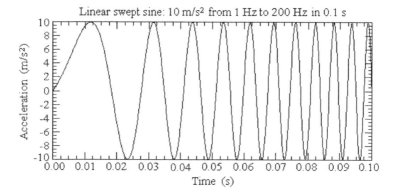

Figure A5.5. *Example of fast swept sine*

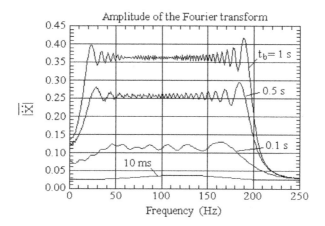

Figure A5.6. *Examples of fast swept sine Fourier transforms*

– sweeping being fast (a few seconds or a fraction of a second, depending on the studied frequency band), the mechanical system responds as to a shock and does not have time to reach the response which it would have in steady state operation or with a slow sweep (Q times the excitation). Accordingly, the dynamics of the necessary instrumentation is less constraining and measurement is taken in a domain where the non-linearities of the structure are less important.

The Fourier transform of the response must be calculated over the whole duration of the response, including the residual signal after the end of sweep.

A5.3. Error of measurement of transfer function using a shock related to signal truncation

With a transient excitation, of shock type or fast swept sine, the transfer function is calculated from the ratio of the Fourier transforms of response and excitation:

$$H(i\Omega) = \frac{Y(i\Omega)}{X(i\Omega)} \qquad [A5.15]$$

where

$$X(i\Omega) = \int_{-\infty}^{+\infty} x(t)\, e^{-i\Omega t}\, dt \qquad [A5.16]$$

$$Y(i\Omega) = \int_{-\infty}^{+\infty} y(t)\, e^{-i\Omega t}\, dt \qquad [A5.17]$$

If $x(t)$ is an impulse unit applied to the time $t = 0$, we have $X(i\Omega) = 1$ whatever the value of Ω and (Volume 1, expression [4.115]):

$$H(i\Omega) = \int_{0}^{\infty} h(t)\, e^{-i\Omega t}\, dt \qquad [A5.18]$$

where $h(t)$ is the impulse response. For a one-degree-of-freedom system of natural frequency f_0 (Volume 1, relationship [4.114]),

$$h(t) = \frac{\omega_0}{\sqrt{1-\xi^2}}\, e^{-\xi \omega_0 t}\, \sin\sqrt{1-\xi^2}\, \omega_0 t \qquad [A5.19]$$

yielding the complex transfer function

$$H(i\Omega) = \frac{1}{\left(1 - \dfrac{\omega^2}{\omega_0^2}\right) + i\, 2\xi\, \dfrac{\omega}{\omega_0}} \qquad [A5.20]$$

Relationship [A5.18] could be used in theory to determine $H(i\Omega)$ from the response to an impulse, but, in practice, a truncation of the response is difficult to avoid, either because the decreasing signal becomes non-measurable or because the time of analysis is limited to a value τ_m [WHI 69]. The effects of truncation have been analyzed by B.L. Clarkson and A.C. Mercer [CLA 65], who showed:

– that the resonance frequency can still be identified from the diagram vector as the frequency to which the rate of variation in the length of arc with frequency, $\frac{ds}{df}$, is maximum;

– that the damping measured from such a diagram (established with a truncated signal) is larger than the true value.

These authors established by theoretical analysis that the error (in %) introduced by truncation is equal to:

$$e(\%) = 100 \left\{ 1 - \left[\frac{1 - e^{-\xi \omega_0 \tau_m} \left(1 + \xi \omega_0 \tau_m + \frac{1}{2} \xi^2 \omega_0^2 \tau_m^2 \right)}{1 - e^{-\xi \omega_0 \tau_m} \left(1 + \xi \omega_0 \tau_m \right)} \right] \right\} \qquad [A5.21]$$

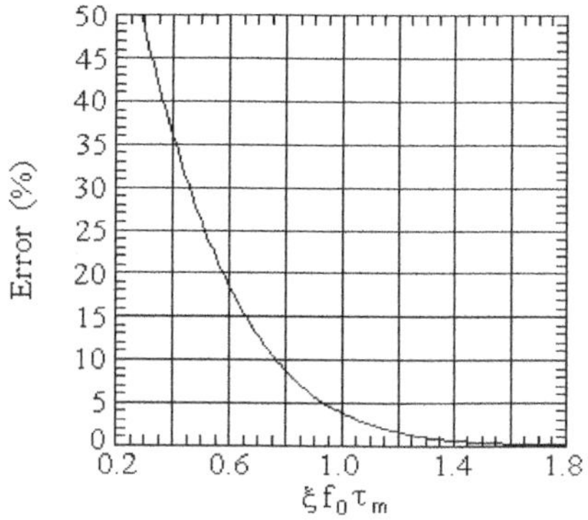

Figure A5.7. *Error of measured value of ξ due to truncation of the signal*

It can be seen that if $\xi \omega_0 \tau_m > 1$, the error of the measured value of ξ is lower than 5%.

It should be noted that we can obtain a very good precision without needing to analyze extremely long records. For example, for $f_0 = 100$ Hz and $\xi = 0.005$, a duration of 2 s led to an error of less than 5% ($\xi \omega_0 \tau_m = 1$).

Appendix 555

A5.4. *Error made during measurement of transfer functions in random vibration*

The function of coherence γ^2 is a measurement of the precision of the calculated value of the transfer function $H(f)$ and is equal to [2.107]:

$$\gamma^2 = \frac{\left[G_{xy}(f)\right]^2}{G_x(f)\,G_y(f)} \qquad [A5.22]$$

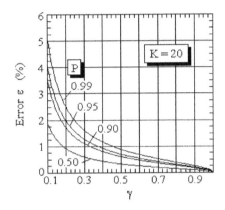

Figure A5.8. *Error of measurement of the transfer function in random excitation versus γ, for $K = 20$*

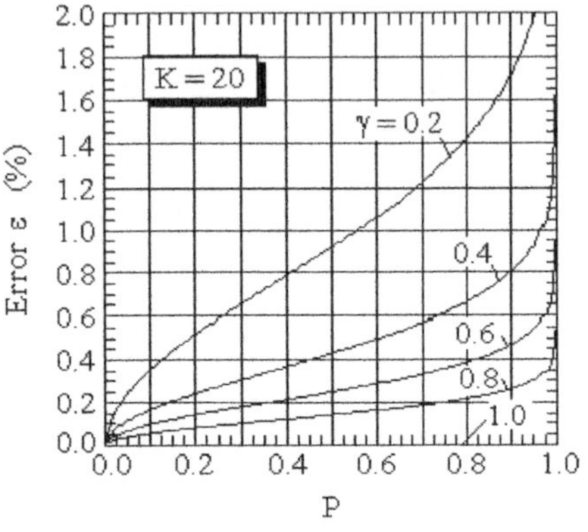

Figure A5.9. *Error of measurement of the transfer function in random excitation versus the probability, for $K = 20$*

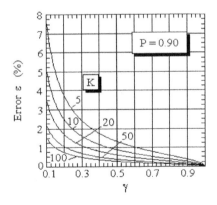
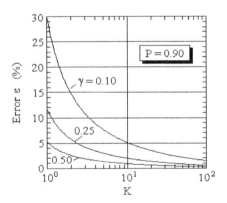

Figure A5.10. *Error of measurement of the transfer function in random excitation versus γ, for $P = 0.90$*

Figure A5.11. *Error of measurement of the transfer function in random excitation versus K, for $P = 0.90$*

If the system is linear and if there is no interference, $\gamma^2 = 1$ and the calculated value of $H(f)$ is correct. If $\gamma^2 < 1$, the error in the estimate of $H(f)$ is provided with a probability P by:

$$\varepsilon = \frac{\Delta |H(f)|}{|H(f)|} \leq \left\{ \left[(1-P)^{-1/K} - 1\right] \left[\frac{1}{\gamma_{xy}^2} - 1\right] \right\}^{\frac{1}{2}} \quad [A5.23]$$

where K is the number of spectra (blocks) used to calculate each PSD [WEL 70].

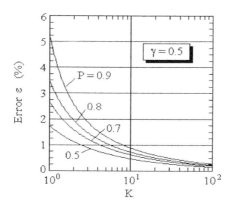

Figure A5.12. *Error of measurement of the transfer function in random excitation versus K, for $\gamma = 0.5$*

A5.5. *Derivative of expression of transfer function of a one-degree-of-freedom linear system*

Let us consider the transfer function:

$$H(\omega) = \frac{1}{\omega_0^2 \left[1 - h^2 + 2i\xi h\right]} \quad [A5.24]$$

where

$$h = \frac{\omega}{\omega_0}$$

using multiplication of the denominator's conjugate quantity, we obtain

$$H(\omega) = \left(1 - h^2\right) A - 2 i \xi h A$$

if we set

$$A = \frac{1}{\omega_0^2 \left[\left(1 - h^2\right)^2 + \left(2 \xi h\right)^2\right]}$$

yielding

$$\frac{dH}{dh} = -2 h A - 2 i \xi A + \left(1 - h^2\right)\frac{dA}{dh} - 2 \xi h i \frac{dA}{dh} \qquad [A5.25]$$

with

$$\frac{dA}{dh} = \frac{4 h \left(1 - 2 \xi^2 - h^2\right)}{\omega_0^2 \left[\left(1 - h^2\right)^2 + \left(2 \xi h\right)^2\right]^2} \qquad [A5.26]$$

$$\frac{dH}{dh} = -2(h + i \xi) A + \left[1 - h^2 - 2 \xi h i\right]\frac{dA}{dh} \qquad [A5.27]$$

A6. Calculation of integrals

A6.1. Integral $I_b(h) = \frac{4\xi}{\pi} \int \frac{h^b}{(1-h)^2 + 4\xi^2 h^2} dh$

$$I_b(h) = \frac{4\xi}{\pi} \frac{h^{b-3}}{b-3} + 2(1 - 2\xi^2) I_{b-2} - I_{b-4} \quad \text{if } b \neq 3$$

$$I_3(h) = \frac{4\xi}{\pi} \ln h + 2(1 - 2\xi^2) I_1 - I_{-1} \qquad \text{if } b = 3 \qquad [A6.1]$$

A6.1.1. *Demonstration*

From [PUL 68]:

$$\left(1-h^2\right)^2 + 4\xi^2 h^2 = h^4 - \beta h^2 + 1$$

if we set $\beta = 2\left(1 - 2\xi^2\right)$. By division, we obtain

$$\frac{h^b}{h^4 - \beta h^2 + 1} = h^{b-4} + \frac{\beta h^{b-2} - h^{b-4}}{h^4 - \beta h^2 + 1} \quad \text{if } b \neq 3$$

yielding recurrence relation [A6.1]. The use of these expressions requires knowledge of $I_b(h)$ for some values from b, for example, $b = 0$, 1 and 2.

A6.1.2. *Calculation of $I_0(h)$ and $I_2(h)$*

The calculation of $I_0(h)$ and $I_2(h)$ can be carried out easily by writing the functions to be integrated respectively in the forms

$$\frac{1}{\left(1-h^2\right)^2 + 4\xi^2 h^2} = \frac{1}{4\sqrt{1-\xi^2}}\left[\frac{h + 2\sqrt{1-\xi^2}}{h^2 + 2h\sqrt{1-\xi^2} + 1} - \frac{h - 2\sqrt{1-\xi^2}}{h^2 - 2h\sqrt{1-\xi^2} + 1}\right]$$

[A6.2]

and

$$\frac{h^2}{\left(1-h^2\right)^2 + 4\xi^2 h^2} = \frac{1}{4\sqrt{1-\xi^2}}\left[\frac{-h}{h^2 + 2h\sqrt{1-\xi^2} + 1} + \frac{h}{h^2 - 2h\sqrt{1-\xi^2} + 1}\right]$$

[A6.3]

Let us set

$$I = \int \frac{h + 2\sqrt{1-\xi^2}}{h^2 + 2h\sqrt{1-\xi^2} + 1} \, dh$$

$$I = \frac{1}{2}\int \frac{2h + 2\sqrt{1-\xi^2}}{h^2 + 2h\sqrt{1-\xi^2} + 1} \, dh + \int \frac{\sqrt{1-\xi^2}}{h^2 + 2h\sqrt{1-\xi^2} + 1} \, dh$$

Knowing what $\delta = h^2 + 2h\sqrt{1-\xi^2} + 1$ can be put in the form

$$\delta = \xi^2\left(\frac{X^2}{\xi^2} + 1\right) \text{ with } X = h + \sqrt{1-\xi^2}$$

it results that

$$I = \frac{1}{2}\ln\left(h^2 + 2h\sqrt{1-\xi^2} + 1\right) + \frac{\sqrt{1-\xi^2}}{\xi} \arctan \frac{h+\sqrt{1-\xi^2}}{\xi}$$

In the same way,

$$J = \int \frac{h - 2\sqrt{1-\xi^2}}{h^2 - 2h\sqrt{1-\xi^2} + 1} dh$$

$$J = \frac{1}{2}\int \frac{2h - 2\sqrt{1-\xi^2}}{h^2 - 2h\sqrt{1-\xi^2} + 1} dh - \int \frac{\sqrt{1-\xi^2}}{h^2 - 2h\sqrt{1-\xi^2} + 1} dh$$

With a change of variable such that:

$$X = h - \sqrt{1-\xi^2}$$

$$h^2 - 2h\sqrt{1-\xi^2} + 1 = \xi^2\left(\frac{X^2}{\xi^2} + 1\right)$$

we obtain

$$J = \frac{1}{2}\ln\left(h^2 - 2h\sqrt{1-\xi^2} + 1\right) - \frac{\sqrt{1-\xi^2}}{\xi} \arctan \frac{h-\sqrt{1-\xi^2}}{\xi}$$

yielding

$$I_0(h) = \frac{\xi}{2\pi\sqrt{1-\xi^2}} \ln \frac{h^2 + 2h\sqrt{1-\xi^2} + 1}{h^2 - 2h\sqrt{1-\xi^2} + 1}$$

$$+ \frac{1}{\pi}\left(\arctan \frac{h+\sqrt{1-\xi^2}}{\xi} + \arctan \frac{h-\sqrt{1-\xi^2}}{\xi}\right) \qquad \text{[A6.4]}$$

NOTES.–

1. *Knowing that*

$$\text{arc tan } x + \text{arc tan } y = \text{arc tan} \frac{x+y}{1-xy} \qquad [A6.5]$$

$I_0(h)$ can be also written

$$I_0(h) = \frac{\xi}{2\pi\sqrt{1-\xi^2}} \ln \frac{h^2 + 2h\sqrt{1-\xi^2} + 1}{h^2 - 2h\sqrt{1-\xi^2} + 1} + \frac{1}{\pi} \text{arc tan} \frac{2h\xi}{1-h^2} \qquad [A6.6]$$

However, expression [A6.6] must be used with caution, [A6.5] being correct only under precise conditions.

Relation [A6.5], established from

$$\tan(a+b) = \frac{\tan a + \tan b}{1 - \tan a \tan b}$$

is exact only if

$$a \neq \frac{\pi}{2} + k\pi$$

$$b \neq \frac{\pi}{2} + k\pi$$

$$a + b \neq \frac{\pi}{2} + k\pi$$

From these first two conditions, if $\tan a = x$ and $\tan b = y$, it is necessary that $-\frac{\pi}{2} < a < \frac{\pi}{2}$ and that $-\frac{\pi}{2} < b < \frac{\pi}{2}$.

Taking into account these inequalities, $a+b$ can thus vary between $-\pi$ and π (limits not included). The second member in addition lays down that $-\frac{\pi}{2} < a+b < \frac{\pi}{2}$. Let us set $\Theta = a+b$. If

$$arc\ tan\ x + arc\ tan\ y > \frac{\pi}{2} \qquad [A6.7]$$

we must thus take as the value $arc\ tan\ \frac{x+y}{1-x\ y}$ of the angle $\Theta - \pi$.

If

$$arc\ tan\ x + arc\ tan\ y < -\frac{\pi}{2} \qquad [A6.8]$$

we consider $\Theta + \pi$.

2. It is easy to show that we are in the situation of an inequality [A6.7] as soon as we have $x+y > 2$. Indeed, let us search for the minimal value of the sum

$$z = tan\ x + tan\left(\frac{\pi}{2} - x\right)$$

i.e. of

$$z = \left(\frac{X^2+1}{X}\right)$$

While setting $X = tan\ x$, we find that the derivative $\frac{dz}{dx}$ is cancelled when $X^2 = 1$, yielding $z = 2$ (condition [A6.7]) or $z = -2$ (condition [A6.8]).

Following what precedes,

$$\int \frac{h}{h^2 + 2h\sqrt{1-\xi^2} + 1} dh = \frac{1}{2} \int \frac{2h + 2\sqrt{1-\xi^2}}{h^2 + 2h\sqrt{1-\xi^2} + 1} dh$$

$$- \sqrt{1-\xi^2} \int \frac{1}{h^2 + 2h\sqrt{1-\xi^2} + 1} dh$$

$$\int \frac{h}{h^2 + 2h\sqrt{1-\xi^2} + 1} dh = \frac{1}{2} \ln\left(h^2 + 2h\sqrt{1-\xi^2} + 1\right) - \frac{\sqrt{1-\xi^2}}{\xi} \operatorname{arc tan} \frac{h + \sqrt{1-\xi^2}}{\xi}$$

and

$$\int \frac{h}{h^2 - 2h\sqrt{1-\xi^2} + 1} dh = \frac{1}{2} \ln\left(h^2 - 2h\sqrt{1-\xi^2} + 1\right) - \frac{\sqrt{1-\xi^2}}{\xi} \operatorname{arc tan} \frac{h - \sqrt{1-\xi^2}}{\xi}$$

yielding

$$I_2(h) = \frac{\xi}{2\pi\sqrt{1-\xi^2}} \ln \frac{h^2 - 2h\sqrt{1-\xi^2} + 1}{h^2 + 2h\sqrt{1-\xi^2} + 1}$$

$$+ \frac{1}{\pi} \left(\operatorname{arc tan} \frac{h + \sqrt{1-\xi^2}}{\xi} + \operatorname{arc tan} \frac{h - \sqrt{1-\xi^2}}{\xi} \right)$$

which can be written, with the same reservations as above,

$$I_2(h) = \frac{\xi}{2\pi\sqrt{1-\xi^2}} \ln \frac{h^2 - 2h\sqrt{1-\xi^2} + 1}{h^2 + 2h\sqrt{1-\xi^2} + 1} + \frac{1}{\pi} \operatorname{arc tan} \frac{2h\xi}{1-h^2} \qquad [A6.9]$$

NOTE.–

When $h \to \infty$, $I_0(h) \to 1$ and $I_2(h) \to 1$.

If $h \to 0$, $I_0(h) \to 0$ yielding, while setting

$$K = \int_0^\infty \frac{G}{(1-h^2)^2 + \frac{h^2}{Q^2}} dh$$

$$(h = \frac{f}{f_0})$$

$$K = G f_0 \int_0^\infty \frac{dh}{\left(1-h^2\right)^2 + \frac{h^2}{Q^2}}$$

$$K = \frac{\pi G f_0 Q}{2} \qquad [A6.10]$$

A6.1.3. *Calculation of $I_1(h)$*

$$I_1(h) = \frac{4\xi}{\pi} \int \frac{h \, dh}{(1-h^2)^2 + 4\xi^2 h^2}$$

$$\frac{h}{(1-h^2)^2 + 4\xi^2 h^2} = \frac{1}{4\sqrt{1-\xi^2}} \left[\frac{1}{h^2 - 2h\sqrt{1-\xi^2} + 1} - \frac{1}{h^2 + 2h\sqrt{1-\xi^2} + 1} \right]$$

yielding

$$I_1(h) = \frac{1}{\pi\sqrt{1-\xi^2}} \left(\arctan \frac{h - \sqrt{1-\xi^2}}{\xi} - \arctan \frac{h + \sqrt{1-\xi^2}}{\xi} \right) \qquad [A6.11]$$

Knowing that

$$\arctan x - \arctan y = \arctan \frac{x-y}{1+xy}$$

$I_1(h)$ can be also written (always with the same reservations):

$$I_1(h) = \frac{1}{\pi\sqrt{1-\xi^2}} \arctan \frac{2\xi\sqrt{1-\xi^2}}{1-2\xi^2 - h^2} \qquad [A6.12]$$

Application

Let us consider the integral

$$L = \int_0^\infty \frac{h \, dh}{(1-h^2) + 4\xi^2 h^2}$$

$$L = \frac{1}{4\xi\sqrt{1-\xi^2}} \left[\pi + \arctan \frac{2\xi\sqrt{1-\xi^2}}{2\xi^2 - 1} \right] \quad \text{if } \xi \leq \frac{1}{\sqrt{2}}$$

$$L = \frac{1}{4\xi\sqrt{1-\xi^2}} \arctan \frac{2\xi\sqrt{1-\xi^2}}{2\xi^2 - 1} \quad \text{if } \xi > \frac{1}{\sqrt{2}} \qquad [A6.13]$$

$$L = \frac{\pi}{4} \quad \text{if } \xi = \frac{1}{\sqrt{2}}$$

The relations below give the expressions of $I_b(h)$ for some values of b [PUL 68]:

$$\alpha = 2\sqrt{1-\xi^2}$$

$$\beta = 2(1 - 2\xi^2)$$

$$I_{-6} = -\frac{4\xi}{\pi}\left[\frac{1}{5h^5} + \frac{\beta}{3h^3} + \frac{\beta^2 - 1}{h}\right] + \beta(\beta^2 - 2)I_0 - (\beta^2 - 1)I_2 \qquad [A6.14]$$

$$I_{-5} = -\frac{4\xi}{\pi}\left[\frac{1}{4h^4} + \frac{\beta}{2h^2} - (\beta^2 - 1)\ln(h)\right] + \beta(\beta^2 - 2)I_1 - (\beta^2 - 1)I_3 \qquad [A6.15]$$

$$I_{-4} = -\frac{4\xi}{\pi}\left[\frac{1}{3h^3} + \frac{\beta}{h}\right] + (\beta^2 - 2)I_0 - \beta I_2 \qquad [A6.16]$$

$$I_{-3} = -\frac{4\xi}{\pi}\left[\frac{1}{2h^2} - \beta \ln(h)\right] + (\beta^2 - 1)I_1 - \beta I_3 \qquad [A6.17]$$

$$I_{-2} = -\frac{4\xi}{\pi h} + \beta I_0 - I_2 \qquad [A6.18]$$

$$I_{-1} = \frac{4\xi}{\pi}\ln(h) + \beta I_1 - I_3 \qquad [A6.19]$$

$$I_0 = \frac{\xi}{\pi\alpha}\ln\left(\frac{h^2+\alpha h+1}{h^2-\alpha h+1}\right) + \frac{1}{\pi}\left[\arctan\left(\frac{2h+\alpha}{2\xi}\right) + \arctan\left(\frac{2h-\alpha}{2\xi}\right)\right] \quad [\text{A6.20}]$$

$$I_1 = \frac{2}{\pi\alpha}\left[\arctan\left(\frac{2h-\alpha}{2\xi}\right) - \arctan\left(\frac{2h+\alpha}{2\xi}\right)\right] \quad [\text{A6.21}]$$

$$I_2 = \frac{\xi}{\pi\alpha}\ln\left(\frac{h^2-\alpha h+1}{h^2+\alpha h+1}\right) + \frac{1}{\pi}\left[\arctan\left(\frac{2h+\alpha}{2\xi}\right) + \arctan\left(\frac{2h-\alpha}{2\xi}\right)\right] \quad [\text{A6.22}]$$

$$I_3 = \frac{\xi}{\pi}\ln\left[(1-h^2)^2 + 4\xi^2 h^2\right] + \frac{\beta}{\pi\alpha}\left[\arctan\left(\frac{2h-\alpha}{2\xi}\right) - \arctan\left(\frac{2h+\alpha}{2\xi}\right)\right] \quad [\text{A6.23}]$$

$$I_4 = \frac{4\xi}{\pi}h + \beta I_2 - I_0 \quad [\text{A6.24}]$$

$$I_5 = \frac{4\xi}{\pi}\frac{h^2}{2} + \beta I_3 - I_1 \quad [\text{A6.25}]$$

$$I_6 = \frac{4\xi}{\pi}\left(\frac{h^3}{3} + \beta h\right) + (\beta^2 - 1)I_2 - \beta I_0 \quad [\text{A6.26}]$$

$$I_7 = \frac{4\xi}{\pi}\left(\frac{h^4}{4} + \frac{\beta h^2}{2}\right) + (\beta^2 - 1)I_3 - \beta I_1 \quad [\text{A6.27}]$$

$$I_8 = \frac{4\xi}{\pi}\left(\frac{h^5}{5} + \frac{\beta h^3}{3} + (\beta^2 - 1)h\right) + \beta(\beta^2 - 2)I_2 - (\beta^2 - 1)I_0 \quad [\text{A6.28}]$$

$$I_9 = \frac{4\xi}{\pi}\left(\frac{h^6}{6} + \frac{\beta h^4}{4} + (\beta^2 - 1)\frac{h^2}{2}\right) + \beta(\beta^2 - 2)I_3 - (\beta^2 - 1)I_1 \quad [\text{A6.29}]$$

Table A6.1. *Integrals $I_b(h)$ for b ranging between –6 and 9*

A6.1.4. Approximations

When the resonance frequency is within the frequency range of the excitation and when the influence of damping is small, it is possible to neglect the term in ξ and to write I_b in the form [MIL 61]:

$$I_b \approx \frac{4\xi}{\pi} \int \frac{h^b \, dh}{\left(1-h^2\right)^2} \qquad [A6.30]$$

provided that $4\xi^2 h^2 \ll \left(1-h^2\right)^2$. The condition is indeed written

$$2\xi h < 1 - h^2$$

or

$$2\xi h < h^2 - 1$$

$$\begin{cases} h^2 + 2\xi h - 1 < 0 \\ h^2 - 2\xi h - 1 > 0 \end{cases}$$

$$h = \frac{-2\xi \pm \sqrt{4\xi^2 + 4}}{2} = -\xi \pm \sqrt{\xi^2 + 1}$$

$$h = \frac{2\xi \pm \sqrt{4\xi^2 + 4}}{2} = \xi \pm \sqrt{\xi^2 + 1}$$

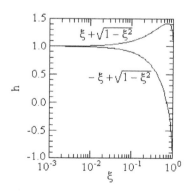

Figure A6.1. *Validity domain of approximation [A6.30]*

yielding

$$-\xi - \sqrt{1+\xi^2} \leq h \leq -\xi + \sqrt{1+\xi^2}$$

$$\xi + \sqrt{1+\xi^2} \leq h \leq \xi - \sqrt{1+\xi^2}$$

Knowing that h is always positive or zero and that ξ varies between 0 and 1, these conditions become

$$h \leq -\xi + \sqrt{1+\xi^2}$$

$$h \geq \xi + \sqrt{1+\xi^2}$$

For $\xi = 0.05$, these conditions lead for example to:

$$\begin{cases} h < 0.9512 \\ h > 1.0512 \end{cases}$$

The smaller the damping, the larger the useful range.

If $\xi = 0.2$

$$\begin{cases} h < 0.82 \\ h > 1.22 \end{cases}$$

We increase, for safety, these values by approximately 40%:

$$h \leq 1.4\left(-\xi + \sqrt{1+\xi^2}\right)$$

$$h \geq 1.4\left(\xi + \sqrt{1+\xi^2}\right)$$

For h small and less than 1:

$$\frac{h^b}{\left(1-h^2\right)^2} \approx h^b \left(1 + 2h^2 + 3h^4 + 4h^6 + \cdots\right) \qquad [A6.31]$$

For h large and higher than 1:

$$\frac{h^b}{\left(1-h^2\right)^2} \approx \frac{h^b}{h^4}\left(1 - \frac{2}{h^2} + \frac{3}{h^4} - \frac{4}{h^6} + \cdots\right) \qquad [A6.32]$$

While replacing $\dfrac{h^b}{\left(1-h^2\right)^2}$ with its expression in I_b and while integrating, it results that

$$I_b \approx \frac{4\xi}{\pi}\left(\frac{h^{b+1}}{b+1} + \frac{2\,h^{b+3}}{b+3} + \frac{3\,h^{b+5}}{b+5} + \cdots\right) \qquad \text{for } h \leq 1$$

$$I_b \approx \frac{4\xi}{\pi}\left(\frac{h^{b-3}}{b-3} - \frac{2\,h^{b-5}}{b-5} + \frac{3\,h^{b-7}}{b-7} - \cdots\right) \qquad \text{for } h \geq 1 \qquad [A6.33]$$

as long as no denominator is zero. If one of them is zero, the corresponding term is to be replaced by a logarithm. Thus, if $b \pm m = 0$, the term

$$\frac{\dfrac{m+1}{2}\,h^{b \pm m}}{b \pm m}$$

must be written

$$\frac{m+1}{2}\ln h$$

A6.1.5. *Particular cases*

$$\int_0^\infty H^2\,h^2\,d\Omega = \frac{\pi}{2}Q\,\omega_0 \qquad [A6.34]$$

$$\int_0^\infty H^2\,d\Omega = \frac{\pi}{2}Q\,\omega_0 \qquad [A6.35]$$

where $h = \dfrac{\Omega}{\omega_0}$ and $H^2 = \dfrac{1}{(1-h)^2 + \dfrac{h^2}{Q^2}}$.

These results can be deduced directly from [A6.22] and [A6.20] [WAT 62].

NOTE.–

More generally, if $b = 0$ [DWI 66]:

$$\int_0^\infty \frac{dx}{a x^4 + 2 b x^2 + c} = \frac{\pi}{2\sqrt{c k}} \qquad [A6.36]$$

where $k = 2 (b+\sqrt{a c})$ and a, c and $k > 0$.

A6.2. Integral $J_b(h) = \displaystyle\int \dfrac{h^b}{(1-h^2)^2} dh$ \hfill [A6.37]

From the above relations, we deduce

$$J_0(h) = -\frac{1}{4} \ln \frac{h-1}{h+1} + \frac{h}{2(1-h^2)} \qquad \text{if } h > 1 \qquad [A6.38]$$

$$J_0(h) = \frac{1}{4} \ln \frac{1+h}{1-h} + \frac{h}{2(1-h^2)} \qquad \text{if } h < 1 \qquad [A6.39]$$

$$J_2(h) = \frac{1}{4} \ln \frac{h-1}{h+1} + \frac{h}{2(1-h^2)} \qquad \text{if } h > 1 \qquad [A6.40]$$

$$J_2(h) = -\frac{1}{4} \ln \frac{1+h}{1-h} + \frac{h}{2(1-h^2)} \qquad \text{if } h < 1 \qquad [A6.41]$$

The expressions of $J_b(h)$ for the other values of b can be calculated starting from these relations using the recurrence formula:

$$\left. \begin{array}{l} J_b(h) = \dfrac{h^{b-3}}{b-3} + 2 J_{b-2} - J_{b-4} \qquad b \neq 3 \\ J_3(h) = \ln h + 2 J_1 - J_{-1} \qquad b = 3 \end{array} \right\} \qquad [A6.42]$$

Bibliography

[ABR 70] ABRAMOWITZ M., STEGUN I., *Handbook of Mathematical Functions*, Dover Publications, Inc., New York, 1970.

[AIT 81] AITCHISON J., BROWN J.A.C., *The Log-normal Distribution with Special Reference to its Uses in Economics*, Cambridge University Press, Monograph 5, 1981.

[ALL 82] ALLIOUD J.P., "Prévision des niveaux vibratoires au moyen de l'analyse statistique dynamique – Application à la maintenance prévisionnelle", *Mécanique-Matériaux-Electricité*, n° 386/387, 2–3, 1982, p. 84–95.

[AND 11] ANDREWS F.J., "Random Vibration – An Overview", http://www.cmsantivibration.co.uk/vibrationpapers_pdf/RandomVibrationOverview.pdf, 2011.

[ANG 61] ANGOT A., "Compléments de mathématiques à l'usage des ingénieurs de l'électrotechnique et des télécommunications", *Editions de la Revue d'Optique*, 4th edition, Collection Scientifique et Technique du CNET, 1961.

[ANT 06] ANTONIOU A., "Digital Signal Processing – Signals, Systems and Filters", Mc Graw-Hill, 2006

[ASP 63] ASPINWALL D.M., "An approximate distribution for maximum response during random vibration", *AIAA Simulation for Aerospace Flight Conference*, Columbus, Ohio, 26–28, , p. 326–330, August 1963.

[BAN 78] BANG B., PETERSEN B.B., MORUP E., *Random Vibration Introduction*, Danish Research Centre for Applied Electronics, Elektronikcentralen, Horsholm, Denmark, September 1978.

[BAR 55] BARTLETT M.S., MEDHI J., "On the efficiency of procedures for smoothing periodograms from time series with continuous spectra", *Biometrika*, Vol. 42, p. 143–150, 1955.

[BAR 61] BARNOSKI R.L., "An investigation of the maximum response of a single degree of freedom system subjected to stationary random noise", *Computer Engineering Associates*, S. 182–5, December 1961.

[BAR 65] BARNOSKI R.L., *The Maximum Response of a Linear Mechanical Oscillator to Stationary and Non-stationary Random Excitation*, NASA-CR-340, December 1965.

[BAR 68a] BARNOSKI R.L., "The maximum response to random excitation of distributed structures with rectangular geometry", *J. Sound Vib.*, Vol. 7, n° 3, p. 333–350, 1968.

[BAR 68b] BARNOSKI R.L., "On the single highest peak response of a dual oscillator to random excitation", *Transactions of the ASME, Journal of Applied Mechanics*, p. 414–416, June 1968,.

[BAR 69] BARNOSKI R.L., MAURER J.R., "Mean-square response of simple mechanical systems to non-stationary random excitation", *Journal of Applied Mechanics*, 36, 1969, p. 221–227.

[BAR 71] BARNOSKI R.L., MAURER J.R., Mean square exceedance characteristics of a single tuned system to amplitude modulated random noise, NASA CR-61352, April 1971.

[BAR 78] BARNOUIN B., THEBAULT J., "Problèmes posés par l'estimation des distributions des contraintes pour le calcul en fatigue des structures tubulaires offshore", *Annales de l'Institut Technique de Bâtiment et des Travaux Publics*, n° 365, Théories et Méthodes de Calcul séries, no. 221, p. 54–72, October 1978,.

[BRA 11] BRANDT A., *Noise and Vibration analysis – Signal Analysis and Experimental Procedures*, Wiley, 2011.

[BEA 72] BEAUCHAMP K.G., PITTEM S.E., WILLIAMSON M.E., "Analysing vibration and shock data – Part II: processing and presentation of results", *Journal of the Society of Environmental Engineers*, Vol. 11-4, Issue 55, p. 26–32, December 1972,.

[BEN 58] BENDAT J.S., *Principles and Applications of Random Noise Theory*, John Wiley & Sons, New York, 1958.

[BEN 61a] BENDAT J.S., "The establishment of test levels from field data – panel session", *The Shock and Vibration Bulletin*, no. 29, Part IV, p. 359–376, 1961.

[BEN 61b] BENDAT J.S., ENOCHSON L.D., KLEIN G.H., PIERSOL A.G., The application of statistics to the flight vehicle vibration problem, ASD Technical Report 61-123, December 1961.

[BEN 62] BENDAT J.S., ENOCHSON L.D., KLEIN G.H., PIERSOL A.G., Advanced concepts of stochastic processes and statistics for flight vehicle vibration estimation and measurement, ASD-TDR-62-973, December 1962.

[BEN 63] BENDAT J.S., ENOCHSON L.D., PIERSOL A.G., Analytical Study of Vibration Data Reduction Methods, The Technical Products Company, Contract NAS8-5093, September 1963.

[BEN 64] BENDAT J.S., Probability Functions for Random Responses: Prediction of Peaks, Fatigue Damage and Catastrophic Failures, NASA CR-33, April 1964.

[BEN 65] BENDAT J.S., ENOCHSON L.D., PIERSOL A.G., Tests for Randomness, Stationarity, Normality and Comparison of Spectra, Air Force Flight Dynamics Laboratory Technical Report, AD0621906, August 1965.

[BEN 71] BENDAT J.S., PIERSOL A.G., *Random Data: Analysis and Measurement Procedures*, Wiley–Interscience, 1971.

[BEN 75] BENNET F.E., "Spectral measurements made from finite duration observations", *Journal of the Society of Environmental Engineers*, vol. 14–1, no 64, p. 11–16, March 1975.

[BEN 78] BENDAT J.S., "Statistical errors in measurement of coherence functions and input/output quantities", *Journal of Sound and Vibration*, 59, 3, p. 405/421, 1978.

[BEN 80] BENDAT J.S., PIERSOL A.G., *Engineering Applications of Correlation and Spectral Analysis*, John Wiley & Sons, New York, 1980.

[BEN 04] BENASCIUTTI D., Fatigue analysis of random loadings, PhD Thesis, University of Ferrera, Italy, 2004.

[BEN 10] BENDAT J., PIERSOL A.G., *Random Data: Analysis and Measurement Procedures*, 4th edn. Wiley Interscience, 2010.

[BER 77] BERNSTEIN M., "Single mode stress required for equivalent damage to multimode stress in sonic feature", *AIAA Dynamics Specialist Conference*, San Diego, California, p. 191–197, 24–25 March 1977.

[BIN 67] BINGHAM C., GODFREY M.D., TUKEY J.W., "Modern techniques of power spectrum estimation", *IEEE Trans. Audio and Electroacoustics*, Vol. AU-15, no. 2, p. 56–66, 1967.

[BLA 58] BLACKMAN R.B., TUKEY J.W., *The Measurement of Power Spectra*, Dover Publications, Inc., New York, 1958, or *Bell System Technical Journal*, Vol. XXXVII, January and March 1958.

[BLA 91] BLANCHET G., PRADO J., *Méthodes numériques pour le traitement du signal*, Masson, 1991.

[BOD 72] BODMER R., "Qu'est ce que la corrélation?", *EMI*, 155/15-4, p. 45–48, 1972.

[BOL 84] BOLOTIN V.U., *Random vibrations of elastic systems*, Martinus Nishoff Publishers, The Hague, Vol. 8, 1984.

[BOO 56] BOOTH G.B., "Random motion", *Product Engineering*, p. 169–176, November 1956.

[BRA 68] BRADLEY J.V., *Distribution-Free Statistical Tests*, Englewood Cliffs, N.J., Prentice-Hall, Inc., 1968.

[BRO 63] BROCH J.T., "Effects of spectrum non-linearities upon the peak distribution of random signals", *Brüel and Kjaer Technical Review*, no. 3, 1963.

[BRO 70] BROCH J.T., *Peak-distribution effects in random-load fatigue*, ASTM-STP 462, p. 105/126, 1970

[BRO 84] BROWNLEE K., *Statistical Theory and Methodology*, Krieger Publishing Company, 1984.

[CAL 69] CALOT G., *Cours de statistique descriptive*, Dunod, 1969.

[CAR 56] CARTWRIGHT D.E., LONGUET-HIGGINS M.S., "Statistical distribution of the maxima of a random function", *Proc. Roy. Soc. A.*, vol. 237, p. 212–232, 1956.

[CAR 68] CARY H., "Facts from figures – the values of amplitude distribution analysis", *IES Proceedings*, p. 49–54, May 1968.

[CAR 73] CARTER G.C., KNAPP C.H., NUTTALL A.H., "Estimation of the magnitude-squared coherence function via overlapped fast Fourier transform processing", *IEEE Trans. on Audio and Electroacoustics*, vol. AU-21, no. 4, p. 337–344, August 1973.

[CAR 80] CARTER G.C., NUTTALL A.H., "On the weighted overlapped segment averaging method for power spectral estimation", *Proceedings of the IEEE*, vol. 68, no. 10, p. 1352–1354, October 1980.

[CAU 61] CAUGHEY T.K., STUMPF H.J., "Transient response of a dynamic system under random excitation", *Transactions of the ASME Journal of Applied Mechanics*, 28, p. 563–566, December 1961.

[CAU 63] CAUGHEY T.K., GRAY A.H., "Discussion of 'Distribution of structural response to earthquakes'", *Journal of Engineering Mechanics Division, ASCE*, vol. 89, no. EM2, p. 159, 1963.

[CHA 72] CHAKRAVORTY M., Transient spectral analysis of linear elastic structures and equipment under random excitation, Department of Civil Engineering, *MIT Research Report* no. R72-18, Structures Publication, no. 340, April 1972.

[CHA 85] CHAUDHURY G.K., DOVER W.D., "Fatigue analysis of offshore platforms subject to sea wave loadings", *International Journal of Fatigue*, vol. 7, no 1, p. 13–19, January 1985.

[CLA 65] CLARKSON B.L., MERCER C.A., "Use of cross-correlation in studying the response of lightly damped structures to random forces", *AIAA Journal*, vol. 12, p. 2287–2291, 1965.

[CLO 03] CLOUGH R.W., PENZIEN J., *Dynamics of Structures*, Third Edition, Computers & Structures, Inc., Berkeley, CA, 2003.

[CLY 64] McCLYMONDS J.C., GANOUNG J.K., "Combined analytical and experimental approach for designing and evaluating structural systems for vibration environments", *The Shock and Vibration Bulletin*, no. 34, Part 2, p. 159–175, December 1964.

[COL 59] COLEMAN J.J., "Reliability of aircraft structures in resisting chance failure", *Journal Operations Research*, vol. 7, no 7, p. 639–645, September/October 1959.

[CON 95] McCONNEL K.G., *Vibration Testing: Theory and Practice*, John Wiley & Sons, New York, 1995.

[COO 65] COOLEY J.W., TUKEY J.W., "An algorithm for the machine calculation of complex Fourier series", *Mathematics of Computation*, vol. 19, no. 90, p. 297–301, April 1965.

[COU 66] COUPRY G., "Une nouvelle méthode d'analyse des phénomènes aléatoires", ONERA, TP no. 330, 1966, *Communication présentée au IVème Symposium International sur l'Instrumentation Aérospatiale*, Cranfield, 21–24 March 1966.

[COU 70] COUPRY G., "Distribution statistique des maximums absolus des processus localement stationnaires et gaussiens à large bande", *La Recherche Aérospatiale*, no. 1, p. 43–48, January/February 1970.

[CRA 58] CRANDALL S.H., *Random vibration*, Technology Press of MIT and John Wiley & Sons, New York, 1958.

[CRA 61] CRANDALL S.H., "Random vibration of a nonlinear system with a set-up spring", *ASME Journal of Applied Mechanics*, Paper no. 61-WA-152, p. 1–6, 1961.

[CRA 63] CRANDALL S.H., MARK W.D., *Random Vibration in Mechanical Systems*, Academic Press, 1963.

[CRA 66a] CRANDALL S.H., CHANDIRAMANI K.L., COOK R.G., "Some first-passage problems in random vibration", *Transactions of the ASME, Journal of Applied* Mechanics, p. 532–538, September 1966,.

[CRA 66b] CRAMER H., "On the intersections between the trajectories of a normal stationary stochastic process and a high level", *Arkiv. Mat.*, 6, p. 337, 1966.

[CRA 67] CRAMER H., LEADBETTER M.R., *Stationary and Related Stochastic Processes*, John Wiley & Sons, New York, 1967.

[CRA 68] CRANDALL S.H., Distribution of maxima in the response of an oscillator to random excitation, MIT Acoust. and Vib. Lab. Rept no. DSR 78867-3, 1968.

[CRA 70] CRANDALL S.H., "First-crossing probabilities of the linear oscillator", *Journal Sound Vib.*, vol. 12, no. 3, 1970, p. 285–299.

[CRA 83] CRANDALL S.H., ZHU W.Q., "Random vibration: a survey of recent developments", *Transactions of the ASME Journal of Applied Mechanics*, vol. 50, no. 4b, p. 953–962, December 1983.

[CRA 92] CRAWFORD A.R., CRAWFORD S., *The simplified handbook pf vibration analysis, Volume 2: Applied Vibration Analysis*, Computational Systems, Inc., Knoxville, TN, 1992.

[CRE 56] CREDE C.E., LUNNEY E.J., Establishment of vibration and shock tests for missile electronics as derived from the measured environment, WADC Technical Report 56-503, ASTIA Document no. AD118133, 1 December 1956.

[CRE 56a] CREDE C.E., "Concepts and trends in simulation", *The Shock and Vibration Bulletin*, 23, p. 1–8, June 1956.

[CUR 64] CURTIS A.J., "Random and complex vibration theory", *IES Proceedings*, p. 35–52, 1964.

[CUR 71] CURTIS A.J., TINLING N.G., ABSTEIN H.T., *Selection and Performance of Vibration Tests*, The Shock and Vibration Information Center, USDD-SVM8, 1971.

[CUR 87] CURTIS A.J., "Applications of digital computers", *Shock and Vibration Handbook*, 27, 3rd edition, C.M. Harris ed., McGraw-Hill, 1987.

[CUR 88] CURTIS A.J., "Concepts in vibration data analysis", *Shock and Vibration Handbook*, p. 22-1–22-28, C.M. Harris ed., McGraw Hill, 1988.

[DAH 74] DAHLQUIST G., BJÖRCK A., *Numerical Methods*, Prentice-Hall, Automatic Computation Series, 1974.

[DAR 72] DART M., Traitement des signaux aléatoires, INSA Lyon, CAST, Stage Etude des Vibrations Aléatoires, 1972.

[DAS 89] DA SILVA PASSOS A., *Méthodes mathématiques du traitement numérique du signal*, Eyrolles, 1989.

[DAV 58] DAVENPORT W.B., ROOT W.L., *Random Signals and Noise*, McGraw-Hill, 1958.

[DAV 64] DAVENPORT A.G., "Note on the distribution of the largest value of a random function with application to gust loading", *Proceedings of Institute of Civil Engineers*, Vol. 28, p. 187–196, 1964.

[DEE 71] DE ESTRADA M.B., Average response spectra for structures excited by filtered white noise, Thesis, Department of Civil Engineering, Massachusetts Institute of Technology, June 1971.

[DER 79] DER KIUREGHIAN A., On response of structures to stationary excitation, Earthquake Engineering Research Center Report no. UCB/EERC 79/32, College of Engineering, University of California, Berkeley, December 1979.

[DER 80] DER KIUREGHIAN A., A response spectrum method for random vibrations, Earthquake Engineering Research Center, Report no. UCB/EERC-80/15, June 1980.

[DEV 62] DE VRIENDT A.B., *La transmission de la chaleur*, Editions Gaëtan Morin, Volumes 1 and 2, 1962.

[DIE 56] DIEDERICH F.W., "The dynamic response of a large airplane to continuous random atmospheric disturbances", *Journal of Aeronautical Sciences*, vol. 23, no.10, pp. 917/930, October 1956.

[DIT 71] DITLEVSEN O., Extremes and first passage times with applications in civil engineering, Doctoral Thesis, Technical University of Denmark, Copenhagen, 23 June 1971.

[DUB 59] DUBOIS W., "Random vibration testing", *Shock, Vibration and Associated Environments Bulletin*, n° 27, Part 2, p. 103–112, June 1959.

[DUR 72] DURRANI T.S., NIGHTINGALE J.M., "Data windows for digital spectral analysis", *Proceedings IEE*, vol. 119, no. 3, p. 342–352, March 1972.

[DWI 66] DWIGHT H.B., *Tables of integrals and others mathematical data*, fourth edition, Macmillan, 1966.

[EIN 05] EINSTEIN, A., "On the Movement of Small Particles Suspended in a Stationary Liquid Demanded by the Molecular Kinetic Theory of Heat," *Annalen der Physik*, V. 17, p. 549. 1905.

[ELD 61] ELDRED K., ROBERTS W.M., WHITE R., Structural vibrations in space vehicles, WADD Technical Report 61-62, December 1961.

[ENO 69] ENOCHSON L.D., "Modern aspects of data processing applied to random vibrations and acoustics", *Proceedings of the Conference on Applications and Methods of Random Data Analysis*, ISCR, University of Southampton, July 1969.

[FAU 69] FAUQUE J.M., BERTHIER D., MAX J., "Analyse spectrale par corrélation", *Journées de l'Automatique dans la Recherche et l'Industrie Nucléaires*, Grenoble, 24–27 June 1969.

[FER 99] FERREE T., Calculating the Power Spectral Density (PSD) in IDL, Electrial Geodesics, Inc., Technical Note, August 23, 1999.

[FIS 28] FISHER R.A., TIPPETT L.H.C., "Limiting forms of the frequency distribution of the largest or smallest member of a sample", *Proceedings of Camb. Phil. Soc.*, 24, 180, 1928.

[FOL 69] FOLEY J.T., "Normal and abnormal dynamic environments encountered in truck transportation", *The Shock and Vibration Bulletin*, n° 40, Part III, p. 31-45, December 1969.

[FOL 72b] FOLEY J.T., Transportation shock and vibration descriptions for package designers. Sandia Laboratories, Albuquerque, SC M 72 0076, July 1972.

[FOR 64] FORLIFER W.R., "The effects of filter bandwith in spectrum analysis of random vibration", *The Shock and Vibration Bulletin*, no. 33, Part II, p. 273–278, 1964.

[FUL 61] FULLER J.R., "Cumulative fatigue damage due to variable-cycle loading", *Noise Control*, July/August 1961, or *The Shock and Vibration Bulletin*, no. 29, Part 4, p. 253–273, 1961.

[FUL 62] FULLER J.R., Research on techniques of establishing random type fatigue curves for broad band sonic loading, ASTIA, The Boeing Co., Final Report No. ASD-TDR-62-501, October 1962, or National Aeronautical Meeting, SAE Paper 671-C, Washington DC, 8–11 April, 1963.

[FUN 53] FUNG Y.C., "Statistical aspects of dynamic loads", *Journal of Aeronautical Sciences*, vol. 20, no. 5, pp. 317–330, May 1953.

[FUN 55] FUNG Y.C., "Tha Analysis of Dynamic Stresses in Aurcraft Structures During Landing as Nonstationary Random Processes", *Journal of Aeronautical Sciences*, vol. 22, no. 4, pp. 449–457, December 1955.

[GAD 87] GADE S., HERLUFSEN H., Use of weighting functions in DFT/FFT analysis, Technical Review, Brüel & Kjaer, Windows to FFT Analysis no. 3 (Part I) and no. 4 (Part II), 1987.

[GAL 57] McGALLEY R.B., Jr., "The evaluation of random-noise integrals", *The Shock and Vibration Bulletin*, Part II, no. 25, p. 243–252, 1957.

[GIL 88] GILBERT R.J., *Vibrations des structures. Interactions avec les fluides. Sources d'excitation aléatoires*, Editions Eyrolles, 1988.

[GIR 04] GIRDHAR P., *Practical Machinery Vibration Analysis and Predictive Maintenance*, edited by SCHEFFER C., Elsevier, 2004.

[GMU 68] GMURMAN V.E., *Fundamentals of Probability Theory and Mathematical Statistics*, ILIFFE Books LTD, London, 1968.

[GOL 53] GOLDMAN S., *Information Theory*, Prentice Hall Inc., New York, 1953.

[GOO 57] GOODMAN N.R., "On the joint estimation of the spectra, cospectrum and quadrature spectrum of a two dimensional stationary Gaussian process", *Scientific Paper*, no. 10, Engineering Statistics Laboratory, New York University, 1957.

[GRA 62] GRABOWSKI S.J., "Vibration loads on a wheeled vehicle", *The Shock and Vibration Bulletin*, no. 30, Part III, p. 45-55, 1962.

[GRA 65] GRAY A.H., Jr., "First-passage time in a random vibrational system", *Journal of Applied Mechanics, Transactions of the ASME*, Paper no. 65, WA/APM-18, p. 1–5, 1965.

[GRA 66] GRAY C.L., "First occurence probabilities for extreme random vibration amplitudes", *The Shock and Vibration Bulletin*, 35, Part IV, p. 99–104, February 1966.

[GRE 61] McGREGOR H.N. et al., "Acoustic problems associated with underground launching of a large missile", *Shock, Vibration and Associated Environments Bulletin*, no. 29, Part IV, p. 317–335, June 1961.

[GRE 81] GREGOIRE R., La fatigue sous charge programmée, CETIM, Technical Note, no. 20, May 1981.

[GUE 80] GUESDON M., GARCIN G., BOCQUET P., "Surveillance des machines par analyse des vibrations", *Mécanique-Matériaux-Electricité*, no. 371–372, p. 395–403, November/December 1980.

[GUM 54] GUMBEL E.J., "Statistical theory of extreme values and some practical applications", *National Bureau of Standards Applied Stress Math*, Series 33, February 1954.

[HAG 62] HAGER R.W., PARTINGTON R.L., LESTIKOW R.J., "Rail transport dynamic environment", *The Shock and Vibration Bulletin*, no. 30, Part III, p. 16–26, 1962.

[HAM 68] HAMMOND J.K., "On the response of single and multidegree of freedom systems to nonstationary random excitations", *Journal of Sound and Vibration* 7, p. 393–416, 1968.

[HAR 78] HARRIS F.J., "On the use of windows for harmonic analysis with the discrete Fourier transform", *Proceedings IEEE*, vol. 66, no. 1, p. 51–83, January 1978.

[HAS 55] HASTINGS C., Jr, *Approximations for Digital Computers*, Princeton University Press, 1955.

[HEA 56] HEAD A.K. and HOOKE F.H., "Random noise fatigue testing", *International Conference on Fatigue of Metals*, Session 3, Paper 9, London, September 1956.

[HER 84a] HERLUFSEN H., Dual channel FFT analysis, Part I, Brüel & Kjaer Technical Review, no.1, 1984.

[HER 84b] HERLUFSEN H., Dual channel FFT analysis, Part II, Brüel & Kjaer Technical Review, no. 2, 1984.

[HIL 70] HILLBERRY B.M., "Fatigue life of 2024-T3 aluminum alloy under narrow- and broad-band random loading", *ASTM.STP*, 462, p. 167–183, 1970.

[HIM 59] HIMELBLAU H., "Random vibration nomograph", *Noise Control*, vol. 5, no. 3, p. 49–50, July 1959.

[HIM 64] HIMELBLAU H., "Graphical method of calculating R.M.S. values for shaped random vibration spectra", *The Shock and Vibration Bulletin*, no. 34, Part II, p. 225–237, 1964.

[HIM 06] HIMELBLAU H., PIERSOL A.G., WISE J.H., GRUNDVIG M.R., Handbook for Dynamic Data Acquisition and Analysis, Institute of Environmental Sciences, IEST-RD-DTE012, 2006.

[HOG 97] HOGG R. V., TANIS E. A., *Probability and statistical inferences*, Saddle River, NJ, Prentice Hall 1997.

[HOR 75] HORBETTE B., Etablissement des spécifications d'essais, Stage ADERA, Théorie des Vibrations et des Chocs, 1975.

[HOW 02] HOWARD R.M., *Principles of random signal analysis and low noise design - the power spectral density and its applications*, John Wiley and Sons, New York, 2002.

[HUS 56] HUSTON W.B., SKOPINSKI T.H., Probability and frequency characteristics of some flight buffet loads, NACA – TN 3733, August 1956.

[ISO 94] ISO 1683, "Acoustics – preferred reference quantities for acoustic levels", first edition, *BS EN 21683. CEN EN 21683*, 1994.

[JAM 47] JAMES H.M., NICHOLS N.B., PHILLIPS R.S., "Theory of Servomechanisms", *MIT Radiation Laboratory Series*, vol. 25, McGraw-Hill, 1947.

[JEN 68] JENKINS G.M., WATTS D.G., *Spectral Analysis and its Applications*, Holden-Day, 1968.

[KAC 76] KACENA W.J. and JONES P.J., "Fatigue prediction for structures subjected to random vibration", *The Shock and Vibration Bulletin*, no. 46, Part III, p. 87–96, August 1976.

[KAT 65] KATZ H., WAYMON G.R., "Utilizing in-flight vibration data to specify design and test criteria for equipment mounted in jet aircraft", *The Shock and Vibration Bulletin*, no. 34, Part 4, p. 137–146, February 1965.

[KAY 81] KAY S.M., MARPLE S.L., "Spectrum analysis – a modern perspective", *Proceedings of the IEEE*, vol. 69, no. 11, p. 1380–1419, November 1981.

[KAZ 70] KAZMIERCZAK F.F., "Objective criteria for comparison of random vibration environments", *The Shock and Vibration Bulletin*, no. 41, Part 3, p. 43–54, December 1970.

[KEL 61] KELLY R.D., "A method for the analysis of short-duration non-stationary random vibration", *Shock, Vibration and Associated Environments Bulletin*, no. 29, Part IV, p. 126–137, June 1961.

[KEL 67] KELLER A.C., "Considerations in the analysis of arbitrary waveforms", *Proceedings IES*, p. 659–671, 1967.

[KEN 61] KENDALL M.G., STUART A., *The Advanced Theory of Statistics, Vol. 2: Inference and Relationship*, Hafner Publishing Company, NY, 1961.

[KNU 98] KNUTH D.E., *The Art of Computer Programming*, Addison-Wesley, 3rd edition, vol. 2, 1998.

[KOR 66] KORN G.A., *Random-process Simulation and Measurements*, McGraw-Hill, 1966.

[KOW 63] KOWALEWSKI J., *Über die Beziehungen zwischen der Lebensdauer von Bauteilen bei unregelmäßig schwankenden und bei geordneten Belastungsfolgen*, DVL-Bericht Nr 249, Porz-Wahn, September 1963.

[KOW 69] KOWALEWSKI J., *Beschreibung regerlloser Vorgänge*, Forstchr. Ber-VDI-Z, Reihe 5, Nr. 7, S7/28, VDI, Verlag GmbH, Dusseldorf, 1969.

[KOZ 64] KOZIN F., Final report on statistical models of cumulative damage, Midwest Applied Science Corp. Report, no. 64 - 17, N 6520969, December 1964.

[KRE 83] KREE P., SOIZE C., *Mécanique aléatoire*, Dunod, 1983.

[LAL 81] LALANNE C., Utilisation de l'environnement réel pour l'établissement des spécifications d'essais, Synthèse sur les méthodes classiques et nouvelles, Cours ADERA, 1981.

[LAL 82] LALANNE C., "Les vibrations sinusoïdales à fréquence balayée", *CESTA/EX no. 803*, 8–6 June 1982.

[LAL 92] LALANNE C., "Fatigue des matériaux", Stage Intespace Environnements Vibratoires Réels – Analyse et Spécifications, March 1992.

[LAL 94] LALANNE C., "Vibrations aléatoires - Dommage par fatigue subi par un système mécanique à un degré de liberté", *CESTA/DT/EX/EC*, no. 1019/94, 1994.

[LAL 95] LALANNE C., "Analyse des vibrations aléatoires", *CESTA/DQS DO 60*, 10 May 1995.

[LAM 76] LAMBERT R.G., "Analysis of fatigue under random vibration", *The Shock and Vibration Bulletin*, no. 46, Part 3, p. 55–72, 1976.

[LAM 80] LAMBERT R.G., *Computation Methods*, General Electric Company, Aerospace Electronic Systems Department, Utica, New York, March 1980.

[LEA 69] LEADBETTER M.R., "Extreme value theory and stochastic processes", *Proceedings of the International Conference on Structural Reliability*, Washington DC, April 1969.

[LEL 73] LELEUX F., Fatigue sous charge aléatoire. Analyse des sollicitations et estimations des durées de vie, CETIM, Technical Note, no. 3, September 1973.

[LEY 65] LEYBOLD H.A., Techniques for examining statistical and power-spectral properties of random time histories, NASA-TND-2714, March 1965 (MS Dissertation, VPI, May 1963).

[LIN 67] LIN Y.K., *Probabilistic Theory of Structural Dynamics*, McGraw-Hill, 1967.

[LIN 70] LIN Y.K., "First-excursion failure of randomly excited structures", *AIAA Journal*, vol. 8, no. 4, p. 720–725, 1970.

[LIN 72] LINSLEY R.C., HILLBERRY B.M., "Random fatigue of 2024-T3 aluminium under two spectra with identical peak-probability density functions", *Probabilistic Aspects of Fatigue*, ASTM STP 511, p. 156–167, 1972.

[LON 52] LONGUET-HIGGINS M.S., "On the statistical distribution of heights of sea waves", *Journal of Marine Research*, no. 3, p. 245–266, 1952.

[LUT 73] LUTES L.D., CHOKSHI, "Maximum response statistics for a linear structure", *Proceedings of the 5th World Conference on Earthquake Engineering*, vol. 8, Series B, p. 358–361, 1973.

[LYO 60] LYON R.H., "On the vibration statistics of a randomly excited hard-spring oscillator", *Journal of the Acoustical Society of America*, Part I, vol. 32, p. 716–719, 1960; Part II, vol. 33, p. 1395/1403, 1961.

[MAG 78] MAGNUSON C.F., Shock and vibration environments for a large shipping container during truck transport, Part II, Sandia Laboratories, Albuquerque, New Mexico, SAND78-0337, May 1978.

[MAR 58] MARSHALL J.T., HARMEN R.A., "A proposed method for assessing the severity of the vibration environment", *The Shock and Vibration Bulletin*, no. 26, Part 2, p. 259–277, December 1958,.

[MAR 66] MARSH K.J., MAC KINNON J.A., Fatigue under random loading, NEL Report, no. 234, July 1966.

[MAR 70] MARK W.D., "Spectral analysis of the convolution and filtering of non-stationary stochastic processes", *Journal of Sound and Vibration*, 11, p. 19–63, 1970.

[MAX 65] MAX J., Les méthodes de corrélation dans le traitement de l'information, CEA/CENG, Technical Note EL/277, May 1965.

[MAX 69] MAX J., FAUQUE J.M., BERTHIER D., "Les corrélateurs dans le traitement des mesures et l'analyse spectrale", *Journées de l'Automatique dans la Recherche et l'Industrie Nucléaires*, Grenoble, 24–27 June 1969.

[MAX 81] MAX J., *Méthodes et techniques de traitement du signal et applications aux mesures physiques*, Volume II, Masson, 1981.

[MAX 86] MAX J., DIOT M., BIGRET R., "Les analyseurs de spectre à FFT et les analyseurs de spectre à corrélation", *Traitement du Signal*, vol. 3, no. 4–5, p. 241–256, 1986.

[MAZ 54] MAZELSKY B., "Extension of power spectral methods of generalized harmonic analysis to determine non-Gaussian probability functions of random input disturbances and output responses of linear systems", *The Journal of the Aeronautical Sciences*, vol. 21, no. 3, p. 145–153, March 1954.

[MEN 86] MENDENHALL W., SCHEAFFER R.L., WACKERLY D.D., *Mathematical Statistics with Applications*, Duxbury Press, CA, 1986.

[MET 03] Metallic Materials and Elements for Aerospace Vehiche Structures, Department of Defense Handbook, MIL-HDBK-5J, 31 January 2003.

[MIL 61] MILES J.W., THOMSON W.T., "Statistical concepts in vibration", in HARRIS C.M., CREDE C.E. (eds), *Shock and Vibration Handbook*, vol. 1-11, McGraw-Hill, 1961.

[MIX 69] MIX D.F., *Random Signal Analysis*, Addison-Wesley, 1969.

[MOO 61] MOODY R.C., "The principles involved in choosing analyzer bandwidth, averaging time, scanning rate and the length of the sample to be analyzed", *The Shock and Vibration Bulletin*, no. 29, Part IV, p. 183–190, 1961.

[MOR 55] MORROW C.T., MUCHMORE R.B., "Shortcomings of present methods of measuring and simulating vibration environments", *Journal of Applied Mechanics*, p. 367–371, September 1955.

[MOR 56] MORROW C.T., "Shock and vibration environments", *Shock and Vibration Instrumentation*, *ASME*, p. 75–86, June 1956.

[MOR 58] MORROW C.T., "Averaging time and data-reduction time for random vibration spectra", *The Journal of Acoustical Society of America*; I – vol. 30, no. 5, p. 456–461, May 1958; II – vol. 30, no. 6, p. 572–578, June 1958.

[MOR 63] MORROW C.T., *Shock and Vibration Engineering*, John Wiley & Sons, New York, vol. 1, 1963.

[MOR 75] MORROW C.T., MUCHMORE R.B., "Shortcomings of present methods of measuring and simulating vibration environments", *Journal of Applied Mechanics*, Vol. 22, no. 3, p. 367–371, September 1975,.

[NAG 06] NAGULAPALLI V.K., GUPTA A., FAN S., Estimation of Fatigue Life of Aluminum Beams subjected to Random Vibration, Department of Mechanical Engineering, Northern Illinois University, 2006.

[NEA 66] MCNEAL R.H., BARNOSKI R.L., BAILIE J.A., "Response of a simple oscillator to nonstationary random noise", *J. Spacecraft*, Vol. 3, no. 3, March 1966, p. 441–443, or Computer Engineering Assoc., Report ES 182-6, March 1962, p. 441–443.

[NEU 70] NEUVILLE J.C., "Erreurs de principe sur les essais sinusoïdaux traités sur ordinateur", *CEA/CESTA/ Z-M EX – DO 449*, 25 March 1970.

[NEW 93] NEWLAND, D.E., *An Introduction to Random Vibrations, Spectral and Wavelet Analysis*, Longman, London, John Wiley & Sons, New York, 1993.

[NUT 71] NUTTALL A.H., Spectral estimation by means of overlapped fast Fourier transform processing of windowed data, NUSC Tech. Rep. 4169, New London, CT, 13 October 1971.

[NUT 76] NUTTALL A.H., Probability distribution of spectral estimates obtained via overlapped FFT processing of windowed data, Naval Underwater Systems Center, Tech. Rep. 5529, New London, CT, 3 December, 1976.

Bibliography   583

[NUT 80] NUTTAL A.H., CARTER G.C., "A generalized framework for power spectral estimation", *IEEE Transactions on Acoustics, Speech, and Signal Processing*, Vol. ASSP-28, no. 3, p. 334–335, June 1980.

[NUT 81] NUTTALL A.H., "Some windows with very good sidelobe behavior", *IEEE Trans. on Acoustics, Speech and Signal Processing*, vol. ASSP-29, no. 1, IEEE, Piscataway, NJ, p. 84–91, February 1981.

[OCH 81] OCHI M.K., "Principles of Extreme Value Statistics and their Application", *Extreme Loads Response Symposium*, Arlington, VA, p.15-30, October 19-20, 1981.

[OSG 69] OSGOOD C.C., "Analysis of random responses for calculation of fatigue damage", *The Shock and Vibration Bulletin*, no. 40, Part II, p. 1–8, December 1969.

[OSG 82] OSGOOD C.C., *Fatigue Design*, Pergamon Press, 1982.

[OST 67] OSTREM F.E., RUMERMAN M.L., Transportation and handling shock and vibration design criteria manual. NASA, George C. Marshall Space Flight Center, Final Report NAS 8-11451, April 1967.

[OST 79] OSTREM F.E., GODSHALL W.D., An assessment of the common carrier shipping environment, Forest Products Laboratory, Forest Service, US Department of Agriculture, Madison, General Technical Report FPL 22, 1979.

[PAE 11] PAEZ T.L., "Random Vibration – History and Overview", *Rotating Machinery, Structural Health Monitoring, Shock and Vibration*, Volume 5, Conference Proceedings of the Society for Experimental Mechanics Series, Volume 8, 105-127, 2011.

[PAP 65] PAPOULIS A., *Probability Random Variables and Stochastic Processes*, McGraw-Hill, 1965.

[PAR 59] PARZEN E., On models for the probability of fatigue failure of a structure, App. Math. and Stat. Lab., Stanford University, Tech. Report, no. 45, 17 April 1959.

[PAR 62] PARIS P.C., The growth of cracks due to variations in loads, PhD Dissertation, Lehigh University, Bethlehem, Pennsylvania, AD63-02629, 1962.

[PAR 64] PARIS P.C., "The fracture mechanics approach to fatigue", *Proceedings 10th Sagamore Army Mater. Res. Conf.*, Syracuse University Press, Syracuse, New York, p. 107–132, 1964.

[PER 74] PERUCHET C., VIMONT P., Résistance à la fatigue des matériaux en contraintes aléatoires, ENICA, 1974.

[PIE 63] PIERSOL Jr. W.J., MOSKOWITZ L., A proposed spectral form for fully developed wind seas based on the similarity theory of S.Z. KITAIGORODSKII, NYU Dept. Meteorology and Oceanography Geophysical Sciences, Lab. Report 63.12, October 1963, or *J. Geophysical Res.*, 69, 5181, 1964.

[PIE 64] PIERSOL A.G., The measurement and interpretation of ordinary power spectra for vibration problems, NASA-CR 90, 1964.

[PIE 70] PIERSOL A.G., MAURER J.R., Investigation on statistical techniques to select optimal test levels for spacecraft vibration tests, Digital Corporation, Report 10909801-F, October 1970.

[PIE 93] PIERSOL A.G., "Optimum resolution bandwidth for spectral analysis of stationary random vibration data", *Shock and Vibration*, John Wiley & Sons, vol. 1, no. 1, p. 33–43, 1993.

[POO 76] POOK L.P., Proposed standard load histories for fatigue testing relevant to offshore structures, National Engineering Laboratory, NEL Report, no. 624, 1976.

[POO 78] POOK L.P., "An approach to practical load histories for fatigue testing relevant to offshore structures", *Journal of the Society of Environmental Engineers*, vol. 17-1, issue 76, p. 22–35, March 1978.

[POT 78] POTTER R., LORTSCHER J., What in the World is Overlapped Processing?, UPDATE, Hewlett-Packard, Santa Clara DSA/Laser Division, no. 1, p. 4–6, September 1978.

[POU 02] POUVIL P., Bruit, Document de Cours, Ecole Nationale Supérieure de l'Electronique et de ses Applications, 2002, http://www.clubeea.org/documents/mediatheque/BRUITCH2.pdf, 2002.

[POW 58] POWELL A., "On the fatigue failure of structures due to vibrations excited by random pressure fields", *The Journal of the Acoustical Society of America*, 30, vol. 12, p. 1130–1135, December 1958.

[PRA 70] PRATT J.S., Random Accelerations: R.M.S. Value, Test Engineering, p.10, March 1970.

[PRE 54] PRESS H., MAZELSKY B., A study of the application of power-spectral methods of generalized harmonic analysis to gust loads on airplanes, National Advisory Committee for Aeronautics, NACA Report 1172, 1954.

[PRE 56a] PRESS H., TUKEY J.W., "Power spectral methods of analysis and their application to problems in airplane dynamics", *Agard Flight Test Manual*, Vol. IV, Part IV-C, June 1956, p. 1–41, or *Bell Telephone System Monograph 26006*, 1957.

[PRE 56b] PRESS H., MEADOWS M.T. and HADLOCK I., A reevaluation of data on atmospheric turbulence and airplane gust loads for application in spectral calculations, NACA Report 1272, 1956.

[PRE 90] PREUMONT A., *Vibrations aléatoires et analyse spectrale*, Presses Polytechniques et Universitaires Romandes, 1990.

[PRE 92] PRESS W. H., TEUKOLSKY S.A., VETTERLING W.T. *et al.*, *Numerical recipes in C: The art of scientific computing*, Cambridge University Press, 1992.

[PRI 67] PRIESTLEY M.B., "Power spectral analysis of nonstationary random processes", *Journal of Sound and Vibration*, 6, p. 86–97, 1967.

[PRI 04] PRIMAK S., KONTOROVICH V., LYANDRES V., *Stochastic Methods and their Applications to Communications: Stochastic Differential Equations Approach*, John Wiley & Sons, New York, 2004

[PUL 68] PULGRANO L.J., ABLAMOWITZ M., "The response of mechanical systems to bands of random excitation", *The Shock and Vibration Bulletin*, no. 39, Part III, p. 73/86, 1968.

[RAC 69] RACICOT R.L., Random vibration analysis – Application to wind loaded structures, Solid Mechanics, Structures and Mechanical Design Division, Report no. 30, Case Western Reserve University, PhD Thesis, 1969.

[RAP 69] RAPIN P., "Applications de la statistique mathématique aux problèmes vibratoires de la mécanique industrielle", *Mécanique*, no. 240, p. 30–39, December 1969.

[RAP 82] RAPIN P., "L'utilisation pratique des machines tournantes – Les comportements anormaux", *Mécanique-Matériaux-Electricité*, no. 386/387, p. 51–57, 1982.

[RAV 70] RAVISHANKAR T.J., Simulation of Random Load Fatigue in Laboratory Testing, Institute for Aerospace Studies, UTIAS Review, no. 29, University of Toronto, March 1970.

[REE 60] REED W.H., HALL A.W., BARKER L.E., Analog techniques for measuring the frequency response of linear physical systems excited by frequency sweep inputs, NASA TN D 508, 1960.

[RIC 39] RICE S.O., "The distribution of the maxima of a random curve", *Amer. J. Math.*, vol. 61, p. 409–416, April 1939.

[RIC 44] RICE S.O., "Mathematical analysis of random noise", *Bell System Technical Journal*, no. 23 (July 1944) and 24 (January 1945).

[RIC 64] RICE J.R., Theoretical prediction of some statistical characteristics of random loadings relevant to fatigue and fracture, PhD Thesis, Lehigh University, 1964.

[RIC 65] RICE J.R., BEER F.P., "On the distribution of rises and falls in a continuous random process", *Trans. of the ASME, Journal of Basic Engineering*, p. 398–404, June 1965.

[ROB 66] ROBERTS J.B., "The response of a single oscillator to band-limited white noise", *J. Sound Vib.*, Vol. 3, no. 2, p. 115–126, 1966.

[ROB 71] ROBERTS J.B., "The covariance response of linear systems to non-stationary random excitation", *Journal of Sound and Vibration*, 14, p. 385–400, 1971.

[ROB 76] ROBERTS J.B., "Probability of first passage failure for lightly damped oscillators", *Stochastics Problems in Dynamics Proceedings*, Southampton, UK, p. 15/1, 15/9, 19–23 July 1976.

[ROS 62] ROSENBLUETH E., BUSTAMENTE J., "Distribution of structural response to earthquakes", *Journal of Engineering Mechanics Division, ASCE*, vol. 88, no. EM3, p.75, 1962.

[ROT 70] ROTH P.R., *Digital Fourier Analysis*, Hewlett-Packard, 1970.

[RUB 64] RUBIN S., "Introduction to dynamics", *Proceedings IES*, p. 3/7, 1964.

[RUD 75] RUDDER F.F., PLUMBEE H.E., "Sonic fatigue design guide for military aircraft", *Technical Report AFFDL-TR-74-112*, May 1975.

[SAN 63] SANDLER I.J., "Techniques of analysis of random and combined random-sinusoidal vibration", *Shock, Vibration and Associated Environments Bulletin*, no. 31, Part 3, p. 211–224, 1963.

[SAN 66] Sandia Corporation Standard Environmental Test Methods, SC-4452 D (M), June 1966.

[SCH 63] SCHIJVE J., "The analysis of random load-time histories with relation to fatigue tests and life calculations", *Fatigue of Aircraft Structures*, Pergamon, p. 115–149, 1963.

[SCH 65] SCHOCK R.W., PAULSON W.E., "A survey of shock and vibration environments in the four major modes of transportation", *The Shock and Vibration Bulletin*, no. 35, Part 5, p. 1-19, 1965.

[SCH 06] SCHARTON T., PANKOW D., SHOLL M., "Extreme Peaks in Random Vibration Testing", *Spacecraft and Launch Vehicles, Dynamic Environments Workshop*, June 27-29, 2006.

[SHA 49] SHANNON C.E., "Communication in the presence of noise", *Proceedings of the IRE*, 37, pp. 10–21, January 1949.

[SHE 83] SHERRAT F., "Vibration and fatigue: basic life estimation methods", *Journal of the Society of Environmental Engineers*, vol. 22-4, no. 99, p. 12–17, December 1983.

[SHE 04] SHESKIN D., *Handbook of Parametric and Nonparametric Statistical Procedures*, 3rd ed., Chapman & Hall, 2004.

[SHI 70a] SHINKLE R.K., STRONG S.A., "A sine of the times, the detection of discrete components in a complex spectrum", *Proceedings IES*, p. 282–292, 1970.

[SHI 70b] SHINOZUKA M., "Random processes with evolutionary power", *Journal of the Engineering Mechanics Division, ASCE 96*, no. EM4, p. 543–545, 1970.

[SHR 95] SHREVE D.H., "Signal Processing for Effective Vibration Analysis", *IRD Mechanalysis, Inc Columbus*, Ohio, pp.1-11, November 1995 (http://www.irdbalancing.com/downloads/SIGCOND2_2.pdf).

[SIE 51] SIEGERT A.J.F., "On the first passage time probability problem", *Physical Review*, vol. 81, no. 4, p. 617, 1951.

[SIE 56] SIEGEL S., *Non parametric Statistics for the Behavioral Sciences*, McGraw-Hill Book Company, 1956.

[SJÖ 61] SJÖSTRÖM S., "On random load analysis, KTH Avh., Thesis no. 156", *Transactions of the Royal Institut of Technology*, Stockholm, Sweden, 1961.

[SKO 59] SKOOG J.A. and SETTERLUND G.G., "Space requirements for equipment items subjected to random excitation", *The Journal of the Acoustical Society of America*, vol. 31, no. 2, p. 227–232, February 1959.

[SLE 61] SLEPIAN D., "Note on the first passage time for a particular Gaussian process", *Annals of Mathematical Statistics*, vol. 32, p. 610, 1961.

[SMI 64] SMITH K.W., "A procedure for translating measured vibration environment into laboratory tests", *Shock, Vibration and Associated Environments Bulletin*, no. 33, Part III, p. 159–177, March 1964.

[SOH 84] SOHANEY R.C., NIETERS J.M., Proper use of Weighting Functions for Impact Testing, Brüel & Kjaer Technical Review, No.4, 1984.

[SPA 84] SPANN F., PATT P., "Component vibration environment predictor", *The Journal of Environmental Sciences*, vol. XXVII, no. 5, p. 19–24, September/October 1984.

[SPA 87] SPANIER J., OLDHAM K.B., *An Atlas of Functions*, Hemisphere Publishing corporation, 1987.

[SPI 74] SPIEGEL M.R., *Formules et tables de mathématiques*, McGraw-Hill, SCHAUM series, 1974.

[STE 39] STEVENS W. L., Distribution of groups in a sequence of alternatives. *Annals of Eugenics*, vol. 9, issue 1, p. 10-17, January 1939.

[STE 67] STERN J., De BARBEYRAC J., POGGI R., *Méthodes pratiques d'étude des fonctions aléatoires*, Dunod, 1967.

[SVE 80] SVETLICKIJ V.A., *Vibrations aléatoires des systèmes mécaniques*, Technique et Documentation, Paris, 1980.

[SWA 63] SWANSON S.R., An Investigation of the Fatigue of Aluminum Alloy Due to Random Loading, Institute of Aerophysics, University of Toronto, UTIA Report no. 84, AD 407071, February 1963.

[SWA 68] SWANSON S.R., "Random load fatigue testing: a state of the art survey", *ASTM Materials Research and Standards*, vol. 8, no. 4, p. 10–44, April 1968.

[SWE 43] SWED F.S., EISENHART C., Tables for testing randomness of grouping in a sequence of alternatives. *The Annals of Mathematical Statistics*, vol. 14, no. 1, p.66–87, 1943.

[SWE 04] SWEITZER K.A., BISHOP N.W.M., GENBERG V.L., "Efficient computation of spectral moments for determination of random response statistics", *International Conference on Noise and Vibration Engineering*, Katholieke Universiteit Leuven, Belgium, ISMA Publications, p.2677-2692, 20-22 September 2004

[TAY 75] TAYLOR H.R., A study of instrumentation systems for the dynamic analysis of machine tools, PhD Thesis, University of Manchester, 1975.

[THO 08] THORBY D., *Structural Dynamics and Vibration in Practice – An Engineering Handbook*, Elsevier Ltd., 2008.

[THR 64] THRALL G.P., *Extreme Values of Random Processes in Seakeeping Applications*, Measurement Analysis Corporation, MAC 307-03, AD 607912, August 1964.

[TIP 25] TIPPET L.H.C., "On the extreme individuals and the range of samples taken from a normal population", *Biometrika*, 17, p. 364–387, 1925.

[TIP 77] TIPTON R., "Noise testing – Part 1. An introduction to random vibration testing", *Test*, p. 12–15, June/July 1977.

[TOL 63] TOLEN J.A., The development of an engineering test standard covering the transportation environment of material. Development and Proof Services, Report no. DPS-1190, December 1963.

[TUS 67] TUSTIN W., "Vibration and shock tests do not duplicate service environment", *Test Engineering*, Vol. 18, no. 2, p. 18–21, August 1967.

[TUS 72] TUSTIN W., *Environmental Vibration and Shock*, Tustin Institute of Technology, 1972.

[TUS 73] TUSTIN W., "Basic considerations for simulation of vibration environments", *Experimental Mechanics*, p. 390–396, September 1973.

[TUS 79] TUSTIN W., "In wide-band random vibration, all frequencies are present", *Test*, Vol. 41, no. 2, p. 12–13, April–May 1979.

[UDW 73] UDWADIA F.E., TRIFUNAC M.D., The Fourier transform, response spectra and their relationship through the statistics of oscillator respons, Earthquake Engineering Research Laboratory, Calif. Institute of Technology, Pasadena, Calif., Report NO-EERL 73-01, April 1973.

[VAN 69] VANMARCKE E.H., First Passage and Other Failure Criteria in Narrow-band Random Vibration: A Discrete State Approach, Research Report R69-68, Department of Civil Engineering, MIT, Cambridge, Mass., October 1969.

[VAN 70] VANMARCKE E.H., "Parameters of the spectral density function: their significance in the time and frequency domains", Department of Civil Engineering, *MIT Research Report n° R70-58*, 1970.

[VAN 72] VANMARCKE E.H., "Properties of spectral moments with applications to random vibration", *Journal of Engineering Mechanics Division, ASCE*, vol. 98, p. 425–446, 1972.

[VAN 75] VANMARCKE E.H., "On the distribution of the first-passage time for normal stationary random processes", *Journal of Applied Mechanics*, vol. 42, p. 215–220, March 1975.

[VAN 79] VANMARCKE E.H., "Some recent developments in random vibration", *Applied Mechanics Reviews*, vol. 32, no. 10, p. 1197–1202, October 1979.

[VIN 72] VINH T., LABOUREAU B., Problème d'identification des corrélogrammes, INSA, CAST, Stage Etude des Vibrations Aléatoires, 1972.

[WAL 40] WALD A., WOLFOWITZ J., "On a test whether two samples are from the same population," *Ann. Math Statist.* 11, p.147–162, June 1940.

[WAL 81] WALKER A.W., "The effects of bandwidth on the accuracy of transfert function measurements of single degree of freedom system response to random excitation", *Journal of Sound and Vibration*, 74, 2, p. 251–263, 1981.

[WAN 45] WANG M.C., UHLENBECK G.E., "On the theory of the Brownian motion II", *Reviews of Modern Physics*, vol. 17, no. 2 and 3, p. 323/342, April–July 1945.

[WAT 62] WATERMAN L.T., "Random versus sinusoidal vibration damage levels", *The Shock and Vibration Bulletin*, n° 30, Part 4, p. 128/139, 1962.

[WEL 67] WELCH P.D., "The use of Fast Fourier Transform for the estimation of power spectra: a method based on time averaging over short, modified periodograms", *IEEE Trans. on Audio and Electroacoustics*, vol. AU-15, no. 2, p. 70–73, June 1967.

[WEL 70] WELLSTEAD P.E., Aspects of real time digital spectral analysis, PhD Thesis, University of Warwick, 1970.

[WHI 69] WHITE R.G., "Use of transient excitation in the dynamic analysis of structures", *The Aeronautical Journal of the Royal Aeronautical Society*, vol. 73, p. 1047–1050, December 1969.

[WHI 72] WHITE R.G., "Spectrally shaped transient forcing functions for frequency response testing", *Journal of Sound and Vibration*, vol. 23, no. 3, 1972, p. 307–318.

[WHI 82] WHITE R.G., PINNINGTON R.J., "Practical application of the rapid frequency sweep technique for structural frequency response measurement", *Aeronautical Journal*, Paper no. 964, May 1982, pp 179–199.

[WIE 30] WIENER N., "Generalized harmonic analysis", *Acta Mathematica*, 55, p. 117–258, 1930.

[WIJ 09] WIJKER J., "Random Vibrations in Spacecraft Structures Design – Theory and Applications", *Solid Mechanics and its Applications*, volume 165, Springer 2009.

[WIR 73] WIRSCHING P.H., HAUGEN E.B., "Probabilistic design for random fatigue loads", *Journal of the Engineering Mechanics Division, ASCE*, 99, no. EM 6, December 1973.

[WIR 81] WIRSCHING P.H., The application of probability design theory to high temperature low cycle fatigue, NASA – CR 165488, N82-14531/9, November 1981.

[WIR 83] WIRSCHING P.H., Probability-based fatigue design criteria for offshore structures, Final Project Report API-PRAC 81-15, The American Petroleum Institute, Dallas, Texas, January 1983.

[YAM 87] YAMAYEE Z.A., KAZIBWE W.E., "Probabilistic modelling of distribution harmonics: data collection and analysis", *Electrical Power & Energy Systems*, vol.9, no. 3, p. 189-192, July 1987.

[YAN 71] YANG J.N., SHINOZUKA M., "On the first excursion probability in stationary narrow-band random vibration", *Journal of Applied Mechanics*, Vol. 38, n0. 4 (Trans. ASME, vol. 93, Series E), December 1971, p. 1017–1022.

[YAN 72] YANG J.N., "Simulation of random envelope processes", *Journal of Sound and Vibrations*, vol. 21, p. 73–85, 1972.

Index

A

acceleration spectral density, 89
acoustic noise, 533
aliasing, 153, 228
amplitude spectral density, 89
analysis time, 216, 219
asymptotic law, 368
autocorrelation, 33, 249, 260
  - coefficient, 188
  - function, 98, 99, 187, 372
autocovariance, 35
average
  - amplitude jump, 339
  - correlation between two successive maxima, 308
  - maximum, 346, 370
  - number of maxima, 326, 328
  - number of maxima per second, 299, 304
  - of highest peaks, 353
  - time interval, 331
  - time interval between two successive maxima, 307
  - total number of maxima, 305
average frequency, 264, 269, 282
  - linear-linear scales, 282
  - linear-logarithmic scales, 284
  - logarithmic-logarithmic scales, 286
  - logarithmic-linear scales, 285
averaging, 167

B, C

band-limited white noise, 95
bandpass filter, 386, 479
bandwidth–time product, 185
barrier
  type B -, 459, 462, 494
  type D -, 459, 497, 504
  type E -, 459, 478, 485, 497
bias error, 233
block(s), 205, 219
  number of -, 229
  number of points per -, 229
block duration, 229
chi-square law, 519
coherence, 555
  - function, 112, 555
coherent output power spectrum, 114
coincident spectral density, 94
collision
  probability of -, 491
confidence
  - interval, 190
  - level, 199

correlation, 176
 - coefficient, 40, 218
 - duration, 41
 - function, 93, 261
 - interval, 446
 - time, 446
correlogram, 35
cospectrum, 94
covariance, 47
crest factor, 33, 101
cross-correlation, 46, 250
 - coefficient, 50
cross-power
 - spectral density, 89
 - spectrum, 91

D, E

damping
 hysteretic -, 394
decibel, 202
degrees of freedom, 183, 187, 217, 249, 442
digitization, 228
Dirac delta function, 98
effective time interval, 423
envelope
 distribution of maxima of -, 379
 - of a process, 374
environment
 real -, 2
ergodicity, 50
error function, 103, 324, 536
Euler's constant, 354, 364, 546
excess kurtosis, 31
expected frequency, 272
exponential law, 516
extreme values, 343

F, G

filter, 159, 232
 bandwidth of -, 231
 - equivalence, 449

first crossing
 -distribution function, 457
 - probability density, 456
first up-crossing distribution function, 463
Fisher's law, 193
Fourier series, 88, 110, 149
Fourier spectrum, 79
Fourier transform, 79, 162
 fast - (FFT), 175
 inverse -, 255
Fourier transformation, 158
Gauss's law, 7, 266, 300, 511
generation of a random signal, 252
Gumbel
 - asymptote, 368
 - law, 528

H, I, K

half-power points, 549
Hanning window, 221
highest peak
 average frequency of -, 381
 average of largest peaks, 363
 expected maximum, 361
 maximum exceeded with risk α, 361
 median, 358
 most probable value, 358
 standard deviation of -, 356
 variance of -, 366
 variation coefficient, 357, 367
instantaneous values
 distribution function, 6
 distribution of -, 5, 259
 probability density,
 threshold crossings, 264
 probability density function, 264
integral I_b, 557
intercorrelation, 48
irregularity factor, 297, 309, 433
kurtosis, 30

L, M, N

Laplace-Gauss law with n variables, 523
largest peaks, 346
lengths of analysis, 200
linear single-degree-of-freedom-system, 233
log-normal law, 513
logarithmic
  - increase, 368
  - rise, 345
Markov process, 486, 493, 494
maxima probability distribution moment, 303
mean
  - clump size, 500
  - number of maxima of the envelope, 479
  - power, 84
  - square bandwidth, 450
  - value, 129
mode, 346, 358, 364
moment, 303
  - of response, 427
narrow band process, 340, 374
non-stationary vibration, 423
noise, 1, 41, 113, 151
  band-limited white -, 269
  narrowband -, 270, 307
normalized
  - autocorrelation, 40
  - covariance, 50
  - cross-correlation function, 50
  - rms error, 180
Nyquist frequency, 157, 228

O, P

octave, 143, 529
  number between two frequencies, 395
  number of dB per -, 396
optimum bandwidth, 240

overlapping, 216
  - weighting function, 221
  - rate, 216, 221
  - statistical error, 218
peak
  - distribution function, 323
  - factor, 101
  - hold spectrum, 250
  - ratio, 101
  - to-average ratio, 33
  - truncation influence on PSD, 104
periodic signal, 149
Poisson
  - law, 460, 471, 517
  - process, 472
positive maximum
  probability of a -, 337
  probability density of -, 337
power-law noise, 97
power spectral density, 81, 262, 371, 528
  constant between two frequencies, 404
  - of a random process, 90
  probability density of maxima, 295
probability distribution of maxima, 295
PSD, 81, 129
  influence of duration, 212
  influence of frequency step, 213
  - of displacement, 131
  - of velocity, 131
  - of periodic signal, 149
  periodic excitation, 177
  variance, 182

Q, R

quadrature spectral density function, 94
quadspectrum, 94
random process, 2
  - autocorrelation function, 16

- autocovariance, 17
- autostationarity, 20
broad-band -, 94, 311
- central moment, 14
- covariance, 17
- cross-correlation function, 16
- cyclostationary -, 22
derivative -, 260
- ensemble average, 12
Gaussian -, 7
- mean, 13
narrow-band -, 96, 311, 325
- quadratic mean, 13
- standard deviation, 15
- stationarity, 17
stationary -, 20
sum of two -, 106
- variance, 14
random variable, 1
random vibration, 550
 - absolute displacement, 409
 - analysis methods, 4
 - autocorrelation, 33
 - autocovariance, 35
 - band-limited noise, 444
 band of -, 395
 - centered moment, 27
 - correlation coefficient, 40
 - correlation duration, 41
 - correlogram, 35
 - covariance, 47
 - crest factor, 33
 - cross-correlation, 46
 - cross-correlation coefficient, 50
 - delay, 34
 - ergodicity, 50
 - excess kurtosis, 31
 - fourth moment, 288
 - intercorrelation, 48
 - kurtosis, 30
 - mean square value, 25
 - mean value, 24
 - moment, 27, 279

narrowband -, 406
- normalized autocorrelation, 40
- normalized covariance, 50
- normalized cross-correlation, 50
- peak-to-average ratio, 33
- quadratic mean, 25
- rms value, 25, 28
- skewness, 29
- standard deviation, 28
- variance, 28
R meter, 313
Rayleigh's law, 7, 300, 322, 520
Rice formula, 265
response
 autocorrelation of -, 445
 average number of peaks, 446
 irregularity factor of -, 431
 P.S.D. of the -, 388
 rms. value of - to white noise, 389
 transitory -, 423
 - to random vibration, 385
 rms absolute acceleration -, 414
 transitory -, 415
 value of the - rms, 387
return period, 344
reverse arrangements test, 58
ripple, 221
rms
 - absolute acceleration, 394
 - absolute displacement, 394
 - absolute velocity, 394
 - displacement, 135
 - relative displacement, 394, 400
 velocity, 61, 401
rms value, 81, 127, 158, 371
 - according to the frequency, 147
 linear-linear scales, 137
 linear-logarithmic scales, 138
 logarithmic- logarithmic scales, 141
 logarithmic-linear scales, 140
runs test, 61

S, T

sampling frequency, 153, 204
Shannon's theorem, 155, 205, 228, 550
shock, 3
skewness, 29
sliding mean, 54
standard deviation, 28, 128
- of maxima, 371
standard error, 184
stationarity, 58
statistical error, 180, 218, 245
   maximum -, 238
   - in decibels, 202
Student law, 527
swept sine, 3
temporal
   - averages, 23
   - step, 229
test
   laboratory -, 3
third octave analysis, 534
threshold, 272, 328, 349
   crossings, 264, 373
   independent - crossings, 460
transfer function, 110, 114, 248, 263, 396, 547
   derivative, 556
   error, 555
   fast swept sine, 551
   random vibration, 551
   shock, 551
   swept sine test, 551
truncation, 105, 553
two-degrees-of-freedom linear system, 321

U, V, W, Z

up-crossing
   - by the envelope of maxima, 472-
   independent envelope peaks, 476
   - of a response level, 455
   maxima independent -, 468
   mean frequency -, 473, 503
   mean time of the first -, 463
   peak independence -, 467
   probability density of first -, 469
   probability of first -, 493
   variance of the first - time, 463
variance, 128, 262
   - of the maximum, 346
Weibull law, 522
white noise, 95, 387, 419, 433, 442, 479
   equivalence with -, 408
wide-band process, 363
Wiener-Khintchine method, 158
window, 216
   Bingham -, 163
   - compensation factor, 164
   Hanning -, 163
   temporal -, 160
   - weighting coefficient, 163
windowing, 160
zero-padding, 170

Summary of Other Volumes in the Series

Summary of Volume 1
Sinusoidal Vibration

Chapter 1. The Need

1.1. The need to carry out studies into vibrations and mechanical shocks
1.2. Some real environments
   1.2.1. Sea transport
   1.2.2. Earthquakes
   1.2.3. Road vibratory environment
   1.2.4. Rail vibratory environment
   1.2.5. Propeller airplanes
   1.2.6. Vibrations caused by jet propulsion airplanes
   1.2.7. Vibrations caused by turbofan aircraft
   1.2.8. Helicopters
1.3. Measuring vibrations and shocks
1.4. Filtering
   1.4.1. Definitions
   1.4.2. Digital filters
1.5. Digitizing the signal
   1.5.1. Signal sampling frequency
   1.5.2. Quantization error
1.6. Reconstructing the sampled signal
1.7. Characterization in the frequency domain
1.8. Elaboration of the specifications
1.9. Vibration test facilities
   1.9.1. Electro-dynamic exciters
   1.9.2. Hydraulic actuators
   1.9.3. Test Fixtures

Chapter 2. Basic Mechanics

2.1. Basic principles of mechanics
   2.1.1. Principle of causality
   2.1.2. Concept of force
   2.1.3. Newton's first law (inertia principle)
   2.1.4. Moment of a force around a point
   2.1.5. Fundamental principle of dynamics (Newton's second law)
   2.1.6. Equality of action and reaction (Newton's third law)
2.2. Static effects/dynamic effects
2.3. Behavior under dynamic load (impact)
2.4. Elements of a mechanical system
   2.4.1. Mass
   2.4.2. Stiffness
   2.4.3. Damping
   2.4.4. Static modulus of elasticity
   2.4.5. Dynamic modulus of elasticity
2.5. Mathematical models
   2.5.1. Mechanical systems
   2.5.2. Lumped parameter systems
   2.5.3. Degrees of freedom
   2.5.4. Mode
   2.5.5. Linear systems
   2.5.6. Linear one-degree-of-freedom mechanical systems
2.6. Setting an equation for n degrees-of-freedom lumped parameter mechanical system
   2.6.1. Lagrange equations
   2.6.2. D'Alembert's principle
   2.6.3. Free-body diagram

Chapter 3. Response of a Linear One-Degree-of-Freedom Mechanical System to an Arbitrary Excitation

   3.1. Definitions and notation
   3.2. Excitation defined by force versus time
   3.3. Excitation defined by acceleration
   3.4. Reduced form
      3.4.1. Excitation defined by a force on a mass or by an acceleration of support
      3.4.2. Excitation defined by velocity or displacement imposed on support
   3.5. Solution of the differential equation of movement
      3.5.1. Methods
      3.5.2. Relative response

3.5.3. Absolute response
3.5.4. Summary of main results
3.6. Natural oscillations of a linear one-degree-of-freedom system
   3.6.1. Damped aperiodic mode
   3.6.2. Critical aperiodic mode
   3.6.3. Damped oscillatory mode

Chapter 4. Impulse and Step Responses

4.1. Response of a mass–spring system to a unit step function (step or indicial response)
   4.1.1. Response defined by relative displacement
   4.1.2. Response defined by absolute displacement, velocity or acceleration
4.2. Response of a mass–spring system to a unit impulse excitation
   4.2.1. Response defined by relative displacement
   4.2.2. Response defined by absolute parameter
4.3. Use of step and impulse responses
4.4. Transfer function of a linear one-degree-of-freedom system
   4.4.1. Definition
   4.4.2. Calculation of $H(h)$ for relative response
   4.4.3. Calculation of $H(h)$ for absolute response
   4.4.4. Other definitions of the transfer function
4.5. Measurement of transfer function

Chapter 5. Sinusoidal Vibration

5.1. Definitions
   5.1.1. Sinusoidal vibration
   5.1.2. Mean value
   5.1.3. Mean square value – rms value
   5.1.4. Periodic vibrations
   5.1.5. Quasi-periodic signals
5.2. Periodic and sinusoidal vibrations in the real environment
5.3. Sinusoidal vibration tests

Chapter 6. Response of a Linear One-Degree-of-Freedom Mechanical System to a Sinusoidal Excitation

6.1. General equations of motion
   6.1.1. Relative response
   6.1.2. Absolute response
   6.1.3. Summary
   6.1.4. Discussion
   6.1.5. Response to periodic excitation

6.1.6. Application to calculation for vehicle suspension response
6.2. Transient response
    6.2.1. Relative response
    6.2.2. Absolute response
6.3. Steady state response
    6.3.1. Relative response
    6.3.2. Absolute response
6.4. Responses $\left|\dfrac{\omega_0 \dot{z}}{\ddot{x}_m}\right|$, $\left|\dfrac{\omega_0 z}{\dot{x}_m}\right|$ and $\dfrac{\sqrt{k m} \dot{z}}{F_m}$
    6.4.1. Amplitude and phase
    6.4.2. Variations of velocity amplitude
    6.4.3. Variations in velocity phase
6.5. Responses $\dfrac{k z}{F_m}$ and $\dfrac{\omega_0^2 z}{\ddot{x}_m}$
    6.5.1. Expression for response
    6.5.2. Variation in response amplitude
    6.5.3. Variations in phase
6.6. Responses $\dfrac{y}{x_m}$, $\dfrac{\dot{y}}{\dot{x}_m}$, $\dfrac{\ddot{y}}{\ddot{x}_m}$ and $\dfrac{F_T}{F_m}$
    6.6.1. Movement transmissibility
    6.6.2. Variations in amplitude
    6.6.3. Variations in phase
6.7. Graphical representation of transfer functions
6.8. Definitions
    6.8.1. Compliance – stiffness
    6.8.2. Mobility – impedance
    6.8.3. Inertance – mass

Chapter 7. Non-viscous Damping

7.1. Damping observed in real structures
7.2. Linearization of non-linear hysteresis loops – equivalent viscous damping
7.3. Main types of damping
    7.3.1. Damping force proportional to the power b of the relative velocity
    7.3.2. Constant damping force
    7.3.3. Damping force proportional to the square of velocity
    7.3.4. Damping force proportional to the square of displacement
    7.3.5. Structural or hysteretic damping
    7.3.6. Combination of several types of damping
    7.3.7. Validity of simplification by equivalent viscous damping
7.4. Measurement of damping of a system

7.4.1. Measurement of amplification factor at resonance
7.4.2. Bandwidth or $\sqrt{2}$ method
7.4.3. Decreased rate method (logarithmic decrement)
7.4.4. Evaluation of energy dissipation under permanent sinusoidal vibration
7.4.5. Other methods
7.5. Non-linear stiffness

Chapter 8. Swept Sine

8.1. Definitions
  8.1.1. Swept sine
  8.1.2. Octave – number of octaves in frequency interval (f_1, f_2)
  8.1.3. Decade
8.2. "Swept sine" vibration in the real environment
8.3. "Swept sine" vibration in tests
8.4. Origin and properties of main types of sweepings
  8.4.1. The problem
  8.4.2. Case 1: sweep where time Δt spent in each interval Δf is constant for all values of f_0
  8.4.3. Case 2: sweep with constant rate
  8.4.4. Case 3: sweep ensuring a number of identical cycles ΔN in all intervals Δf (delimited by the half-power points) for all values of f_0

Chapter 9. Response of a Linear One-Degree-of-Freedom System to a Swept Sine Vibration

9.1. Influence of sweep rate
9.2. Response of a linear one-degree-of-freedom system to a swept sine excitation
  9.2.1. Methods used for obtaining response
  9.2.2. Convolution integral (or Duhamel's integral)
  9.2.3. Response of a linear one-degree-of freedom system to a linear swept sine excitation
  9.2.4. Response of a linear one-degree-of-freedom system to a logarithmic swept sine
9.3. Choice of duration of swept sine test
9.4. Choice of amplitude
9.5. Choice of sweep mode

Appendix

Vibration Tests: a Brief Historical Background

Bibliography

Index

Summary of Volume 2
Mechanical Shock

Chapter 1. Shock Analysis

1.1. Definitions
   1.1.1. Shock
   1.1.2. Transient signal
   1.1.3. Jerk
   1.1.4. Simple (or perfect) shock
   1.1.5. Half-sine shock
   1.1.6. Versed sine (or haversine) shock
   1.1.7. Terminal peak sawtooth (TPS) shock (or final peak sawtooth (FPS))
   1.1.8. Initial peak sawtooth (IPS) shock
   1.1.9. Square shock
   1.1.10. Trapezoidal shock
   1.1.11. Decaying sinusoidal pulse
   1.1.12. Bump test
   1.1.13. Pyroshock
1.2. Analysis in the time domain
1.3. Temporal moments
1.4. Fourier transform
   1.4.1. Definition
   1.4.2. Reduced Fourier transform
   1.4.3. Fourier transforms of simple shocks
   1.4.4. What represents the Fourier transform of a shock?
   1.4.5. Importance of the Fourier transform
1.5. Energy spectrum
   1.5.1. Energy according to frequency
   1.5.2. Average energy spectrum

1.6. Practical calculations of the Fourier transform
　1.6.1. General
　1.6.2. Case: signal not yet digitized
　1.6.3. Case: signal already digitized
　1.6.4. Adding zeros to the shock signal before the calculation of its Fourier transform
　1.6.5. Windowing
1.7. The interest of time-frequency analysis
　1.7.1. Limit of the Fourier transform
　1.7.2. Short term Fourier transform (STFT)
　1.7.3. Wavelet transform

Chapter 2. Shock Response Spectrum

2.1. Main principles
2.2. Response of a linear one-degree-of-freedom system
　2.2.1. Shock defined by a force
　2.2.2. Shock defined by an acceleration
　2.2.3. Generalization
　2.2.4. Response of a one-degree-of-freedom system to simple shocks
2.3. Definitions
　2.3.1. Response spectrum
　2.3.2. Absolute acceleration SRS
　2.3.3. Relative displacement shock spectrum
　2.3.4. Primary (or initial) positive SRS
　2.3.5. Primary (or initial) negative SRS
　2.3.6. Secondary (or residual) SRS
　2.3.7. Positive (or maximum positive) SRS
　2.3.8. Negative (or maximum negative) SRS
　2.3.9. Maximax SRS
2.4. Standardized response spectra
　2.4.1. Definition
　2.4.2. Half-sine pulse
　2.4.3. Versed sine pulse
　2.4.4. Terminal peak sawtooth pulse
　2.4.5. Initial peak sawtooth pulse
　2.4.6. Square pulse
　2.4.7. Trapezoidal pulse
2.5. Choice of the type of SRS
2.6. Comparison of the SRS of the usual simple shapes
2.7. SRS of a shock defined by an absolute displacement of the support
2.8. Influence of the amplitude and the duration of the shock on its SRS
2.9. Difference between SRS and extreme response spectrum (ERS)

2.10. Algorithms for calculation of the SRS
2.11. Subroutine for the calculation of the SRS
2.12. Choice of the sampling frequency of the signal
2.13. Example of use of the SRS
2.14. Use of SRS for the study of systems with several degrees of freedom
2.15. Damage boundary curve

Chapter 3. Properties of Shock Response Spectra

3.1. Shock response spectra domains
3.2. Properties of SRS at low frequencies
   3.2.1. General properties
   3.2.2. Shocks with zero velocity change
   3.2.3. Shocks with $\Delta V = 0$ and $\Delta D \neq 0$ at the end of a pulse
   3.2.4. Shocks with $\Delta V = 0$ and $\Delta D = 0$ at the end of a pulse
   3.2.5. Notes on residual spectrum
3.3. Properties of SRS at high frequencies
3.4. Damping influence
3.5. Choice of damping
3.6. Choice of frequency range
3.7. Choice of the number of points and their distribution
3.8. Charts
3.9. Relation of SRS with Fourier spectrum
   3.9.1. Primary SRS and Fourier transform
   3.9.2. Residual SRS and Fourier transform
   3.9.3. Comparison of the relative severity of several shocks using their Fourier spectra and their shock response spectra
3.10. Care to be taken in the calculation of the spectra
   3.10.1. Main sources of errors
   3.10.2. Influence of background noise of the measuring equipment
   3.10.3. Influence of zero shift
3.11. Specific case of pyroshocks
   3.11.1. Acquisition of the measurements
   3.11.2. Examination of the signal before calculation of the SRS
   3.11.3. Examination of the SRS
3.12. Pseudo-velocity shock spectrum
   3.12.1. Hunt's relationship
   3.12.2. Interest of PVSS
3.13. Use of the SRS for pyroshocks
3.14. Other propositions of spectra
   3.14.1. Pseudo-velocity calculated from the energy transmitted
   3.14.2. Pseudo-velocity from the "input" energy at the end of a shock

3.14.3. Pseudo-velocity from the unit "input" energy
3.14.4. SRS of the "total" energy

Chapter 4. Development of Shock Test Specifications

4.1. Introduction
4.2. Simplification of the measured signal
4.3. Use of shock response spectra
   4.3.1. Synthesis of spectra
   4.3.2. Nature of the specification
   4.3.3. Choice of shape
   4.3.4. Amplitude
   4.3.5. Duration
   4.3.6. Difficulties
4.4. Other methods
   4.4.1. Use of a swept sine
   4.4.2. Simulation of SRS using a fast swept sine
   4.4.3. Simulation by modulated random noise
   4.4.4. Simulation of a shock using random vibration
   4.4.5. Least favorable response technique
   4.4.6. Restitution of an SRS by a series of modulated sine pulses
4.5. Interest behind simulation of shocks on shaker using a shock spectrum

Chapter 5. Kinematics of Simple Shocks

5.1. Introduction
5.2. Half-sine pulse
   5.2.1. General expressions of the shock motion
   5.2.2. Impulse mode
   5.2.3. Impact mode
5.3. Versed sine pulse
5.4. Square pulse
5.5. Terminal peak sawtooth pulse
5.6. Initial peak sawtooth pulse

Chapter 6. Standard Shock Machines

6.1. Main types
6.2. Impact shock machines
6.3. High impact shock machines
   6.3.1. Lightweight high impact shock machine
   6.3.2. Medium weight high impact shock machine
6.4. Pneumatic machines
6.5. Specific testing facilities

6.6. Programmers
    6.6.1. Half-sine pulse
    6.6.2. TPS shock pulse
    6.6.3. Square pulse – trapezoidal pulse
    6.6.4. Universal shock programmer

Chapter 7. Generation of Shocks Using Shakers

7.1. Principle behind the generation of a signal with a simple shape versus time
7.2. Main advantages of the generation of shock using shakers
7.3. Limitations of electrodynamic shakers
    7.3.1. Mechanical limitations
    7.3.2. Electronic limitations
7.4. Remarks on the use of electrohydraulic shakers
7.5. Pre- and post-shocks
    7.5.1. Requirements
    7.5.2. Pre-shock or post-shock
    7.5.3. Kinematics of the movement for symmetric pre- and post-shock
    7.5.4. Kinematics of the movement for a pre-shock or a post-shock alone
    7.5.5. Abacuses
    7.5.6. Influence of the shape of pre- and post-pulses
    7.5.7. Optimized pre- and post-shocks
7.6. Incidence of pre- and post-shocks on the quality of simulation
    7.6.1. General
    7.6.2. Influence of the pre- and post-shocks on the time history response of a one-degree-of-freedom system
    7.6.3. Incidence on the shock response spectrum

Chapter 8. Control of a Shaker Using a Shock Response Spectrum

8.1. Principle of control using a shock response spectrum
    8.1.1. Problems
    8.1.2. Parallel filter method
    8.1.3. Current numerical methods
8.2. Decaying sinusoid
    8.2.1. Definition
    8.2.2. Response spectrum
    8.2.3. Velocity and displacement
    8.2.4. Constitution of the total signal
    8.2.5. Methods of signal compensation
    8.2.6. Iterations
8.3. D.L. Kern and C.D. Hayes' function
    8.3.1. Definition

8.3.2. Velocity and displacement
8.4. ZERD function
  8.4.1. Definition
  8.4.2. Velocity and displacement
  8.4.3. Comparison of ZERD waveform with standard decaying sinusoid
  8.4.4. Reduced response spectra
8.5. WAVSIN waveform
  8.5.1. Definition
  8.5.2. Velocity and displacement
  8.5.3. Response of a one-degree-of-freedom system
  8.5.4. Response spectrum
  8.5.5. Time history synthesis from shock spectrum
8.6. SHOC waveform
  8.6.1. Definition
  8.6.2. Velocity and displacement
  8.6.3. Response spectrum
  8.6.4. Time history synthesis from shock spectrum
8.7. Comparison of WAVSIN, SHOC waveforms and decaying sinusoid
8.8. Waveforms based on the $\cos^m(x)$ window
8.9. Use of a fast swept sine
8.10. Problems encountered during the synthesis of the waveforms
8.11. Criticism of control by SRS
8.12. Possible improvements
  8.12.1. IES proposal
  8.12.2. Specification of a complementary parameter
  8.12.3. Remarks on the properties of the response spectrum
8.13. Estimate of the feasibility of a shock specified by its SRS
  8.13.1. C.D. Robbins and E.P. Vaughan's method
  8.13.2. Evaluation of the necessary force, power and stroke

Chapter 9. Simulation of Pyroshocks

9.1. Simulations using pyrotechnic facilities
9.2. Simulation using metal to metal impact
9.3. Simulation using electrodynamic shakers
9.4. Simulation using conventional shock machines

Appendix

Mechanical Shock Tests: A Brief Historical Background

Bibliography

Index

Summary of Volume 4
Fatigue Damage

Chapter 1. Concepts of Material Fatigue

1.1. Introduction
    1.1.1. Reminders on the strength of materials
    1.1.2. Fatigue
1.2. Types of dynamic loads (or stresses)
    1.2.1. Cyclic stress
    1.2.2. Alternating stress
    1.2.3. Repeated stress
    1.2.4. Combined steady and cyclic stress
    1.2.5. Skewed alternating stress
    1.2.6. Random and transitory stresses
1.3. Damage arising from fatigue
1.4. Characterization of endurance of materials
    1.4.1. S-N curve
    1.4.2. Influence of the average stress on the S-N curve
    1.4.3. Statistical aspect
    1.4.4. Distribution laws of endurance
    1.4.5. Distribution laws of fatigue strength
    1.4.6. Relation between fatigue limit and static properties of materials
    1.4.7. Analytical representations of S-N curve
1.5. Factors of influence
    1.5.1. General
    1.5.2. Scale
    1.5.3. Overloads
    1.5.4. Frequency of stresses
    1.5.5. Types of stresses
    1.5.6. Non-zero mean stress

1.6. Other representations of S-N curves
    1.6.1. Haigh diagram
    1.6.2. Statistical representation of Haigh diagram
1.7. Prediction of fatigue life of complex structures
1.8. Fatigue in composite materials

Chapter 2. Accumulation of Fatigue Damage

2.1. Evolution of fatigue damage
2.2. Classification of various laws of accumulation
2.3. Miner's method
    2.3.1. Miner's rule
    2.3.2. Scatter of damage to failure as evaluated by Miner
    2.3.3. Validity of Miner's law of accumulation of damage in case of random stress
2.4. Modified Miner's theory
    2.4.1. Principle
    2.4.2. Accumulation of damage using modified Miner's rule
2.5. Henry's method
2.6. Modified Henry's method
2.7. Corten and Dolan's method
2.8. Other theories

Chapter 3. Counting Methods for Analyzing Random Time History

3.1. General
3.2. Peak count method
    3.2.1. Presentation of method
    3.2.2. Derived methods
    3.2.3. Range-restricted peak count method
    3.2.4. Level-restricted peak count method
3.3. Peak between mean-crossing count method
    3.3.1. Presentation of method
    3.3.2. Elimination of small variations
3.4. Range count method
    3.4.1. Presentation of method
    3.4.2. Elimination of small variations
3.5. Range-mean count method
    3.5.1. Presentation of method
    3.5.2. Elimination of small variations
3.6. Range-pair count method
3.7. Hayes' counting method
3.8. Ordered overall range counting method
3.9. Level-crossing count method

3.10. Peak valley peak counting method
3.11. Fatigue-meter counting method
3.12. Rainflow counting method
  3.12.1. Principle of method
  3.12.2. Subroutine for rainflow counting
3.13. NRL (National Luchtvaart Laboratorium) counting method
3.14. Evaluation of time spent at a given level
3.15. Influence of levels of load below fatigue limit on fatigue life
3.16. Test acceleration
3.17. Presentation of fatigue curves determined by random vibration tests

Chapter 4. Fatigue Damage by One-degree-of-freedom Mechanical System

4.1. Introduction
4.2. Calculation of fatigue damage due to signal versus time
4.3. Calculation of fatigue damage due to acceleration spectral density
  4.3.1. General case
  4.3.2. Particular case of a wideband response, e.g. at the limit $r = 0$
  4.3.3. Particular case of narrowband response
  4.3.4. Rms response to narrowband noise G_0 of width Δf when $G_0 \, \Delta f$ = constant
  4.3.5. Steinberg approach
4.4. Equivalent narrowband noise
  4.4.1. Use of relation established for narrowband response
  4.4.2. Alternative: use of mean number of maxima per second
4.5. Calculation of damage from the modified Rice distribution of peaks
  4.5.1. Approximation to real maxima distribution using a modified Rayleigh distribution
  4.5.2. Wirsching and Light's approach
  4.5.3. Chaudhury and Dover's approach
  4.5.4. Approximate expression of the probability density of peaks
4.6. Other approaches
4.7. Calculation of fatigue damage from rainflow domains
  4.7.1. Wirsching's approach
  4.7.2. Tunna's approach
  4.7.3. Ortiz-Chen's method
  4.7.4. Hancock's approach
  4.7.5. Abdo and Rackwitz's approach
  4.7.6. Kam and Dover's approach
  4.7.7. Larsen and Lutes ("single moment") method
  4.7.8. Jiao-Moan's method

4.7.9. Dirlik's probability density
4.7.10. Madsen's approach
4.7.11. Zhao and Baker model
4.7.12. Tovo and Benasciutti method
4.8. Comparison of S-N curves established under sinusoidal and random loads
4.9. Comparison of theory and experiment
4.10. Influence of shape of power spectral density and value of irregularity factor
4.11. Effects of peak truncation
4.12. Truncation of stress peaks
    4.12.1. Particular case of a narrowband noise
    4.12.2. Layout of the S-N curve for a truncated distribution

Chapter 5. Standard Deviation of Fatigue Damage

5.1. Calculation of standard deviation of damage: Bendat's method
5.2. Calculation of standard deviation of damage: Mark's method
5.3. Comparison of Mark and Bendat's results
5.4. Standard deviation of the fatigue life
    5.4.1. Narrowband vibration
    5.4.2. Wideband vibration
5.5. Statistical S-N curves
    5.5.1. Definition of statistical curves
    5.5.2. Bendat's formulation
    5.5.3. Mark's formulation

Chapter 6. Fatigue Damage using Other Calculation Assumptions

6.1. S-N curve represented by two segments of a straight line on logarithmic scales (taking into account fatigue limit)
6.2. S-N curve defined by two segments of straight line on log-lin scales
6.3. Hypothesis of non-linear accumulation of damage
    6.3.1. Corten-Dolan's accumulation law
    6.3.2. Morrow's accumulation model
6.4. Random vibration with non-zero mean: use of modified Goodman diagram
6.5. Non-Gaussian distribution of instantaneous values of signal
    6.5.1. Influence of distribution law of instantaneous values
    6.5.2. Influence of peak distribution
    6.5.3. Calculation of damage using Weibull distribution
    6.5.4. Comparison of Rayleigh assumption/peak counting
6.6. Non-linear mechanical system

Chapter 7. Low-cycle Fatigue

7.1. Overview
7.2. Definitions
    7.2.1. Baushinger effect
    7.2.2. Cyclic strain hardening
    7.2.3. Properties of cyclic stress–strain curves
    7.2.4. Stress–strain curve
    7.2.5. Hysteresis and fracture by fatigue
    7.2.6. Significant factors influencing hysteresis and fracture by fatigue
    7.2.7. Cyclic stress–strain curve (or cyclic consolidation curve)
7.3. Behavior of materials experiencing strains in the oligocyclic domain
    7.3.1. Types of behaviors
    7.3.2. Cyclic strain hardening
    7.3.3. Cyclic strain softening
    7.3.4. Cyclically stable metals
    7.3.5. Mixed behavior
7.4. Influence of the level application sequence
7.5. Development of the cyclic stress–strain curve
7.6. Total strain
7.7. Fatigue strength curve
7.8. Relation between plastic strain and number of cycles to fracture
    7.8.1. Orowan relation
    7.8.2. Manson relation
    7.8.3. Coffin relation
    7.8.4. Shanley relation
    7.8.5. Gerberich relation
    7.8.6. Sachs, Gerberich, Weiss and Latorre relation
    7.8.7. Martin relation
    7.8.8. Tavernelli and Coffin relation
    7.8.9. Manson relation
    7.8.10. Ohji *et al.* relation
    7.8.11. Bui-Quoc *et al.* relation
7.9. Influence of the frequency and temperature in the plastic field
    7.9.1. Overview
    7.9.2. Influence of frequency
    7.9.3. Influence of temperature and frequency
    7.9.4. Effect of frequency on plastic strain range
    7.9.5. Equation of generalized fatigue
7.10. Laws of damage accumulation
    7.10.1. Miner rule
    7.10.2. Yao and Munse relation
    7.10.3. Use of the Manson–Coffin relation

7.11. Influence of an average strain or stress
7.12. Low-cycle fatigue of composite material

Chapter 8. Fracture Mechanics

8.1. Overview
8.2. Fracture mechanism
   8.2.1. Major phases
   8.2.2. Initiation of cracks
   8.2.3. Slow propagation of cracks
8.3. Critical size: strength to fracture
8.4. Modes of stress application
8.5. Stress intensity factor
   8.5.1. Stress in crack root
   8.5.2. Mode I
   8.5.3. Mode II
   8.5.4. Mode III
   8.5.5. Field of equation use
   8.5.6. Plastic zone
   8.5.7. Other form of stress expressions
   8.5.8. General form
   8.5.9. Widening of crack opening
8.6. Fracture toughness: critical K value
8.7. Calculation of the stress intensity factor
8.8. Stress ratio
8.9. Expansion of cracks: Griffith criterion
8.10. Factors affecting the initiation of cracks
8.11. Factors affecting the propagation of cracks
   8.11.1. Mechanical factors
   8.11.2. Geometric factors
   8.11.3. Metallurgical factors
   8.11.4. Factors linked to the environment
8.12. Speed of propagation of cracks
8.13. Effect of a non-zero mean stress
8.14. Laws of crack propagation
   8.14.1. Head law
   8.14.2. Modified Head law
   8.14.3. Frost and Dugsdale
   8.14.4. McEvily and Illg
   8.14.5. Paris and Erdogan
8.15. Stress intensity factor
8.16. Dispersion of results
8.17. Sample tests: extrapolation to a structure

8.18. Determination of the propagation threshold K_S

8.19. Propagation of cracks in the domain of low-cycle fatigue

8.20. Integral J

8.21. Overload effect: fatigue crack retardation

8.22. Fatigue crack closure

8.23. Rules of similarity

8.24. Calculation of a useful lifetime

8.25. Propagation of cracks under random load
   8.25.1. Rms approach
   8.25.2. Narrowband random loads
   8.25.3. Calculation from a load collective

Appendix

Bibliography

Index

Summary of Volume 5
Specification Development

Chapter 1. Extreme Response Spectrum of a Sinusoidal Vibration

1.1. The effects of vibration
1.2. Extreme response spectrum of a sinusoidal vibration
   1.2.1. Definition
   1.2.2. Case of a single sinusoid
   1.2.3. General case
   1.2.4. Case of a periodic signal
   1.2.5. Case of n harmonic sinusoids
   1.2.6. Influence of the dephasing between the sinusoids
1.3. Extreme response spectrum of a swept sine vibration
   1.3.1. Sinusoid of constant amplitude throughout the sweeping process
   1.3.2. Swept sine composed of several constant levels

Chapter 2. Extreme Response Spectrum of a Random Vibration

2.1. Unspecified vibratory signal
2.2. Gaussian stationary random signal
   2.2.1. Calculation from peak distribution
   2.2.2. Use of the largest peak distribution law
   2.2.3. Response spectrum defined by k times the rms response
   2.2.4. Other ERS calculation methods
2.3. Limit of the ERS at the high frequencies
2.4. Response spectrum with up-crossing risk
   2.4.1. Complete expression
   2.4.2. Approximate relation
   2.4.3. Approximate relation URS – PSD
   2.4.4. Calculation in a hypothesis of independence of threshold overshoot

2.4.5. Use of URS
2.5. Comparison of the various formulae
2.6. Effects of peak truncation on the acceleration time history
    2.6.1. Extreme response spectra calculated from the time history signal .
    2.6.2. Extreme response spectra calculated from the power spectral densities
    2.6.3. Comparison of extreme response spectra calculated from time history signals and power spectral densities
2.7. Sinusoidal vibration superimposed on a broadband random vibration
    2.7.1. Real environment
    2.7.2. Case of a single sinusoid superimposed to a wideband noise
    2.7.3. Case of several sinusoidal lines superimposed on a broadband random vibration
2.8. Swept sine superimposed on a broadband random vibration
    2.8.1. Real environment
    2.8.2. Case of a single swept sine superimposed to a wideband noise
    2.8.3. Case of several swept sines superimposed on a broadband random vibration
2.9. Swept narrowbands on a wideband random vibration
    2.9.1. Real environment
    2.9.2. Extreme response spectrum

Chapter 3. Fatigue Damage Spectrum of a Sinusoidal Vibration

3.1. Fatigue damage spectrum definition
3.2. Fatigue damage spectrum of a single sinusoid
3.3. Fatigue damage spectrum of a periodic signal
3.4. General expression for the damage
3.5. Fatigue damage with other assumptions on the S–N curve
    3.5.1. Taking account of fatigue limit
    3.5.2. Cases where the S–N curve is approximated by a straight line in log–lin scales
    3.5.3. Comparison of the damage when the S–N curves are linear in either log–log or log–lin scales
3.6. Fatigue damage generated by a swept sine vibration on a single-degree-of-freedom linear system
    3.6.1. General case
    3.6.2. Linear sweep
    3.6.3. Logarithmic sweep
    3.6.4. Hyperbolic sweep
    3.6.5. General expressions for fatigue damage
3.7. Reduction of test time
    3.7.1. Fatigue damage equivalence in the case of a linear system

3.7.2. Method based on fatigue damage equivalence according
to Basquin's relationship
3.8. Notes on the design assumptions of the ERS and FDS

Chapter 4. Fatigue Damage Spectrum of a Random Vibration

4.1. Fatigue damage spectrum from the signal as function of time
4.2. Fatigue damage spectrum derived from a power spectral density
4.3. Simplified hypothesis of Rayleigh's law
4.4. Calculation of the fatigue damage spectrum with Dirlik's probability density
4.5. Up-crossing risk fatigue damage spectrum
4.6. Reduction of test time
  4.6.1. Fatigue damage equivalence in the case of a linear system
  4.6.2. Method based on a fatigue damage equivalence according to
  Basquin's relationship taking account of variation of natural damping
  as a function of stress level
4.7. Truncation ofthe peaks of the "input" acceleration signal
  4.7.1. Fatigue damage spectra calculated from a signal as a function of time
  4.7.2. Fatigue damage spectra calculated from power spectral densities
  4.7.3. Comparison of fatigue damage spectra calculated from signals
  as a function of time and power spectral densities
4.8. Sinusoidal vibration superimposed on a broadband randomvibration
  4.8.1. Case of a single sinusoidal vibration superimposed on broadband
  random vibration
  4.8.2. Case of several sinusoidal vibrations superimposed on a
  broadband random vibration
4.9. Swept sine superimposed on a broadband random vibration
  4.9.1. Case of one swept sine superimposed on a broadband random vibration
  4.9.2. Case of several swept sines superimposed on a broadband
  random vibration
4.10. Swept narrowbands on a broadband random vibration

Chapter 5. Fatigue Damage Spectrum of a Shock

5.1. General relationship of fatigue damage
5.2. Use of shock response spectrum in the impulse zone
5.3. Damage created by simple shocks in static zone of the response spectrum

Chapter 6. Influence of Calculation Conditions of ERSs and FDSs

6.1. Variation of the ERS with amplitude and vibration duration
6.2. Variation of the FDS with amplitude and duration of vibration
6.3. Should ERSs and FDSs be drawn with a linear or logarithmic
frequency step?

6.4. With how many points must ERSs and FDSs be calculated?
6.5. Difference between ERSs and FDSs calculated from a vibratory signal according to time and from its PSD
6.6. Influence of the number of PSD calculation points on ERS and FDS
6.7. Influence of the PSD statistical error on ERS and FDS
6.8. Influence of the sampling frequency during ERS and FDS calculation from a signal based on time
6.9. Influence of the peak counting method
6.10. Influence of a non-zero mean stress on FDS

Chapter 7. Tests and Standards

7.1. Definitions
  7.1.1. Standard
  7.1.2. Specification
7.2. Types of tests
  7.2.1. Characterization test
  7.2.2. Identification test
  7.2.3. Evaluation test
  7.2.4. Final adjustment/development test
  7.2.5. Prototype test
  7.2.6. Pre-qualification (or evaluation) test
  7.2.7. Qualification
  7.2.8. Qualification test
  7.2.9. Certification
  7.2.10. Certification test
  7.2.11. Stress screening test
  7.2.12. Acceptance or reception
  7.2.13. Reception test
  7.2.14. Qualification/acceptance test
  7.2.15. Series test
  7.2.16. Sampling test
  7.2.17. Reliability test
7.3. What can be expected from a test specification?
7.4. Specification types
  7.4.1. Specification requiring *in situ* testing
  7.4.2. Specifications derived from standards
  7.4.3. Current trend
  7.4.4. Specifications based on real environment data
7.5. Standards specifying test tailoring
  7.5.1. The MIL–STD 810 standard
  7.5.2. The GAM EG 13 standard
  7.5.3. STANAG 4370
  7.5.4. The AFNOR X50–410 standard

Chapter 8. Uncertainty Factor

8.1. Need – definitions
8.2. Sources of uncertainty
8.3. Statistical aspect of the real environment and of material strength
   8.3.1. Real environment
   8.3.2. Material strength
8.4. Statistical uncertainty factor
   8.4.1. Definitions
   8.4.2. Calculation of uncertainty factor
   8.4.3. Calculation of an uncertainty factor when the real environment is only characterized by a single value

Chapter 9. Aging Factor

9.1. Purpose of the aging factor
9.2. Aging functions used in reliability
9.3. Method for calculating the aging factor
9.4. Influence of the aging law's standard deviation
9.5. Influence of the aging law mean

Chapter 10. Test Factor

10.1. Philosophy
10.2. Normal distributions
   10.2.1. Calculation of test factor from the estimation of the confidence interval of the mean
   10.2.2. Calculation of test factor from the estimation of the probability density of the mean strength with a sample of size n
10.3. Log–normal distributions
   10.3.1. Calculation of test factor from the estimation of the confidence interval of the average
   10.3.2. Calculation of test factor from the estimation of the probability density of the mean of the strength with a sample of size n
10.4. Weibull distributions
10.5. Choice of confidence level

Chapter 11. Specification Development

11.1. Test tailoring
11.2. Step 1: analysis of the life-cycle profile. Review of the situations
11.3. Step 2: determination of the real environmental data associated with each situation
11.4. Step 3: determination of the environment to be simulated
   11.4.1. Need

11.4.2. Synthesis methods
11.4.3. The need for a reliable method
11.4.4. Synthesis method using PSD envelope
11.4.5. Equivalence method of extreme response and fatigue damage
11.4.6. Synthesis of the real environment associated with an event (or sub-situation)
11.4.7. Synthesis of a situation
11.4.8. Synthesis of all life profile situations
11.4.9. Search for a random vibration of equal severity
11.4.10. Validation of duration reduction
11.5. Step 4: establishment of the test program
11.5.1. Application of a test factor
11.5.2. Choice of the test chronology
11.6. Applying this method to the example of the "round robin" comparative study
11.7. Taking environment into account in project management

Chapter 12. Influence of Calculation Conditions of Specification

12.1. Choice of the number of points in the specification (PSD)
12.2. Influence of the Q factor on specification (outside of time reduction)
12.3. Influence of the Q factor on specification when duration is reduced
12.4. Validity of a specification established for a Q factor equal to 10 when the real structure has another value
12.5. Advantage in the consideration of a variable Q factor for the calculation of ERSs and FDSs
12.6. Influence of the value of parameter b on the specification
12.6.1. Case where test duration is equal to real environment duration
12.6.2. Case where duration is reduced
12.7. Choice of the value of parameter b in the case of material maden up of several components
12.8. Influence of temperature on parameter b and constant C
12.9. Importance of a factor of 10 between the specification FDS and the reference FDS (real environment) in a small frequency band
12.10. Validity of a specification established by reference to a one-degree-of-freedom system when real structures are multi-degree-of-freedom systems

Chapter 13. Other Uses of Extreme Response, Up-Crossing Risk and Fatigue Damage Spectra

13.1. Comparisons of the severity of different vibrations
13.1.1. Comparisons of the relative severity of several real environments
13.1.2. Comparison of the severity of several standards

    13.1.3. Comparison of earthquake severity
  13.2. Swept sine excitation – random vibration transformation
  13.3. Definition of a random vibration with the same severity as a series of shocks
  13.4. Writing a specification only from an ERS (or an URS)
    13.4.1. Matrixinversion method
    13.4.2. Method by iteration
  13.5. Establishment of a swept sine vibration specification

Appendix

Formulae

Bibliography

Index

Printed and bound by CPI Group (UK) Ltd, Croydon, CR0 4YY
27/02/2023
03195010-0005